The RAH-66 Comanche Helicopter

Technical Accomplishment, Program Frustration

The RAH-66 Comanche Helicopter

Technical Accomplishment, Program Frustration

Arthur W. Linden and the Comanche Team

LIBRARY OF FLIGHT

Paul Park, Editor-in-Chief
University of Texas at Arlington
Arlington, Texas

Published by
American Institute of Aeronautics and Astronautics, Inc.
12700 Sunrise Valley Drive, Reston, VA 20191-5807

American Institute of Aeronautics and Astronautics, Inc., Reston, Virginia

1 2 3 4 5

Library of Congress Cataloging-in-Publication Data

Names: Linden, Arthur W., author.
Title: The RAH-66 Comanche helicopter: technical accomplishment, program frustration/Arthur W. Linden and the Comanche Team.
Description: Reston, VA : The American Institute of Aeronautics and Astronautics, Inc., [2021] | Series: Library of flight | Includes bibliographical references and index.
Identifiers: LCCN 2021061084 (print) | LCCN 2021061085 (ebook) | ISBN 9781624106248 (hardcover) | ISBN 9781624106255 (ebook)
Subjects: LCSH: Comanche (Reconnaissance/attack helicopter)—History. | Attack helicopters—United States—Design and construction—History.
Classification: LCC UG1232.A88 L56 2021 (print) | LCC UG1232.A88 (ebook) | DDC 623.74/60472–dc23/eng/20211217
LC record available at https://lccn.loc.gov/2021061084
LC ebook record available at https://lccn.loc.gov/2021061085

ISBN: 978-1-62410-624-8

Ebook ISBN: 978-1-62410-625-5

All photos are courtesy of the I.I. Sikorsky Historical Archives, except where noted. Proceeds from the sale of this book will be donated to the Igor I. Sikorsky Historical Archives, Inc. of Stratford Connecticut.

CONTENTS

PROLOGUE

The Igor I. Sikorsky Historical Archives started a project a number of years ago to put summary descriptions of all the aircraft the company had produced on our website. These started out as a few pages each with a number of illustrations. However, some authors expanded their description to many pages.

I was asked to write about the RAH-66 Comanche, a project that consumed the last 13 years of my 37-year career at Sikorsky. I was assigned as Sikorsky Comanche engineering manager in 1987. I then went on to be the Sikorsky program manager and finally the Boeing Sikorsky program director from 1996 to June 2000.

As I started to write, I quickly exceeded 60 pages and I hadn't really gotten started. I realized that I had a great quantity of material that should be properly documented, and perhaps a book was in order.

I also realized that to properly do a book I would need the help of other ex-Comanche folks. We convened a meeting with a number of the key personnel to see if there was an interest in writing a book. Included were Bill Harper, Bruce Kay, Ken Rosen, Tom Sheehy, George Sipprell, Rus Stiles, and myself. Their titles and Comanche positions are detailed in the Acknowledgements section later in this book. Enthusiasm for the project was clear, and so we launched the effort. Others would join us as we moved forward.

A key decision was whom we should involve in telling this history. After the first few years of conceptual design, Sikorsky was teamed with Boeing. The work was split 50/50, meaning Boeing had a major part of the story to tell. And the Army, of course, played a major role in how the program evolved. But our group decided that only by restraining the story to Sikorsky people could we get to a reasonable picture of the history of the program in any kind of practical time period. We would try to be careful to tell Boeing's and the Army's parts as objectively as we could.

We also quickly recognized that there were really two separate stories to be told. First was all of the challenging technical issues and their development. Then we started a simple chapter oh the programmatic issues, including all the program delays and five restructures in 13 years, and how and why the program was canceled in 2004. This soon expanded to multiple chapters. It was then decided to do the book in two parts—technical and programmatic.

On the technical side, we all had a large stash of material and many stories to tell. We also knew that much of the Comanche story was already on the Internet. We could find material on almost any subject. Some of it, however, was not correct. So our objective became to put the complete technical history together in one location with the accuracy that comes from the original participants in the program.

If a particular chapter was primarily written by one or two people, they are identified under the chapter title. If no authors are identified for a chapter, it simply means that the chapter was compiled from the contributions of numerous people.

A difference between the authors is that some chose to tell personal stories about their coworkers. These are interesting and add an element of detail new to the reader. But mentioning specific individuals can be difficult. By definition, this leaves out many others who made substantial contributions to the program. This should not be inferred to be a judgment about the importance of these individuals to the program compared to many others who are not mentioned. Those not mentioned should know that we do also recognize their dedicated work on the program.

A conscious effort was made not to assign any undue credit or blame to any individual or group for anything that happened on the program. Over almost 20 years, hundreds of people worked on the program. In hindsight, some things could have been done better, but virtually everyone was working in a positive direction as they saw it at the time. This book gives the whole Boeing Sikorsky/Army team the credit for all the program successes and the blame for whatever went wrong.

Art Linden

INTRODUCTION

On December 5, 2003, Lt. Gen Richard A. Cody, Army Deputy Chief of Staff and an experienced aviator, flew the RAH-66 Comanche. According to the January 2004 issue of *National Defense*

> "he gave the aircraft a stellar review. ...Cody said that the Comanche had 'exceeded his expectations' and was particularly impressed by the rotor system, engine response and control. Cody also reportedly said the aircraft was 'easy to fly' and should be put into the hands of soldiers as soon as possible." [1]

Twelve weeks later the program was cancelled.

Mike Hirschberg, executive director of the Vertical Flight Society (then AHS International), wrote in the January/February 2017 issue of *Vertiflite* magazine:

> "The RAH-66 Comanche was the most advanced helicopter ever built. First, inadequate R&D funding levels during the Clinton administration stretched out its development over two decades. Then with only two prototypes completed, the Comanche had to be sacrificed in 2003 to field upgrades to the existing U.S. Army rotorcraft fleet so it would remain viable in Afghanistan and Iraq.
>
> "... This terrible choice was necessitated by the lack of sufficient funds for vertical lift aviation." [2]

The contracted Comanche program, initiated in 1991, lasted 13 years. By late 2003 the Boeing Sikorsky team was riding high. (At least that was the perception we were putting out to the public.) For Christmas 2003 we created and passed out a Christmas tree ornament celebrating the program moving into the manufacturing stage. Note that this was a "first in a series." We intended to do this every year.

In 1985, the U.S. Army initiated the Light Helicopter Experimental (LHX) program with contracts to industry for risk-reduction activities. LHX was to replace the Army's Vietnam-era light helicopter fleet - the Hughes/McDonnell Douglas OH-6, and the Bell OH-58 and UH/AH-1 scout, utility, and attack aircraft. The program represented a 4500 aircraft production run to replace the existing 7000 light-fleet helicopters.

Because of the large size of the projected production contract, the Army directed the four major U.S. helicopter companies to form two two-company teams. In 1985 Sikorsky teamed with Boeing calling ourselves the "First Team" to compete against the McDonnell Douglas-Bell "SuperTeam".

An extensive Development/Validation phase followed with aircraft design studies and risk reduction testing of numerous design concepts and aircraft systems.

Much new technology was available to advance the LHX beyond the UH-60 Black Hawk utility and AH-64 Apache attack helicopters developed in the early 1970s. Most important of these on the airframe side were the ability to achieve low radar and infrared signatures, a new emphasis on low acoustic signature, and new structural technologies in composite and advanced metallic materials. Great advances had been made in electronic flight controls and mission avionics—new high-resolution sensors were emerging with longer-range day and night targeting capability. Onboard computing power was advancing at a rapid rate.

The breakthrough advantage of the LHX would be its ability to "see without being seen." (Fig I.1 and I.2) With low observables to make it difficult to detect, the latest technology in sighting and targeting systems to extend stand-off range, and the ability to communicate data to all on the battlefield, LHX would be the first helicopter that could detect targets without being detected itself. This was a long way from the Vietnam era when the enemy knew a helicopter was approaching long before it got to the fight simply from its acoustic signature.

The original requirement for low observables came from the threat of a large tank invasion through the Fulda gap in Germany. That was the major threat when the LH requirements were defined. Although we believed, and stated, that low observables would be a big advantage in other types of conflict,

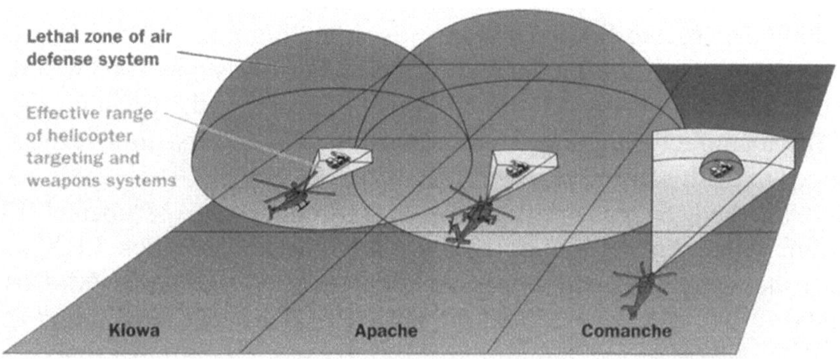

Fig. I.1 Seeing Without Being Seen.

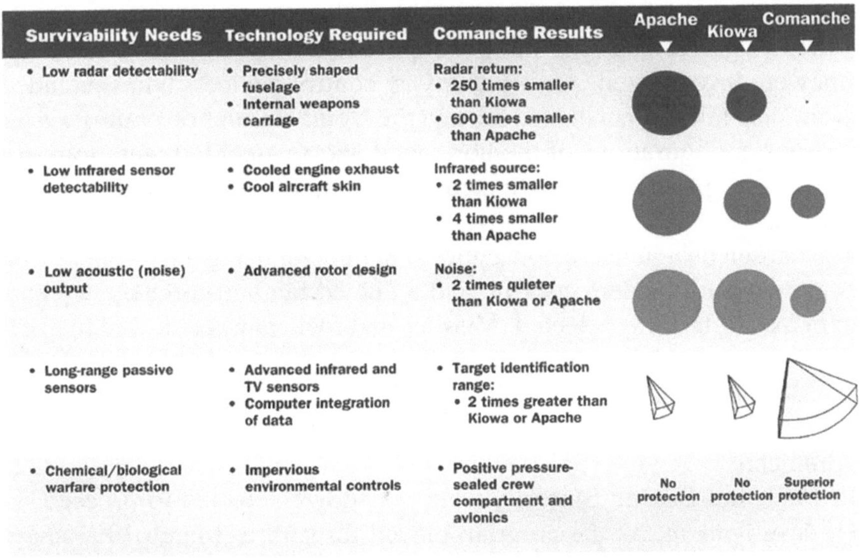

Fig. I.2 The Comanche dramatically reduced aircraft detectability and increased targeting range.

there were those detractors over the years that tried to define us as only a "cold war " weapon system after the cold war was over.

The Army also generated a new focus on air superiority with new requirements for high levels of maneuverability across all aircraft operations to improve its survivability. Developments in fly-by-wire and automated flight controls were proving able to lower pilot workload while allowing more precise flight path control and improved safety.

New multifunction cockpit displays, helmet-mounted sighting systems, and multi-spectral sensors all enhanced target detection range and accuracy. Weapons systems were being improved in parallel, and the LHX was to be compatible with air-to-ground and air-to-air missiles, rockets, and an accurate new gun.

The Comanche also required breakthroughs in supportability to simplify maintenance and reduce life-cycle costs. The design needed electronic subsystem fault detection/isolation, interpretation/prognostics, and two-way audio/video and data communication to the maintainer and support center. A Portable Maintenance Aid (PMA)/electronic maintainer logbook would be a brand new concept. We needed Interactive Electronic Technical Manuals (IETM). The flight line tool kit would have only 49 hand tools. All this was unheard of in the 80's and 90's.

The utility version, which was to replace the UH-1 Hueys, was abandoned by the Army early on to save development costs.

The aircraft configuration discussions stated with looking at a number of available options: conventional helicopters, tilt rotors and the Advancing

Blade Concept, which used two rigid counter-rotating rotors. Studies included single versus dual-pilot cockpits. Four, five and six-bladed main rotors were investigated. Anti-torque/yaw control device studies included a conventional tail rotor, a ducted fan and the Notar circulation control system. Numerous configurations of the horizontal and vertical tail were looked at. Various concepts were investigated for radar, IR and acoustic signature reduction.

The chosen design was a conventional helicopter using a five-bladed main rotor, a two-pilot tandem cockpit, and a ducted fan anti-torque/yaw control system. A "T" tail was selected. Missiles and rockets were stowed internally and only exposed when being fired. The nose-mounted gun swiveled 180 degrees to stow in a low radar signature position.

We felt that it was a well-integrated design that best met the Army's requirements.

In 1991, the Boeing Sikorsky team was down-selected to proceed with LHX development. As the program moved forward to March 1993 and the design was firmed up, the Army Chief of Staff, Gen. Gordon R. Sullivan, declared the LHX (by then re-designated as the RAH-66 Comanche) to be the "quarterback of the digital battlefield."

During the thirteen years of the contracted program, support from DoD and the congress vacillated. Initially support was strong and the funding was approved in 1991 for the Demonstration/Validation Full Scale Development phase. However a few years later the yearly funding was reduced, requiring a stretch out of the schedule and a full program restructuring. This happened two more times before the Full Scale Development Program was successfully concluded in 2000. Through these restructures the original schedule for Low Rate Initial Production in 1996 was extended to 2005. The final restructure in 2002 had a phased schedule with LRIPs from 2009 to 2011. In a number of these restructures additional requirements were added, or the mission equipment systems were redefined to incorporated further advances in technology. These changes were introduces to keep the program current and/or to make it more attractive to the Army user. But they also usually increased aircraft weight and cost.

The YRAH-66 Demonstration-Validation prototype aircraft flew for the first time in January 1996 and completed envelope expansion in 1999.

In June 2000 the U.S. Army awarded the Boeing Sikorsky team the RAH-66 Engineering Manufacturing and Development (EMD) contract.

By late 2003 the Boeing Sikorsky team was riding high. (At least that was the perception we were putting out to the public.) The 2003 configuration of the aircraft is shown in Fig. I.3.

For Christmas 2003 we created and passed out a Christmas tree ornament celebrating the program moving into the manufacturing stage, fig I.4

Note that this was a "first in a series." We intended to do this every year.

Fig. I.3 The final aircraft configuration.

During the thirteen years of the contracted program, support from DoD and the congress vacillated. Initially support was strong and the funding was approved in 1991 for the Demonstration/Validation Full Scale Development phase. However a few years later the yearly funding was reduced, requiring a stretch out of the schedule and a full program restructuring. This happened two more times before the Full Scale Development Program was successfully concluded in 2000. Through these restructures the original schedule for Low Rate Initial Production in 1996 was extended to 2005. The final restructure in 2002 had a phased

Fig. I.4 The 2003 Christmas Ornament and Packaging.

Comanche (RAH-66)
Presents
"From Vision to Reality"

History will record 2003 as a year that includes many accomplishments for our U. S. Armed Forces and the Comanche program. On March 19, U.S. Army aviation assets participated in the beginning of Operation Iraqi Freedom (OIF) along with Joint and multi-national Coalition forces. The "vision" of Comanche took a significant step forward when the Engineering Manufacturing and Development (EMD) contract was signed on June 30, 2000. The "reality" of Comanche took shape on March 31, 2003 when the "manufacturing" phase began with major assembly of the first EMD aft fuselage in Philadelphia, PA. Shortly afterwards, on August 11, 2003, the forward fuselage major assembly control point was loaded in Bridgeport, CT. Army aviation has always played a significant role in combat operations dating back to the Korean War. Comanche will provide the next leap ahead capability for Army aviation.

It will be the decisive edge for the twenty-first century Army Objective Force. It is built for stealth, agility, rapid response, lethality, supportability, and deployability. Coupled with its advanced mission equipment suite and full complement of weapon systems, it will be an unmatched scout with clout!

This year's "first in series" ornament design was created to commemorate the Boeing Sikorsky Joint Venture (JV) beginning manufacture of the RAH-66 Comanche armed reconnaissance helicopter in conjunction with the JV First Team partners. The spinning coin captures the profile of the new Comanche against old glory in the background on one side and documents the significant 2003 Comanche historical milestones on the opposing side. This symbol represents the renewed sense of patriotism displayed by many of our countrymen in 2003 and the support they have displayed for all our deployed Forces. The dates benchmark the Comanche that will be in their future.

Fig. I.4 ***(Continued)*.**

schedule with LRIPs from 2009 to 2011. In a number of these restructures additional requirements were added, or the mission equipment systems were redefined to incorporated further advances in technology. These changes were introduced to keep the program current and/or to make it more attractive to the Army user. But they also usually increased aircraft weight and cost.

But, in 2004 the program was cancelled.

Twelve weeks after Lt. Gen Cody's very positive remarks after having flown the aircraft, on February 23, 2004 the Department of Defense announced that it was terminating the program. Even at that time Lt. Gen Cody said

> "...I think I'm the only general officer that's flown the Comanche. And I will tell you up front, and pass on to my friends at Boeing and Sikorsky that are on this team, they built a tremendous aircraft. It is the most flexible, most agile aircraft that we have produced in this country, and the people that built it ought to be very, very proud of it. Tremendous flying characteristics and leap-ahead technology that's going to help us as we move forward." [3]

The Comanche, the Army's only new helicopter program in over 20 years, with the ability to "see without being seen", was meant to be a game-changer on the battlefield. Now it was cancelled.

So, what happened? How did Comanche become expendable? How did a program that was originally scheduled to enter Low Rate Initial Production in 1996 get delayed to 2011? How did an aircraft originally designed to replace the Army's light scout/attack fleet grow to become a virtual replacement for the much larger, heavier AH-64 Apache?

Comanche was leap-ahead technology when it was envisioned. But developmental issues including slow technical maturity and Army funding reductions caused four restructures of the program.

Why did the Boeing Sikorsky/Army program management team decide that a sixth program restructure was necessary, requiring more time and more money, when we had just recently signed a $3.1 billion EMD contract?

Eventually development costs grew to where the Army could no longer afford it while fighting two wars and facing reduced DoD budgets.

This is a story of great accomplishment and much frustration. It is the story of a program that never stayed stable for more than a few years at a time. It is the story of a program that enjoyed strong support at the most senior levels of the Army and DoD—until it didn't.

This is the Comanche story, as told by some of the key members of the Sikorsky Comanche development team.

REFERENCES

[1] Sandra I. Erwin, "Comanche Review Shrouded in Secrecy," *National Defense*, January 2004.

[2] Mike Hirschberg, Defense Priorities for the New Administration, VFS Executive Director, "Vertiflite, the magazine of the Vertical Flight Society," January/February 2017, p. 4.

[3] U.S. Army Press Conference, Briefing on the Restructure and Revitalization of Army Aviation, Acting Army Secretary Les Brownlee and other Senior Army officials, 23 February 2004.

Part One

THE COMANCHE WEAPONS SYSTEM

Chapter 1

LHX Concept Development (1979–1988)

Arthur W. Linden

In the early 1980s, the Army had two new medium-size helicopters in or near production. Sikorsky was building the UH-60 Black Hawk for the utility mission, and McDonnell Douglas was building the AH-64 Apache for the attack role. The fact that a new light helicopter program was on the horizon was very obvious. And as early as 1981, the Army initiated conceptual design studies that envisioned one new dynamic system (engines, transmission system, controls, and rotors) that could be applied to both utility and scout/attack aircraft.

Army combat doctrine was transitioning from a heavy mechanized army to a more air-mobile concept, which stressed maneuver warfare and made Army aviation assets an even more important integral element of combat operations. This military strategy included the armed/reconnaissance helicopter as an integral element of the ground force maneuver and a valuable combat asset that would increase the ground force survivability and operational effectiveness.

A 1982 Army Aviation Mission Area Analysis (AAMAA) modeled a mid-to-late-1990s Warsaw Pact enemy with sophisticated, technically advanced combat threat systems in Europe and Southwest Asia. Assumptions included dispersed, networked command and control; mobile threat systems; nuclear, chemical, biological, and electronic warfare environments; and high-density threat levels.

The AAMAA identified 77 deficiencies and 260 shortfalls in the existing light helicopter fleet, and it recommended the development of a new scout, attack, and utility helicopter. In March 1982, an Army Aviation Systems Program Review endorsed the recommendations of the AAMAA. The Army wanted a new light helicopter.

A buzz of related activity started within the Army technical centers. The Army believed that lack of helicopter air-to-air combat capability was a serious deficiency that needed to be addressed in any new acquisition. And because many new technologies were becoming available, the Army's Aviation Applied Technology Directorates (AATDs) at both Moffett Field, California, and Ft. Eustis, Virginia, were engaged in research, study, and test programs across many of these advanced helicopter design technologies expected in a new scout-attack aircraft.

AIR-TO-AIR COMBAT TESTS (AACTS)

In 1983, the AATD at Ft. Eustis initiated a series of one-on-one simulated air-to-air combat tests (AACTs) involving several helicopters. The objective was to gather real data regarding what maneuvering capability meant in a simulated air-to-air battle. The first two phases evaluated the Army's current fleet of light helicopters—the OH-58, UH-1, and AH-1.

Sikorsky participated in AACT Phase III tests with an S-76A flying against the UH-60A.

The AACT was set up in a restricted airspace around the Naval Air Warfare Center at Patuxent River, Maryland, in order to adequately capture the data using special tracking tools. The task was to engage in air-to-air combat and use simple aircraft maneuvering performance to gain the tactical advantage. The pilots used whatever maneuvering capability was available to engage the other aircraft and assumed a fixed weapons system.

What the data typically showed was that the UH-60A power margins gave the aircraft the ability to climb above the S-76 and gain a tactical advantage in some cases. It also showed that when the S-76 used a significant side-kick maneuver to point and shoot, it gained a huge tactical advantage. This rapid side-kick maneuver ultimately made its way into the required specification for the LHX.

The AACT program quantified the significance of how maneuvering performance changed the survival equation. The data from all phases of these AACT tests were documented in U.S. Army Aviation Systems Command (AVSCOM) technical reports.

MANEUVERABILITY AND AGILITY

Another key technology effort was a study program again initiated by the Army Aviation Applied Technology Directorate at Ft. Eustis to identify the key design drivers for helicopter maneuverability and agility. Sikorsky was selected to conduct the Maneuverability and Agility (M/A) Design Sensitivity Analysis. The analysis used first principles but also employed analysis based on a typical simulation of a 10,000-lb generic helicopter (essentially varying parameters around an S-76 baseline). The study was conducted over 18 months and resulted in analysis of virtually every helicopter design parameter while the aircraft was accelerating and decelerating, climbing and descending, making steady and transient turns, through pullups and pushovers, in hover turns, and in lateral and sideslip flight. The results quantified how helicopter maneuvering capability was affected or how the agility of the aircraft changed when these parameters were changed.

The study also made a serious attempt to identify the impact that each parameter had on the overall vehicle weight. The experience of going through such a study turned out to be very useful in sensitizing the Sikorsky design team on the cost-benefit of LHX design choices.

As a follow-on to this Maneuverability and Agility study, a Phase II contract was awarded to Sikorsky. This study was conducted to determine the characteristics of eight different helicopters that flew in the Army's AACTs and included a correlation of the two efforts. The results validated previously developed Phase I design study parameter sensitivities. The helicopters evaluated were the UH-60A, S-76A, AH-64A, AH-1S, OH-58A, and SA-365N-1. The CH-53E and Soviet Mi-28 were also analyzed, although no AACT data were available for them. The characteristics of these vehicles were assessed by flying detailed computer models through a series of nine discrete maneuvers.

Selected flight test data from AACT II, III, and IV were analyzed using a workstation to replay the data and generate synthetic out-of-cockpit views. A method for scoring the simulated maneuvers and the AACT engagements was devised. M/A ratings for all of the subject aircraft were determined using the simulated maneuver scores. Correlation of the ratings to the AACT scores was very good, with the winning air combat aircraft having the highest M/A rating in four of the five engagement pairs analyzed. This study also produced a unique database employing a common simulation technology and evaluation methodology for all of the subject helicopters. The study confirmed that we clearly understood the maneuverability and agility attributes and the design parameters that allowed for air combat success. Now we could quantify their effects.

Sikorsky profited significantly as a result of our direct participation in both the AACTs and the M/A study, enabling us to write a more credible and focused LHX design proposal to the Army and also to better understand several of the design choices and tradeoffs through the design process.

The ARTI Program

In December 1983, the Army signed Advanced Rotorcraft Technology Integration (ARTI) contracts with Bell Helicopter Textron, Boeing Vertol, McDonnell Douglas, and Sikorsky.

With the initial requirement of several thousand aircraft, the Army was concerned it might not have the manpower to fill cockpit seats. To reduce the number of pilots required, the ARTI program challenged industry to come up with a viable single-pilot cockpit approach. The objective was to demonstrate with simulation experiments the technical feasibility of the single-pilot scout/attack (SCAT) aircraft and reduce risk through the development of an integrated automated cockpit. The intent was to prove that single-pilot operations were feasible and supported by available technologies of an integrated cockpit, automated navigation, digital map, automatic targeting, interactive voice controls, sensor fusion, wide-field-of-view helmet displays, and low workload automated flight control system.

Sikorsky's ARTI partners were selected to provide technology substantiation for our design. They included TRW for communications [integrated communications, navigation, and identification avionics (ICNIA)] and aircraft survivability [integrated electronic warfare system (INEWS)]; Northrop Grumman for survivability techniques, targeting, and day/night imaging systems; Martin Marietta for target detection and day/night imaging systems; Kaiser for helmet display and helmet tracking systems; and Hamilton Standard for flight controls, inertial sensors, helmet displays, and general helicopter electronics.

A 1985 contract modification added development of the mission computer, electro-optical target acquisition and designation system (EOTADS) and the night vision pilotage system (NVPS). There was an ongoing cooperative effort between the Army Aviation Systems Command (AVSCOM) and the Air Force Aerospace Medical Research Laboratory (AFAMRAL) to demonstrate a helmet-mounted display (HMD).

Also in 1985, Sikorsky teamed with Boeing to form the "First Team." The two companies integrated their unique approaches into one design.

Because the ARTI focus was on single-pilot operation and the desired technology to achieve the enhanced mission, the cockpit needed to manage an information-rich environment. Controls also had to be easy to operate. That necessitated automated decision making and simple command structures. The LHX mission equipment package (MEP) requirements evolved into a software-intensive "assistant pilot" requirement and the need for impressive threat detection, threat protection, communication, and control/display capabilities.

By the middle of the ARTI program, our design featured a control/display approach that minimized button selections for critical mission tasks by keeping menu hierarchies broad and shallow. The goal was that the pilot should need no more than three button selections for all critical tasks.

Adaptive Fuel Control Program

Another area of technical interest back in this pre-LHX timeframe was the concern over rotor speed transients during maneuvering flight. Whether a maneuver is performed by changing power directly or the power change results from a change to the inflow through the main or tail rotor, it can become a high-workload piloting task to control the resulting rotor speed transients from a nominal setting. That transient can result from several maneuver scenarios such as a rapid acceleration or deceleration, a change from level flight to a rapid climb or descent, or an angle-of-attack change to the rotor, such as in a pull-up.

Besides potentially limiting pilot aggressiveness, these transients make it difficult to terminate maneuvers precisely. Imagine a scenario where a pilot of an attack helicopter is trying to turn toward a target, point a fixed weapon, and fire the weapon. The pointing accuracy would be affected by these rotor speed transients. In an adversarial situation, the combatant that can turn and aim the weapon to fire first is going to survive, and rotor speed transient control is a life-or-death factor.

Engineers of the U.S. Army Technical Directorate at Ft. Eustis were very aware of this issue because of a vast amount of feedback from fleet operations on several helicopters. Its technical experts wondered to what extent modifications to engine fuel control systems could be used to help reduce these transients. It eventually awarded a contract to Chandler Evans Co. (CECO), an engine fuel control manufacturer, to study this issue and to research the potential and feasibility of a practical adaptive fuel control system. CECO selected Sikorsky as a study partner.

The Adaptive Fuel Control Program spanned a few years in a couple of phases and ended with a test program that was conducted at the Sikorsky West Palm Beach, Florida, test facility. CECO and Sikorsky conducted the test, which reviewed and measured the effect of a prototype controller on an S-76A with Allison engines and a standard fuel control system. The tests revealed some very interesting results. A laser gun was mounted on the side of the aircraft, and several maneuvers were flown that simulated maneuvering to a ground target and taking shots at the target. Comparisons were made of the accuracy following the maneuver. This was a way of demonstrating just how effective controlling the rotor speed transients would be in a very likely battlefield scenario.

As it turned out, both a roll anticipator and yaw anticipator were eventually added to the control architecture. Because there were noticeable power shifts and rotor speed transients during aggressive roll and yaw maneuvers, by modifying the fuel control triggered by pilot inputs we were able to reduce these transients to a reasonable level. The net effect was more aggressive and efficient maneuvering capability. If we had not been through the study program, it is likely that we would have gone through the test program making no adjustments, and Comanche would have had ended up with poorer engine transient speed performance, resulting in less precision capability.

A full report of this Adaptive Fuel Control study is documented and was presented at an American Helicopter Society (now the Vertical Flight Society)

(AHS) Forum in the 1990s; the intent here is to point out that Sikorsky was very sensitive to this issue and applied similar fuel control adaptive features on the LHX program.

SIMULATION

For decades, flight simulation was a major tool in helicopter design, being used in the requirements development of flight controls and flying qualities. It was also a tool used in the design of aircraft structures by projecting component structural loads. The ARTI program brought an entirely new focus towards simulation: pilot-in-the-loop full mission simulation. This mandated new tools and innovative approaches to simulating hardware and software of the entire cockpit.

Because the military specifications were not fully defined at that time, part task, part mission, and full mission experiments were used as design tools to fully develop and validate the cockpit. By providing a full mission environment, operational Army aviators were able to offer ideas and criticism very early in the development process. The Army completely endorsed the concept of using simulation for this purpose.

The LHX was envisioned to have a "glass cockpit" with advanced controls and digital displays including the helmet-mounted display, a multiaxis sidearm controller, controller grips with multifunction switches, and voice-activated controls. The mission equipment package (MEP) would include systems used for targeting and identification, navigation, communications, and night pilotage.

Weapons and fire control systems modeling was needed. In addition, modeling of threat forces was required as well as additional models of friendly/support forces.

Because it was too early in the design process to have a representative aerodynamic model of the LHX, a decision was made to utilize the model of the S-76. At this time, Sikorsky was planning a flight controls test vehicle, grafting a single-pilot cockpit onto the nose of the S-76. This aircraft was named the Shadow. This one aerodynamic simulation model could thus be used for both the ground and Shadow simulations. This allowed us to leverage that development and not require added costs to develop a separate aircraft simulation model.

The approach taken for simulating the MEP and armament systems was significantly simpler. Little was available on the commercial market in the form of software tools that would help in creating graphics. This was particularly true for symbology such as overlays for a map or helmet-mounted display (HMD).

To create symbology, the ARTI simulation utilized the on-board processor selected for the Shadow in-flight simulator mission processor. Although this solved the immediate problem, it was designed as an airborne unit and as such introduced several problems related to updates needed for simulation purposes as well as the ability to make changes quickly and efficiently.

The HMD integrated into the ARTI simulation was a prototype of a helmet system developed by Hamilton Standard. Its primary requirement was to provide a wide field of view. The prototype utilized a common motorcycle helmet to house the electronics and lens. Head movements were sensed through a magnetic head tracker. The prototype HMD provided a full 80-deg wide field of view, but was quite heavy, which resulted in uncomfortable loads on a pilot's head and neck. To reduce the apparent weight of the unit and the resultant pilot fatigue, a bungee arrangement attached the helmet to a pulley in the cockpit ceiling and a counterweight.

In the Boeing Sikorsky teaming agreement, Sikorsky was responsible for the pilot–vehicle interface. A reconfigurable pedestal upon which a variety of display configurations could be mounted was installed. The simulator cockpit accommodated a single crewmember with a multiaxis sidearm controller. Initially, the simulator was a fixed base facility with a 120-deg horizontal projection screen (Fig. 1.1), but eventually migrated to a full-motion dome with 180-deg projection for out-the-window visuals (Fig. 1.2).

Simulation cockpit hardware relied on commercial off-the-shelf components where possible to reduce costs. These included cathode-ray tube (CRT) displays instead of flat panels. New design concept input controls were included such as touchscreens, programmable display pushbuttons, and voice commands. New grip switches were included such as five-way slew controllers to act as a joystick-type control for several MEP and armament systems.

Fig. 1.1 First-generation Sikorsky simulation.

The armament system was simplified by having weapons (missiles, rockets, and a gun) coupled to a trigger switch. The simulator's visual system graphics models created the fly-out effect. No actual flight dynamics were calculated for the weapons fly-out. When a weapon was fired and the event occurred after satisfying constraints, a hit or kill was established based on probability. No account was taken of miss distance in this calculation.

No aircraft systems were simulated, so there was no need for cockpit gauges depicting health of engines, transmissions, and so forth. It was envisioned that much of the typical LHX systems monitoring would be automated.

Prior to the integration of audio warnings, announcements of threat detections were provided by simulator operators speaking scripted messages as the aircraft progressed through the scenarios.

Boeing was responsible for the flight controls and constructed a motion-based simulator with a 200-deg horizontal out-the-window projection capability at its facility in Pennsylvania (Fig. 1.3). It had the ability to fly the simulation model controlled by the actual flight control computers. Reconfigurable, nonflightworthy controllers were installed so the mechanical characteristics could be investigated. Flight control laws were developed by flying through the high fidelity visual simulation environment projected on the wide screen in front of the cockpit. It also had the ability to insert a threat aircraft into the scene, to permit evaluation of air-to-air capabilities.

Boeing also utilized a fixed base cockpit laboratory to evaluate software implementation of the mission equipment package.

Martin Marietta, being responsible for the pilotage and targeting sensors, utilized a terrain board simulator at its Orlando facility. Scale models of trees, obstacles, buildings, and targets were placed on the board, which incorporated

Fig. 1.2 Second-generation Sikorsky simulation.

Fig. 1.3 Final LHX simulator.

realistic terrain relief, creating a three-dimensional world of "real" objects. High resolution cameras were mounted on a three-degrees-of-freedom (*x*-*y*-*z*) moving gantry to simulate sensor video.

In June and July 1990, the Army conducted an evaluation of our LHX design using the full mission simulator at Sikorsky. Full battle simulation was conducted, including air-to-air combat, air-to-ground combat, and day and night operations. By then, dozens of Army aviators had flown the simulators, providing valuable feedback. A group of independent evaluators flew missions and provided workload data to be analyzed for inclusion into the ARTI final report.

ADVANCED COMPOSITE AIRFRAME PROGRAM (ACAP)

Beyond the improved flight dynamic and single pilotage requirements that the LHX would demand, new airframe structural concepts were going to be demanded to reduce weight and to improve crashworthiness. Fiber-reinforced resin composite materials promised more crashworthy airframe structures free of metallic corrosion and more tolerant of battle damage. They also promised to reduce the radar cross section of low observable aircraft. In 1981, the Army's Aviation Applied Technology Directorate (AATD) at Ft. Eustis contracted both Bell Helicopter and Sikorsky Aircraft to design, fabricate, and crash-test representative LHX-sized, low observable composite fuselages. The effort was called the Advanced Composite Airframe Program (ACAP). Sikorsky's design used the dynamic components and flight controls from the S-76 helicopter and was designated internally as the Sikorsky S-75. Bell based its D292 demonstrator on its commercial Model 222 dynamics.

Both the S-75 and D292 incorporated blow-out panels to relieve overpressure from internal High Explosive Incendiary (HEI) detonations. Otherwise, the structure could rupture catastrophically.

The fuselage shape was our original attempt at reducing radar signature.

Sikorsky flew its ACAP demonstrator in 1984 (Fig. 1.4). Bell flew its D292 a year later.

FIRST FLIGHT

In 1986, the S-75 static test article was successfully drop-tested in the Impact Dynamics Research Facility (IDRF) at the NASA Langley Research Center in Virginia. The test simulated a crash impact with a sink speed of 38 fps and 10 deg of roll and pitch. The crash was deemed survivable. Cabin volume was maintained, and anthropomorphic dummies in the cockpit and cabin did not sustain excessive deceleration forces.

In June 1999, at the end of the ACAP program, the Sikorsky team suggested that the S-75 flight test aircraft be placed in an aviation museum. NASA had a different plan, and crash-tested the flight article with 38 fps vertical velocity and 32.5 fps horizontal velocity (Fig. 1.5). This second drop test provided similar results and demonstrated the ability of three-dimensional finite element crash simulation codes to predict airframe structural response.

Most importantly, ACAP demonstrated that composite airframes could meet Army requirements and be lighter in weight and lower in cost than comparable metal structures. Demonstrations and tests showed that military requirements for crashworthiness, ballistic survivability, radar signature, reliability, and maintainability could be met with pragmatic design solutions.

Fig. 1.4 Sikorsky S-75 ACAP technology demonstrator helicopter.

Fig. 1.5 ACAP crash impact test conducted at NASA, Langley Drop Test Facility (NASA Photo).

PRELIMINARY DESIGN STUDIES

Sikorsky started looking at the light helicopter requirement as far back as 1979, but a serious effort was initiated in 1981 in anticipation of an Army request for information (RFI) for potential rotary-wing aircraft technologies from industry. At that time the requirements were:

- 4292 aircraft in both scout/attack (SCAT) and utility configurations to replace 7000 existing aircraft
- Single pilot
- 8000-lb SCAT mission gross weight
- 70% commonality between SCAT and utility versions
- 50% of the maintenance manhours per flight hour of the AH-1 and UH-1
- 25% specific fuel consumption reduction compared to the T-53 engine
- SCAT aircraft engineering and manufacturing development (EMD) fly-off competition
- Production goal of FY92

Other important requirements were air transportability load time; self-deployability range; reliability, availability, and maintainability (RAM) attributes; and life cycle costs.

Sikorsky conceptual design studies evaluated the coaxial advancing blade concept (ABC), a tilt rotor, a compound helicopter, and a conventional helicopter. The XH-59A ABC concept demonstrator was nearing the end of its flight test program and had received a glowing report from an Army evaluation team for its speed, maneuverability/agility, and compactness. There were some at Sikorsky who thought it was the obvious answer for the new Army light helicopter. Others were enthusiastic about the tilt rotor. Still others felt strongly that a new, advanced technology conventional helicopter was the best solution.

The level of technology assumed for the helicopter configuration at that time was the use of an integrated technology (articulated) rotor (ITR); a conventional technology planetary gearbox with advanced gearing materials, tooth shapes, and free wheel units; an advanced digital-optical control system (ADOCS)-type flight control system; and higher harmonic control.

The rotor for the helicopter design used four blades and had a diameter of 38 ft, giving it a SCAT disc loading of 7.4 psf. The antitorque rotor was a 39-in. fan-in-fin with 11 blades. Two 1200-hp Intermediate Rated Power at Sea Level Standard conditions (IRP at SLS) engines were used. The design gross weight and the SCAT mission weight were 8385 lb, somewhat exceeding the 8000-lb goal. The utility mission weight was 10,060 lb.

Meanwhile, in January 1983, an Army LHX Special Working Group (SWG) developed the framework for the LHX program, and in May 1983, a Justification for a Major System New Start (JMSNS) was approved, authorizing initiation of concept exploration for the LHX.

In 1985, the Army gave out $900,000 contracts to the four major U.S. helicopter companies—Bell, Boeing Vertol, Hughes, and Sikorsky. These tasked each company to come up with its LHX best technical approach (BTA).

Sikorsky again looked at a tilt rotor, a helicopter, an ABC, and a NOTAR (no tail rotor) helicopter.

The specification for the LHX included many requirements that were new and unique. Detectability specifications covered radar, infrared, and aural signatures. The helicopter had to operate and survive in environments with nuclear, chemical, and biological warfare threats. These requirements had to be understood and meaningful metrics created, and the metrics had to be traded off against other aircraft attributes.

To accomplish the tradeoffs, technology specialists developed elaborate three-dimensional carpet plots that quantitatively showed the relationship of the various specified requirements to other aircraft attributes; for example, radar signature was plotted against detectability measured in distance and probability of detection, with a third variable of rate of climb, because shaping the fuselage changed the aerodynamic characteristics. It was one thing for the technologists to create these data, but quite another feat for the design teams to digest it, understand it, and make meaningful decisions.

An approximation of the required mission gross weight was first made. Based on UH-60 and AH-64 technology, a helicopter rudimentary weight empty fraction (WE/GW) at that time was approximately 0.55. Applying technologies such as composite aircraft structures, this WE fraction was projected to be reduced to approximately 0.45. Militarization features then needed to be considered before the final WE fraction was established to calculate the potential mission gross weight.

Based on the mission gross weight, the rotor diameters, power required, fuel volume, and so forth were established, and configuration layouts were created. From these configuration layouts, the weight and subsystem sizes were adjusted. This process was repeated until the mission requirements and platform definition closed.

Fig. 1.6 The initial helicopter had a *V* tail and a ducted fan antitorque system.

The resulting conventional helicopter baseline design at this time had a 40-ft main rotor diameter. The tilt rotor had 23.8-ft-diameter prop-rotors on each side, and the coaxial helicopter (ABC) had a 35-ft main rotor diameter. The conventional helicopter dash speed was 180 kt. The tilt rotor promised 250 kt and the ABC 230 kt.

Figures 1.6 to 1.8 are illustrations of a few of the early concepts. These paintings were done by Andy Whyte, a Sikorsky preliminary design engineer who was also an accomplished artist.

Fig. 1.7 The utility configuration being used as an air ambulance.

Fig. 1.8 A high-speed concept using the ABC coaxial rigid rotor.

The ABC and the tilt rotor were not pursued any further because their higher speed was not required.

One interesting trade study was the location of the EOTADS sensor. We looked at a mast-mounted sensor, a roof-mounted sensor, and a nose-mounted sensor. Elevating the sensor above the rotor was thought to reduce the exposure of the aircraft to the threat systems during target acquisition and target designation. However, at close examination of terrain features, it was concluded that the terrain features were not representative of looking over a wall but more like looking over a curved surface with depth. In simulations it was concluded that the aircraft did have to be exposed significantly to get complete area coverage with the sensor above the rotor, similar to the aircraft exposure required with the sensor on the nose of the aircraft. Although others had concluded differently, we concluded that there was no advantage of placing the sensor above the rotor.

When the BTAs from this study were presented to the Army, three of the manufacturers suggested conventional helicopters. Bell's BTA was a tilt rotor.

TEAMING WITH BOEING

Sikorsky and Boeing teamed up in 1985 and we called ourselves the Boeing Sikorsky First Team. Bell and McDonnell Douglas then formed the competing SuperTeam.

The original Army plan was to have the two teamed companies compete with each other in production. This would require duplicate tooling for two production lines. Eventually, when the production quantity was reduced to 1292, this idea was eliminated, because the cost of duplicate composite tooling became too high.

In the Boeing Sikorsky teaming agreement, Sikorsky was the lead on the airframe, air vehicle integration, simulation, and final vehicle assembly. Boeing was the lead on the mission equipment package (MEP) and integrated logistics support (ILS). However, each company would take a portion of the other's lead responsibilities. Boeing was assigned to develop the flight control system, the main rotor blades, and the tail of the aircraft. Sikorsky developed the crew interface requirements and had responsibility for the helmet sighting and tracking system of the MEP.

We joined forces prior to the decision that the LHX would be a conventional helicopter. In initial discussions, Sikorsky exposed Boeing to the details of the recently completed Sikorsky XH-59A ABC program. Boeing had recently teamed up with Bell for the development of the large USMC V-22 tilt rotor aircraft and discussed the possibility of a tilt rotor. The issue was resolved once the Army decided that speeds higher than conventional helicopters were not necessary.

At this point in the program, designers had great ambitions and great unknowns. Furthermore, the government was evaluating engine proposals from two competing teams. This led to tradeoff studies for many distinct LHX variants: scout and utility, single and dual pilot, and conventional and low observable helicopters, all evaluated with two competing engines.

Also at this time, personal computers were being introduced into the design process. Not everyone at Sikorsky had embraced the use of personal computers. Some senior engineers close to retirement were actually counting on retiring before they had to learn how to use them. That attitude quickly vanished when program leadership started to exclusively use Microsoft Outlook to schedule meetings and use email for communications. From then on, everyone including the dinosaurs joined the digital age.

Preliminary design studies were conducted using a traditional design organizational structure. There was a small joint engineering team co-located in Philadelphia to establish a conceptual helicopter configuration; however, subsystem preliminary design and trade studies were conducted by parallel design teams at Boeing in Philadelphia and Sikorsky in Stratford, Connecticut.

Frequent coordination meetings were held at both sites. There was a local commuter airline that flew between Bridgeport and Philadelphia several times a day, with flight times slightly over an hour, so it was very convenient to hold face-to-face meetings. In fact, Sikorsky team members usually got home earlier from Philadelphia than from Stratford. This was because the last flight from Philadelphia left at 5:45 pm, arriving in Bridgeport at around 7 pm. That was earlier than normal workdays during the intense days leading up to, during, and after the LHX proposal submittal.

A series of technology interchange meetings were conducted with both companies. It was understood that these were very important trades. Because the terminology was almost a foreign language, much time was spent learning the nuances of the various metrics. The process was very rigorous and thorough, and resulted in numerous design iterations. Each iteration was designated as a BTA and given a number. There were at least eight BTAs before the design was frozen.

Early on, having two diverse design teams trading concepts was a good thing, because it introduced new thinking and ideas that might not have otherwise been considered. In some instances there was intransigence due to company traditions and culture. It was often joked that midway on the New Jersey Turnpike between Philadelphia and Stratford, a rest stop sign declared: North of Here, Sikorsky Processes at Work; South of Here, Boeing Processes in Effect.

Sikorsky was exposed to Boeing's "lifting tail-rotor" concept in June 1985 when Boeing and Sikorsky first teamed. Like the Sikorsky UH-60 aircraft, the LHX utility configuration required a large center of gravity range. And like the UH-60, one option was to tilt the tail rotor to provide some aft lift to accommodate the aft cg range. Tilting the tail rotor also reduced the broadside radar return from the tail rotor, but it did not reduce the leading edge returns from the blades or the main rotor/tail rotor interaction noise if the rotor tilt was limited to the UH-60 tilt. Boeing's solution to these issues was to tilt the rotor to almost 90 deg and increase the chord of the tail rotor blades to provide enough real estate between the leading edge of the blade and the blade spar to accommodate radar cross-section treatment material. Putting the tail rotor near the same plane as the main rotor reduced the main rotor vortex interaction with the tail rotor. As speed increased, the main rotor vortex tended to go below the plane of the tail rotor, and that reduced the aircraft's noise signature.

Although the LO characteristics led Boeing toward the lifting tail-rotor concept, the LHX had a number of maneuver requirements that required the platform to tolerate operating at negative *g* to reduce exposure to ground threats. The issue with the lifting tail-rotor design was the loss of yaw control during and transition to a negative *g* maneuver.

A team from Sikorsky was moved to Philadelphia to evaluate the Boeing concepts. Figure 1.9 shows a Boeing concept for the LHX utility version. A six-person committee was commissioned at the companies to adjudicate the trades. This committee consisted of the chief of design, technology manager, and system engineering leader from each company, and was called the CO-6. Although most of the trades were amicably settled, some were bitterly contested.

The Boeing team then came to Stratford to evaluate the Sikorsky Fantail configuration. We worked together to establish which BTA the team would take forward. Once it was realized that the lifting tail rotor could not conduct the negative *g* maneuver efficiently, the concept was dropped. In October 1985, the Army released the LHX request for proposal (RFP), and the Boeing Sikorsky team proposed the Fantail configuration.

Fig. 1.9 Early LHX Boeing utility concept (Boeing Photo).

Back to the Design Process

In 1985, when Boeing Sikorsky received the contract from the Army for concept formulation and preliminary design, both scout/attack (SCAT) and utility configurations were required. Thirteen SCAT and 10 utility missions were defined by the Army. Additional LHX program goals now included:

- Significant combat capability and survivability improvements over existing fleet aircraft
- Control power for nap-of-the-earth (NOE) flight, air-to-air (agility), and power margin
- Dash speed of 180 kt
- 8000+500 lb primary mission gross weight (PMGW) for SCAT version
- Unit flyaway cost not to exceed $6 million for SCAT and $4 million for the utility aircraft (in constant 1984 dollars)
- Transportable in C-17, C-141B, and C-5A aircraft
- Crashworthy
- Low noise

A competitive development and production program was defined with the two teams, utilizing pilot production tooling. The full-scale development (FSD) program included industry participation in the requirements development process, with aircraft growth margins built-in.

Phase I included detailed design of hardware and software, brass-board demonstration of mission equipment package (MEP), and development of plans to achieve RAM/ILS, MANPRINT/training, pilot production, and production competition goals. Risk reduction was intended to enhance the ability to meet production and operating cost guarantees when the Phase II proposals for FSD were submitted. These were also to be firm-fixed-price proposals, with no government-funded facilitation.

An interesting addition to the LHX requirements was the Department of Defense's direction that both the air vehicle and the engines were to be designed in the metric system of units. This was fashionable at the time because only the United States, Burma, and Liberia did not use the metric system, relying solely on English measurements. Thus, the thinking was that by using the power and influence of the Department of Defense, this program would be the vanguard to lead the country to become compliant with what was thought to be a global imperative. Some studies were conducted, and it was deemed to be feasible with a nominal impact.

This led the designers into unfamiliar territory, with discussions as to where that rule was to apply—initially everywhere, eventually not quite.

The eventual LHX weight-empty requirement was usually stated as 7500 lb. At some point someone embarrassed the Army by pointing out that that was not metric. Then it was restated as 3402 kg. But in design discussions, it was almost always stated as 7500 lb.

Metrification was accomplished in a fairly straightforward manner with just a few minor problems, which were primarily cultural. Length and weight conversions were readily adapted, but it was difficult to develop "feels" for other quantities such as stress and density. In the English system these quantities are expressed as pounds per square foot and pounds per cubic inch, respectively. The metric equivalent units were pascals and kilograms per cubic meter. The natural sense of how big or small these units were took time to learn. For convenience, some analyses were conducted using English measurements and then converted to metric for formal documentation. It took a while for the designers to get comfortable with the new system. Some never did. Other minor conflicts arose around whether to use hard or soft conversions for certain materials and hardware. Although hardware such as bolts and screws could be procured in metric increments, this was not always the case for other off-the-shelf commodities; for example, aluminum sheet thicknesses were standardized in English units and when converted to metric were rounded to uneven numbers, so the English system was maintained for these conditions.

Some years later the requirement for the use of the metric system was rescinded, and the aircraft was redesigned in conventional units, at a substantial additional program cost. The engines were also to be metric, and it is believed that they still are.

In 1986 the companies presented their ARTI results to the Army. Boeing Sikorsky concluded that a single-pilot LHX SCAT was technically feasible, including the basic tasks of flight control, navigation, and communication, and could be flown effectively with acceptable workload. The primary issues were the inability of a single pilot to continue nap-of-the-earth flight while simultaneously engaging in off-axis self-defense or conducting reconnaissance digital communications.

Some of the other companies at the meeting concluded that the single-pilot approach was completely feasible. We thought their conclusion might have been politically motivated—telling the Army what it wanted to hear. Boeing Sikorsky was the first to present the bad news that two crewmembers were needed for the full mission. The Army seemed quite uncomfortable to hear that we disagreed with the other presenters, and we were challenged to explain our position.

In the end, the Army agreed with our analysis. The LHX would be a dual-pilot aircraft with single-pilot operational capability.

In March 1986, the Army completed another update of the LHX BTA. This included:

- A composite airframe
- An advanced technology rotor system
- High-efficiency engines
- A lightweight drive system
- Very high speed integrated circuits (VHSIC) mission computers
- Advanced displays
- Voice-interactive subsystems
- Integrated aircraft survivability equipment
- Digital avionics
- Fly-by-light/fly-by-wire flight controls
- Cockpit integration/automation

In July 1986, the competitive fly-off approach was eliminated because its costs exceeded the available funding. Instead, in September 1986, the Army released risk reduction contracts for the higher-than-medium-risk technologies that had been identified. Two contracts were signed with the teams for continuation of risk reduction work on mission computers, EOTADS, HMD, mission computer high speed processing for navigation equipment, night pilotage system (night and adverse weather combat missions), plus survivability equipment and aided target acquisition system.

The EOTADS was viewed as an evolution of the Apache and Army Helicopter Improvement Program (AHIP) systems. Also included were a second-generation infrared sensor, a low-light-level TV, and a laser rangefinder/designator with automatic target detection and recognition and improved range.

The Army's unusual requirement that the 7500-lb empty weight took precedence over meeting all the technical requirements created a further complexity in developing the design. We quickly determined that all the initial needs defined in the required operational capability (ROC) and the Comanche system configuration could not, in fact, be met within the weight and cost restrictions. Therefore, an optimization process was required to get the "best" solution.

This also meant that our design and the SuperTeam's design would be different, depending on the choices made as to which requirements to compromise. Thus, the Army evaluators would be judging two concepts that met different requirements. Just one more thing to make the LHX competition challenging.

The Boeing Sikorsky design was evolving. The weapons pod configuration was the original packaging concept for the weapons. The pods lasted until we teamed with Boeing, and the Boeing structural lead designer came up with the idea to use the center box beam as the primary airframe structure. This allowed the weapons to be mounted inside the external skin but outside the primary structure.

The V-tail was originally designed for radar signature reduction, but when we did the analysis the physical size got very large. The design was then changed to the canted fan and vertical tail and a high-mounted horizontal tail (Fig. 1.10).

Fig. 1.10 Near final configuration.

The analysis was continued to evaluate what additional tailoring of the requirements might be needed if additional weight saving was required in the future. Also provided was a prioritized list of attributes and design features that should be added to the configuration if at all possible.

In 1987, an Army LHX analysis of alternatives (AOA) was completed. This again included the tilt rotor that was still being advocated by some people within the Army. The AOA verified that the existing Army aircraft could not be modernized to be effective on the future battlefield. The advanced technology helicopter and the tilt rotor both demonstrated substantial improvements in survivability, supportability, and cost. It was the higher cost and technical feasibility of the tilt rotor that led the AOA to recommend the advanced technology helicopter.

The tilt rotor advocates were a strong group and still wanted it considered again, if not for the attack role then at least for the utility role. This led the Office of the Secretary of Defense (OSD) to stop the program and contract with the RAND Corporation and the Institute of Defense Analyses (IDA) to conduct independent studies to confirm the AOA results. These IDA/RAND studies delayed the program by about one year.

IDA/RAND Study: Helicopters vs Tilt Rotors

So, the subject was studied again to examine the effectiveness of a tilt-rotor helicopter vs a single-rotor helicopter. Boeing Sikorsky felt that the study team had a bias in favor of the tilt rotor. Sikorsky pilots and handling qualities personnel attended the study debrief to industry at the RAND headquarters in Santa Monica, California. We voiced our opinion of the study results, which we felt had a strong and misguided bias toward the tilt rotor.

The same IDA/RAND study group conducted a separate simulation event at the NASA Ames Research Center. We were allowed to participate in and question the study and simulation findings. The fact that the tilt rotor could trim at any desired pitch attitude, whereas a helicopter could not, they felt gave the tilt rotor a distinct air combat advantage for weapons delivery. It was a cornerstone capability that they centered their study on. We questioned the conclusions reached and the usefulness and practicality of that concept for the role. We also pointed out that the helicopter had several other advantages like being able to turn quicker using large sideslip, higher load factors, and quickness.

RAND also conducted a shadow cost and operational effectiveness analysis (COEA) in 1990 because the tilt-rotor proponents were still criticizing the helicopter decision. It eventually concluded again that the tilt rotor was not appropriate for LHX. The tilt-rotor configuration never competed again for the LHX for a number of practical reasons, most related to size and required gross weight as well as concerns with maneuverability.

REVISED REQUIREMENTS (AGAIN!)

In June 1988, the Army released the next revisions to the requirements for the LHX. This was for a SCAT configuration only and required that it have an empty weight of 7500 lb or less and a maximum average production cost of $7.5 million per aircraft or less.

This was an odd requirement. Some simplistic analysis had concluded that helicopters cost $1000 per lb, and the way to hold down cost was to hold down weight. This made little sense because much money is spent in helicopter programs specifically to reduce weight with advanced materials and higher-risk design approaches. These drive up production costs rather than reduce them. A weight-is-cost rationale may make sense when the design keeps within conventional practices; however, it does not work when extraordinary measures and new technologies are required to reduce aircraft weight.

At this point, the Army was ready to contract with the two competing teams in a Demonstration/Validation program to further develop the technology and reduce the risk of the program.

Chapter 2

THE DEMONSTRATION/VALIDATION PROGRAM (1988 TO 1991)

ARTHUR W. LINDEN

The Boeing Sikorsky team was awarded the initial Demonstration/Validation (Dem/Val) contract in September 1988. Now under contract, we continued to conduct trade studies and experimental testing to further the design. The Army's strict requirements for aircraft empty weight became the controlling factor in establishing what systems made up the aircraft and what level of the requirements the aircraft could meet. The criterion was operational effectiveness. Every attribute and every subsystem had to earn its way onto the aircraft.

The proposed design now included a bearingless main rotor (BMR) system, a fly-by-wire (FBW) flight control system, advanced control laws, a new glass cockpit with sidearm controllers, a fan-in-fin antitorque system, a new all-composite airframe with a low radar signature, a new engine exhaust system to reduce infrared (IR) signature, a split-torque main gearbox, and an advanced mission equipment package (MEP) and avionics and helmet-mounted displays. On paper it looked like a solid design, but could we make it all come together and work as advertised?

The design drawings were prepared using the new 3-D CATIA system. After all the drawings were completed, a 3-D model check was performed to identify any interferences. Packaging was extremely dense. Catching interferences early in the design phase saved a lot of time and expense before parts and tools were made. However, as important as interface management was, vulnerability to changes was still a significant problem. It was not practical to design everything simultaneously, so inevitably as details were fleshed out, the interfaces were sometimes affected, which sometimes necessitated design changes. This was an ongoing process, particularly with subcontracted components, such as the environmental control system (ECS) and MEP, which lagged the basic airframe design.

Operations and sensitivity analyses were used to assess the influence of the design options on operational effectiveness in terms of mission capability, survivability, and availability. This structured process ensured that all the attributes were evaluated, all the attributes impacted were assessed by the same matrix, and all the parameters were in the correct categories for the evaluation.

The heavy emphasis on supportability/availability was found to decrease the Comanche's specific operational and support costs by 40%, resulting in a large reduction in maintenance manhours compared to the then-current fleet.

The overall result of these studies was a five to one improvement in combat effectiveness when compared to the existing light helicopter fleet.

Over the early years of the program and the three years of the Dem/Val contract, the Boeing Sikorsky team accomplished the following:

- 80 documented and Army-reviewed trade studies
- Over 67 risks-reducing demonstrations, including a total MEP system integration demonstration and a complete combat missions simulation
- Over 10,000 hours of wind tunnel occupancy including testing of:
  - Airframe/drag
  - Rotor–airframe interaction
  - Fantail/empennage
  - Engine installation and exhaust
  - Aeromechanical stability
- Full-scale wind tunnel testing of the bearingless main rotor in excess of 200 kt and 19,000 lb of lift
- Three surrogate aircraft used to develop and demonstrate major systems:
  - Target acquisition system (TAS) in CH-47
  - Fantail on S-76
  - Cockpit, MEP, and flight control laws on Shadow
- 4500 hours of radar cross-section testing
- 3800 hours of second-generation forward-looking infrared (FLIR) test

Requirements tailoring would be an important discriminator in the competition with the SuperTeam. Our solution to close on the weight included reducing the crashworthiness level to 38 fps from the 42 fps requirement, providing the frag barrier between the two cockpits only when intel indicated that it was needed based on the threat, radar provisions only for potential growth in the future, a radar cross-section (RCS) reduction kit for when the operational scenario required it, a 12.7/14.5-mm self-sealing fuel system vs. 14.5-mm, reduced area armor protection (no leg armor), no active countermeasures, image intensification tubes on the helmet vs in the Electro Optical Sensor System (EOSS) for backup pilotage, and a max continuous gearbox rating vs. a max power rating.

Sikorsky had two major flight test efforts underway with the Shadow and Fantail demonstrator aircraft.

Fig. 2.1 The experimental S-76 Shadow test aircraft.

The Shadow had a single-pilot cockpit grafted onto the front end of a commercial S-76 airframe (see Fig. 2.1). The research pilot in the front seat could be isolated from the main cockpit, which accommodated a safety crew. The objectives of the program were to risk reduce some of the night vision system design elements and to help resolve some single-pilot questions. Additional objectives were to risk reduce the FBW cockpit elements and some of the basic low-speed, landing, and takeoff concerns regarding control switching on ground contact. Shadow was a very critical risk reduction tool for Comanche. The Shadow program is described in detail in Chapter 6, Handling Qualities.

Early on, Boeing Sikorsky had opted to use a fan-in-fin antitorque system (see Fig. 2.2). Sikorsky had some prototype experience with a fan on the larger S-67 helicopter several years earlier but decided it was necessary to conduct a risk reduction prototype program to get some LHX-specific performance,

Fig. 2.2 S-76B fantail demo aircraft.

Fig. 2.3 A "family" photo: the S-76B fantail, the LHX mockup, and the S-76B Shadow.

acoustics, and handling data, and to get more experience with the build and integration concerns. The test vehicle chosen was again an S-76B. This program is described in detail in Chapter 12, The Fantail. Figure 2.3 Shows the Fantail demonstrator, the Comanche mockup and the Shadow together for a "family" photo.

There was a short-lived effort at this time to get European content into the program, the idea being to expand the aircraft market and get additional funding. This included a few visits to Europe and a number of discussions with their companies and militaries, but nothing came of this effort.

A Commercial LHX?

Also at this time there was a preliminary design study at Sikorsky for a commercial LHX, combining the advanced LHX technologies with the S-76 (see Fig. 2.4).

The takeoff weight would grow from the S-76B's 11,100 lb to 13,100 lb, and the useful load for a North Sea mission would grow from 3095 lb to 5520 lb.

Advantages claimed were:

- U.S. government commonality
- Smaller overall size
- Lower vibration with five blades
- Safer operation with the Fantail
- Greater reliability and reduced maintenance due to advanced technology

There was one article in the press on this concept in *Helicopter World*, the October–December 1990 issue [1], but this approach did not go beyond the preliminary design phase.

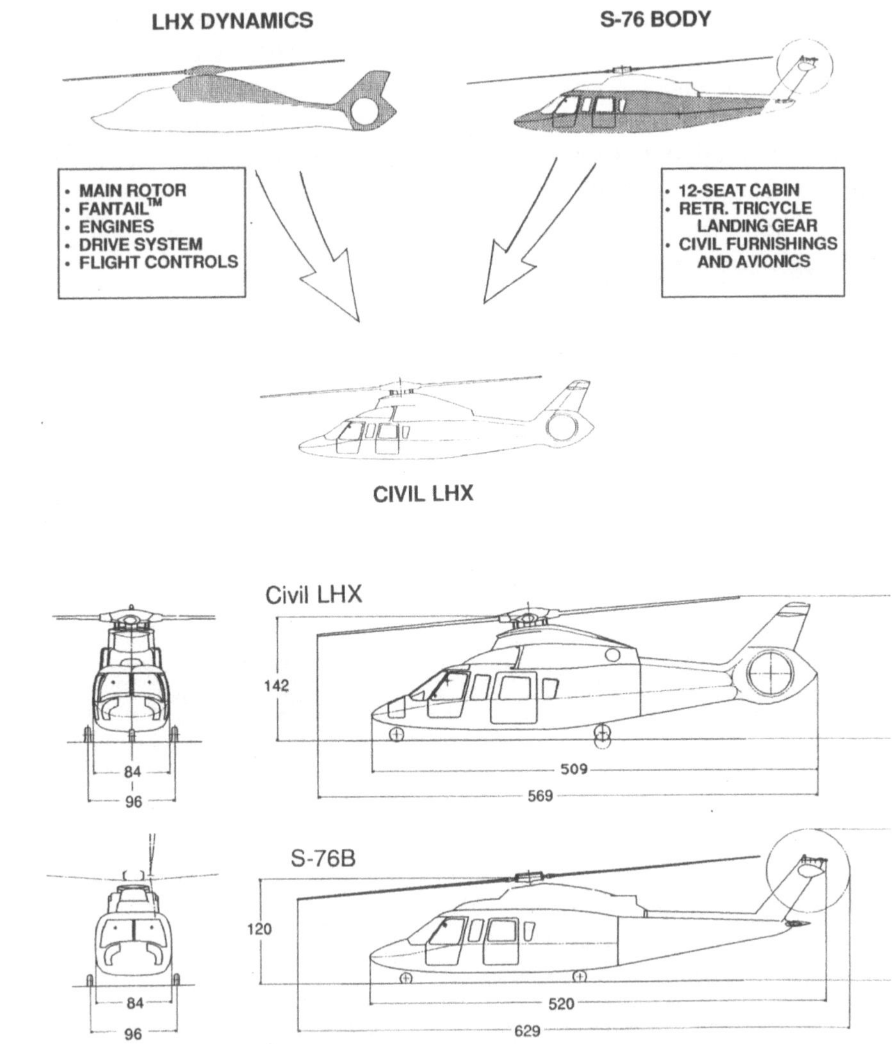

Fig. 2.4 A concept for a civil helicopter based on LHX dynamic components.

BACK TO THE PROPOSAL TO THE ARMY

The Boeing Sikorsky joint cost proposal was a huge effort because costs had to be broken down into a very detailed work breakdown structure (WBS) and address all of the thousands of discrete requirements contained in the government's specification. Aggressive pricing was necessary because this was a very competitive and important program. Estimates were compared to historical YUH-60A Utility Tactical Transport Aircraft System (UTTAS) and V-22 prototype data. Not surprisingly, the initial estimates were overly conservative and significantly higher than pricing targets.

To whittle the estimates down, a series of grueling cost reviews were conducted in Philadelphia, chaired by the Sikorsky and Boeing engineering vice presidents. In the end, a reasonable cost proposal estimate was achieved.

On 31 August 1990, the proposal was submitted to the Army. A full trailer truck was required to deliver all the documentation. The cover letter is shown in Fig. 2.5. The proposed aircraft is shown in Fig. 2.6.

BOEING SIKORSKY
LIGHT HELICOPTER FIRST TEAM

31 August 1990

Major General Donald R. Williamson
Commanding General
U.S. Army Aviation Systems Command
4300 Goodfellow Boulevard
St. Louis, MO 63120-1798

Dear General Williamson:

On behalf of the Boeing Sikorsky First Team, we are pleased to provide this offer in response to your Light Helicopter (LH) Full Scale Development (FSD) Request for Proposal.

Working in concert with Army combat and materiel developers, as well as field users, the Boeing Sikorsky First Team has generated an optimum, operationally effective and suitable LH design solution. Our approach to LH balances significant leap-ahead warfighting capability with affordable acquisition and ownership costs at the lowest possible risk to the Army. Integral to our optimization process have been the First Team Product Development Teams who, in addition to utilizing traditional criteria, have made full use of Demonstration/Validation test results and trade study conclusions as well as lessons learned from fielded Army systems.

Our offer complies with your requirements, employs minimal tailoring, and underscores our fundamental commitment to cost, technical, and schedule credibility. LH, as proposed by the Boeing Sikorsky First team, is realistically achievable on a timely basis within our proposed FSD pricing arrangement. In addition to offering a compliant proposal we have, in our cost volume, identified several deferrals that could reduce FSD cost.

We believe that linking profitability to meeting Army requirements is the best way to demonstrate contractor commitment. Accordingly, we have proposed subjective award fees for FSD based on the $7.5 million design-to-cost objective, the 3402 kg (7500 lb) weight goal, the $500 per flight-hour support cost target, achievement of key specification requirements, and schedule performance. Similarly, we have proposed an incentive fee appropriate to FSD, which underscores our commitment to cost realism and cost performance. Both the incentive provisions and award fee elements are shared by the entire First Team. This fee arrangement illustrates the First Team concept that each member's ultimate success is dependent on a combination of individual effort and team performance as a whole.

The First Team offer includes a firm commitment to link profitability throughout the entire production phase to meeting DTC, O&S, and weight objectives. Additionally, the First Team has effectively blended make/buy, co-production, and competition elements into a total acquisition plan that will provide the Army the best value over the program life cycle.

Through the commitments outlined above, we pledge total Boeing Company and United Technologies Corporation support to deliver a Full Scale Development program that ensures producing and fielding the LH on time and with leap-ahead warfighting capabilities.

Sincerely,

F. Shrontz
Chairman and Chief Executive Officer
The Boeing Company

R. Daniell
Chairman and Chief Executive Officer
United Technologies Corporation

BOEING SIKORSKY LH PROGRAM OFFICE
Scott Plaza II, Suite 635
Philadelphia, PA 19113

Fig. 2.5 Cover letter for the 1990 proposal.

Fig. 2.6 The proposed LHX aircraft.

The following data describe the LH as it was proposed against the Army requirements:

	Requirement	First Team Design
Dash speed	≥170 kt	177 kt
Vertical rate of climb	≥500 ft/min	1225 ft/min
Load factor	+3.5 g	+3.5 g; −1.0 g
Hover turn to target	5 s	4.6 s
Masking	≤2 s	1.6 s
Constant altitude 90-deg turn	≤6 s	5.5 s
0 to 85 kt accelerating turn	≤12 s	11.7 s
Primary mission gross weight		10,088 lb
Alternate maximum gross weight		12,813 lb
Empty weight	7500 lb	7499 lb

The table shown in Fig. 2.7, from a Boeing Sikorsky press release, presents more detail. (There are some minor differences compared to the previous table.)

GENERAL ARRANGEMENT AND DETAILED DRAWINGS

Figures 2.8 through 2.11 show detailed drawings of the proposed design. Ad Fig. 2.12 describes the various systems.

Proposal submittal was followed by a protracted period of responding to errors, omissions, and clarifications (EOCs). The team strove to make sure that we were fully compliant with all of the Army's requirements and that the contract price was competitive.

Boeing Sikorsky RAH-66 Comanche

Engines	
Manufacturer	LHTEC (Allison and Garrett)
Model	Two T800-LHT-800
Max at 4,000 ft, 95° F, kW (shp)	690 (925)
Transmission	
Max kW (shp)	1,532 (2,054)
Bearingless main rotor system	
Blades	5
Construction	composite
Diameter, m (ft)	11.90 (39.04)
Disc area, m² (ft²)	111.21 (1,197.04)
FANTAIL anti-torque rotor system	
Blades	8
Length, cm (in)	11.43 (4.50)
Construction	composite
Chord, cm (in)	17.0 (6.69)
Diameter, m (ft)	1.37 (4.50)
Performance [at 4,000 ft, 95° F]	
Dash speed, km/h (kts)	> 328 (> 177)
Vert. rate of climb, m/m (fpm)	360 (1,182)
Masking	1.6 sec
180° hover turn to target	4.7 sec
Snap turn to target at 80 kts	3.0 sec
Load factor	+3.5g to -1.0g
Range, self-deploy, km (nm)	2,335 (1,260)
Accommodation	
Cockpit crew seats	2
Q9 *Dimensions, external*	
Length, fuselage, m (ft)	13.22 (43.37)
Length, rotors turn, m (ft)	14.28 (46.85))
Width, landing gear, m (ft)	2.31 (7.58)
Width, horizontal stabilizer, m (ft)	2.82 (9.25)
Width, fuselage, m (ft)	2.29 (7.51)
Height, top of horiz. stabilizer, m (ft)	3.39 (11.12)

Weights	
Self-deploy, kg (lbs)	7,790 (17,174)
Primary mission gross, kg (lbs)	4,587 (10,112)
Empty, kg (lbs)	3,402 (7,500)
Useful load, kg (lbs)	1,185 (2,612)
Fuel capacity	
Internal, liters (gals)	984 (260)
Self-deploy, liters (gals)	4,542 (1,200)
Missile armament (reconnaissance mission)	
Internal	4 AGM-114 Hellfire
	2 AIM-92 Stinger
20-mm ammunition	500 rounds
Missile armament (maximum)	
Internal and external	14 AGM-114 Hellfire
	or 18 AIM-92 Stinger
	or 62 70 mm (2.75 in) rockets
Turreted 20-mm Gatling gun	
Basic ammunition load	500 rounds
Rates of fire	750 or 1,500 rounds/min
Targeting system	
2nd generation focal-plane-array FLIR	
Low-light-level TV	
Laser range-finder/designator	
Aided target detection/classification	
Night-vision/adverse-weather pilotage system	
2nd generation focal-plane-array FLIR	
Image intensifiers	
Wide-field-of-view helmet-mounted display	
Low observable characteristics	
Reduced radar cross-section	
Integrated infrared suppression	
Low acoustic/visual signatures	
Missions	
Armed reconnaissance	
Light attack	
Air combat	

Fig. 2.7 Comanche Technical Data for the 1990 Proposal.

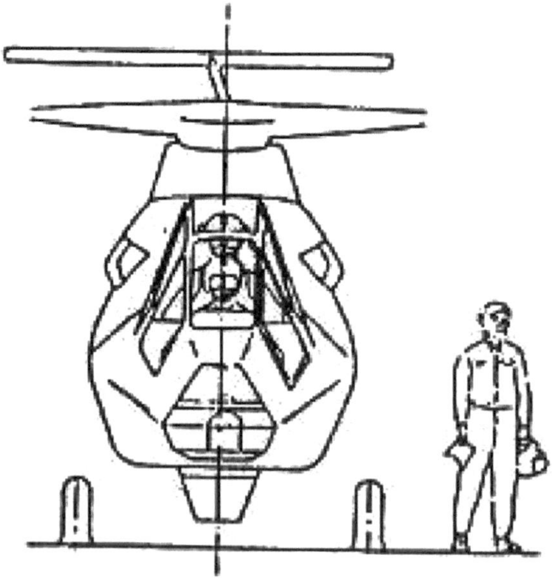

Fig. 2.8

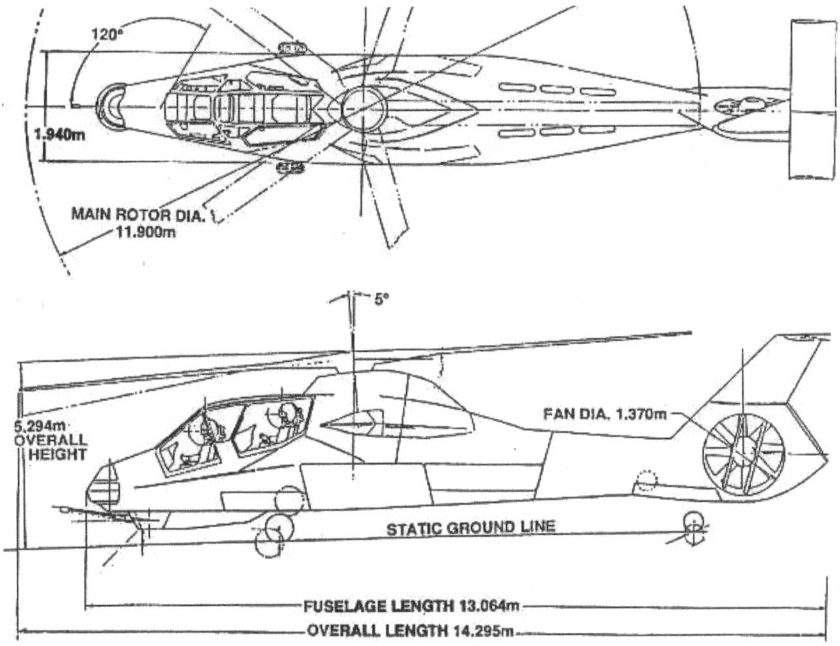

Fig. 2.9 Comanche Three-View Drawing.

Then silence ensued as the source selection board deliberated.

The next program event was very intensive detailed negotiations with the Army at an old military facility in Granite City, Illinois, near its headquarters. These negotiations went on for several months, during which each statement of work line item and cost estimate was examined and negotiated by the Army's technical subject matter experts. The result was a further trimming of any excess budgets. Everyone eventually sighed with relief that the task was over, and though the budgets were challenging, they could be achieved.

When we were negotiating the contract, our view was that the SuperTeam's NOTAR was the Army's preferred antitorque solution. We didn't think we were in a good position. Boeing and Sikorsky also were viewed as utility helicopter companies, never having designed or built an armed reconnaissance helicopter. There were many reasons in 1991 not to select the Boeing Sikorsky team.

During contract negotiations, however, it became clear that the SuperTeam's proposed aircraft didn't meet the 7500-lb constraint. We learned later that they were told to modify their proposal to meet the empty weight requirement, and they reduced their EOSS size and capability—not good. They also had supportability and air transportability problems—their aircraft had to be put on its side to fit in the C-130.

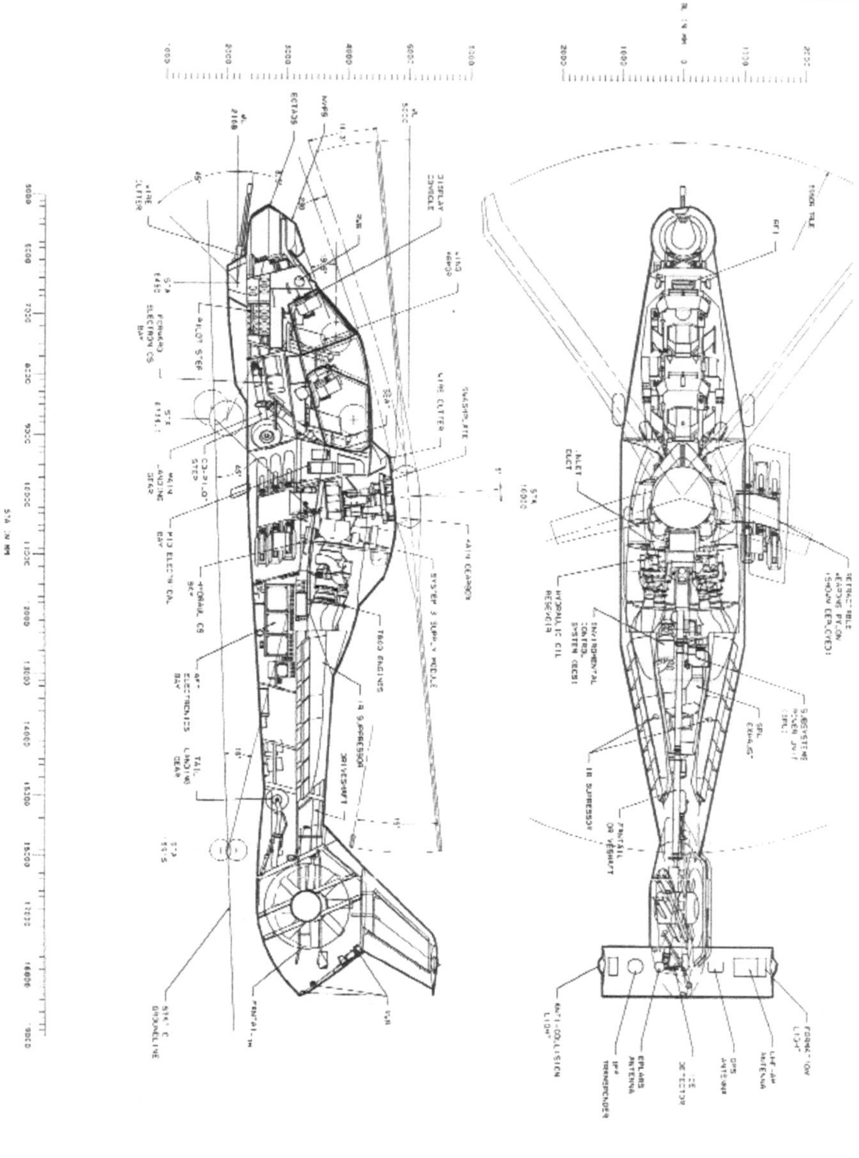

Fig. 2.10 Comanche Inboard Profile Drawing. part 1.

Naming The Aircraft The "Comanche"

As is its custom, in April 1991 the Army named the program after a Native American tribe, in this case the Comanche Nation of Oklahoma. The Comanche Nation enthusiastically supported this (Fig. 2.13). Its chairman, Wallace Coffey, spoke at the Comanche rollout ceremony. The nation also

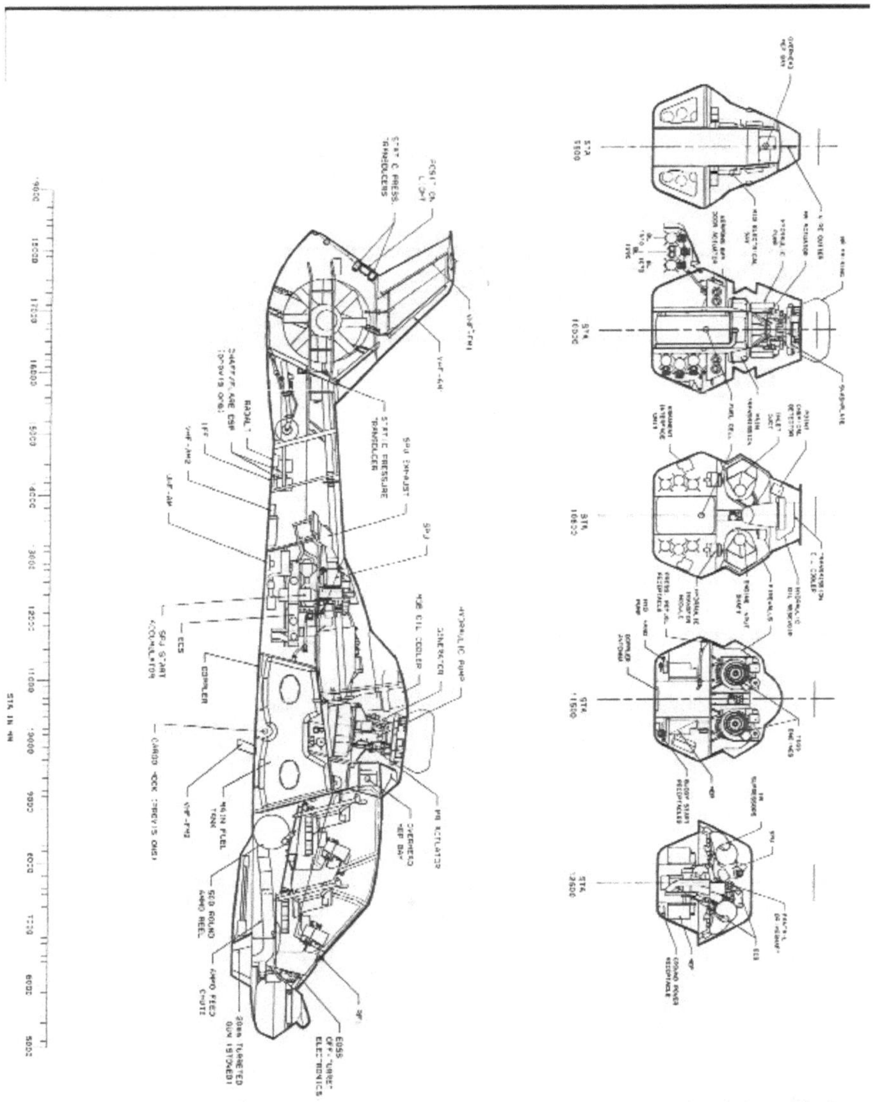

Fig. 2.11 Comanche Inboard Profile Drawing, part 2.

extended an award that it used to honor outstanding individuals of the tribe to those working on Comanche. This was the Pu-Ku-Tsi award, named after members of the tribe who were outstanding in their bravery and fought to the death for their cause. The first to receive the Pu-Ku-Tsi award was Army Chief of Staff General Gordon L. Sullivan, who received it at the rollout ceremony.

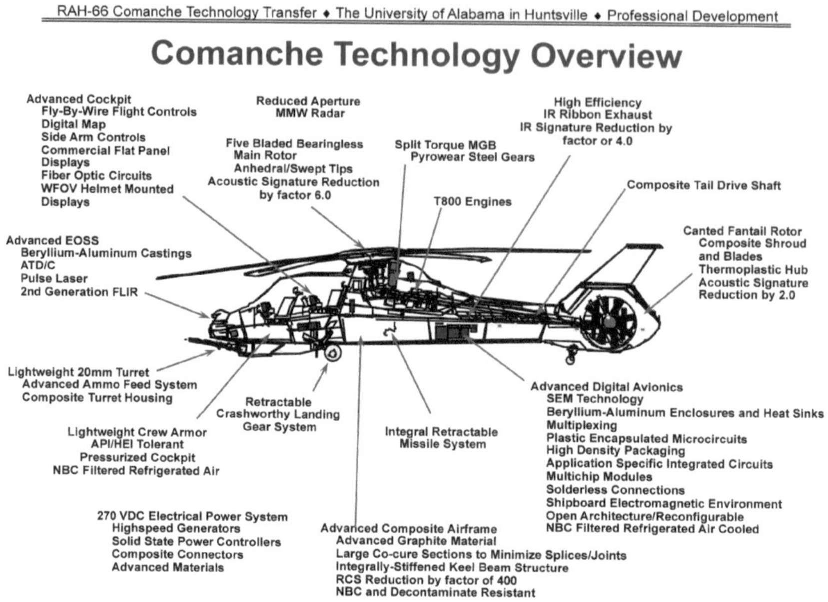

Fig. 2.12 The various Comanche systems.

Fig. 2.13 The Comanche Nation fully supported the program.

CONTRACT AWARD

Finally, on Friday 8 April 1991, there was to be an announcement after the stock market closed. Sikorsky employees were told to gather in the factory high bay area to await the announcement. The mood was very tense. Several hundred people milled about waiting for someone to come out and break the news. As time went on, finally the beginning of the president's entourage started to make its way to the platform where company president Gene Buckley was to speak. Everyone was on pins and needles. Gene climbed up to the stage and said that he had been informed by Senator Joe Lieberman that "We have won LH!" The jubilant crowd went crazy. Although he amplified the victory announcement with some details, he was drowned out by the roar of the crowd.

REFERENCE

[1] LH technologies studied for civil '76, *Helicopter World*, October–December 1990, p. 44.

Chapter 3

The Mission Equipment Package

Frank Schrader with contributions from, Raj Kaushik, Rus Stiles and Bill Harper

The mission equipment package (MEP) was incorporated into a wide array of leading-edge technology components. The purpose of the MEP was to support the pilot (potentially single) in the complex scout/light attack mission. Its role commenced with the initial planning phases for the mission, it was key to the execution and completion of the mission, and it provided data for debriefs and repairs.

The MEP managed navigation, communication, night pilotage, target acquisition, fire control, and aircraft systems control. It provided control of all non-flight-critical subsystems and provided the displays for all the information needed by the pilots. System housekeeping and failure detection added a dimension of automation not seen in any other tactical helicopter and was a major contributor to pilot workload reduction.

The cockpit needed to provide an information-rich environment and had to be easy to operate. That necessitated automated decision making and simple command structures. Thus, the MEP requirements evolved into a software-intensive "assistant pilot" requirement and the need for impressive threat detection, threat protection, communication, and control/display requirements. This software complexity remained, even with the second cockpit.

The Comanche MEP was a significant advancement over the electronic systems installed on the Black Hawk, Apache, and Kiowa Warrior systems that the Army had fielded in the 1970s. No other helicopter or fixed wing aircraft at the time had the level of technology planned for Comanche, although the F-22 and B-1 were considering some of the electronic elements.

In the 1970s, most electronics had started to use transistors. Sikorsky flew its first digital navigation computer in the early 1970s. As predicted by Boyle's Law, computer speeds were increasing along a log scale, doubling every 3 years, and transistors were being replaced by smaller and smaller integrated circuits.

The weights engineers optimistically drew the conclusion that weights would drop significantly with each coming processor improvement, but greed prevailed. The extra processing speeds and memory capacities did not result in a smaller package to do the same old job; instead, the capabilities of the equipment multiplied to accomplish automation, resulting in approximately the same package size, weight, and heat dissipation but doing a significantly more exotic job.

Comanche had an integrated avionics architecture as opposed to a federated architecture. In a federated system, each function was implemented by distributed hardware-based subsystems consisting of dedicated functions and physical communication channels or networks. The federated architecture featured clear boundaries of responsibility and limited sharing of hardware and communications resources for various functions. Faults associated with a specific function could easily be identified and located to the assigned hardware simply through the loss of the function.

An integrated architecture reduces the hardware required and wiring points by sharing and multitasking assets. This increases reliability, reduces weight and cost, and simplifies fault tolerance by providing the option to reassign functions to different hardware assets and/or networks; however, it introduces the potential of fault propagation, which increases the complexity of diagnostics required.

The MEP provided highly automated, gracefully degrading, redundant functionality for communications, navigation, controls and displays, weapons, aircraft survivability equipment, and other aircraft systems. It also provided a data interface to the flight control system and the on-board integrated fault detection/fault isolation function.

At the heart of the MEP were two mission computer clusters (MCCs). Each MCC was capable of providing full MEP processing functionality and employed multiple and physically separate modules, which enhanced fault tolerance.

Additionally, the MEP was responsible for monitoring the health of aircraft systems such as hydraulics, electrical, lubrication, and fuel. This reduced the administrative burden for the pilot and let him then focus on mission accomplishment. There was some concern that this requirement forced the MEP to be flight critical, but it was decided that because the flight control computers—which were flight critical—controlled their system reconfiguration, failure of the MEP only affected pilot workload, making the MEP mission critical. The capability for the MEP to significantly reduce pilot workload was unique and required extensive integration of system sensors (pressures, temperatures, flows, etc.). Combined with advanced software algorithms, the MEP effectively eliminated the need to display traditional system health gauges—a capability not yet achieved in other aircraft. Indeed, the MEP was far ahead of its time.

The scout/attack mission required certain basic, but specialized, MEP capabilities:

- Threat avoidance through intelligent mission planning and in-flight threat detection
- Continuous receipt and update of target/threat intelligence from:
  - Passive long-range target acquisition system (TAS) and aircraft survivability equipment (ASE) sensors operating in smart automated search modes
  - Army Tactical Command and Control System (ATCCS), enhancing battle execution
- Secure digital and voice communications in a nap-of-the-earth flight regime
- Accurate, covert (nonradiating) navigation for localization and reporting of threats
- Pilot night vision capability to allow covert nap-of-the-earth flight even at night and in bad weather
- Integrated display of the horizontal tactical situation overlaid on maps for quick pilot consumption
- Integrated heads-up displays of the real world and tactically significant elements to allow aggressive piloting (minimizing the pilot's need to look inside the cockpit)
- Timely assimilation and display of all information to minimize pilot workload
- Easy access to system capabilities to minimize control actions that would take away from the pilot's capability to fly aggressively
- Automated weapons selection and aiming

This need for centralized and integrated information led to the use of a central mission computer to sort and display relevant information as needed in a dynamic environment.

In the same timeframe as LHX, the U.S. Air Force Systems Lab at Wright-Patterson Air Force Base was developing an architectural concept called PAVE PILLAR. It mirrored the LHX architecture in many ways. The goal was to use common module designs across many functions and use software to customize the functions as needed. The Army and Navy adopted the outputs from PAVE PILLAR as well. The primary program outputs were a standardized MIL-STD-1750A computer module; the integrated communications, navigation, and identification avionics (ICNIA) reprogrammable very-high-speed integrated circuit (VHSIC) modules and the integrated electronic warfare system (INEWS) survivability suite modules, all designed to a standard module packaging envelope (the SEM-E module format).

The LHX/Comanche systems and sensors were connected to the central mission computer via high-speed data buses (HSDB) and displayed video was transmitted over a high-speed video data distribution network (VDDN). Both

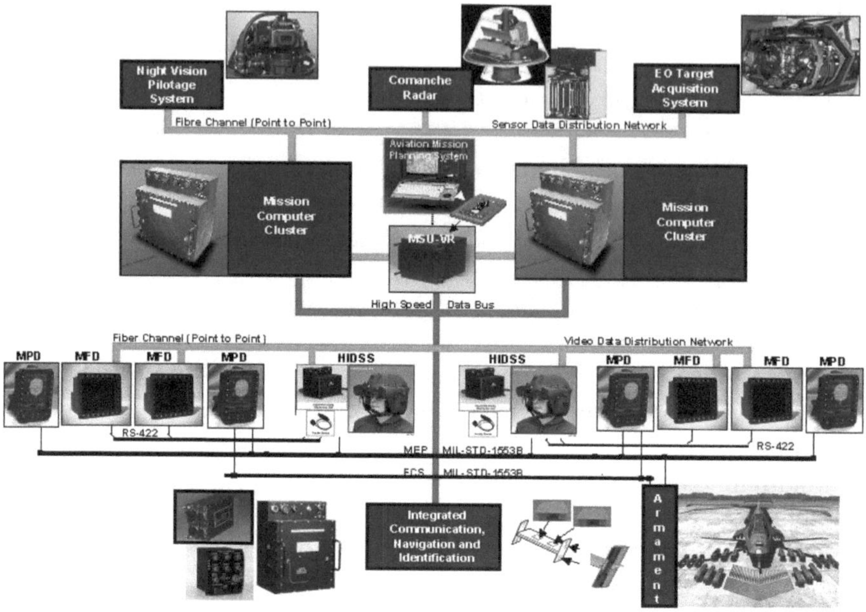

Fig. 3.1 Comanche's MEP architecture.

of these buses used fiber optic lines to attain the digital data rates required. Simpler, low data rate systems were connected using the MIL-STD-1553 wire transmission lines. The helmet integrated display and sighting system (HIDSS), through its enhanced display electronics unit, controlled all the analog cockpit switches and sensors and sent their state data and diagnostics onto the 1553 bus. Similarly, the dual air vehicle interface control system (AVICS) controlled the sensors, switches, and simple analog systems throughout the rest of the aircraft and sent their state data and diagnostics onto the 1553 bus.

Figure 3.1 shows what the MEP architecture looked like.

The cockpit was designed to put all controls and displays within easy reach of the pilot, but the requirement to bring hands and eyes inside was kept to a minimum.

The Mission Equipment Package Components

As the program entered the EMD phase in 2000, the MEP consisted of the following components:

- Dual mission computers
- A mass storage system for mission data entry, digital map storage, mission computer programming, and a mission log, which also logged maintenance events

- Dual 10 × 12 head-down LCD color multifunction displays (MFDs) in each cockpit
- Dual 6 × 8 LCD color multipurpose displays (MPDs) in each cockpit
- A target acquisition and designation system mounted on the nose
- A night vision pilotage system (NVPS) mounted on the nose
- A HIDSS in each cockpit, with provisions for incorporating night vision goggle imagery
- An ICNIA suite
- An aircraft survivability suite consisting of radar and laser warning systems and a chaff dispenser (It turned out that the INEWS originally proposed was primarily jammers, which Comanche resisted.)
- A terrain mapping and display system
- Dual air vehicle interface and control system (AVICS) to provide a data bus collector for simple aircraft sensors
- A rotary air data system to provide omnidirectional airspeed and altitude information
- Provisions for the LONGBOW radar
- Interfaces with the armament system (guns, rockets, and missiles), the diagnostics system, and the 270-V power system.

THE JOINT INTEGRATED AVIONICS WORKING GROUP (JIAWG)

The JIAWG was started in 1985 as an attempt at commonality between Air Force and Navy advanced fighter programs and had been meeting for two years before the light helicopter (LH) joined the working group. The objective was to lower costs by using common modules or equipment across all three platforms, and by 1987 the JIAWG was talking about software commonality as well.

The JIAWG did establish a common avionics baseline that focused on common modules. These modules flowed to the mission processor, ICNIA, and INEWS. All used digital processing, Ada software documentation, and reprogrammable functions. The ICNIA and INEWS programs were "developing technology" that was not program specific, but based on the common modules.

When the LH joined JIAWG, we reviewed all our planned equipment as candidates for commonality.

At the function level, we all had mission computers, comm/nav/identification equipment, survivability systems, and controls and displays. We also wanted elements of the PAVE PILLAR architecture, including fault tolerance, built-in test, modular construction, and reconfiguration capability. And we were designing around the SEM-E modules favored by PAVE PILLAR, so commonality looked promising.

But the devil is in the details.

For instance, we all wanted inertial systems linked to the global positioning system (GPS); however, LHX wanted three lightweight systems to support flight controls and had limited (compared to the other services) navigation accuracy requirements in the short mission, nap-of-the-earth environment. Most of the navigation was well within GPS limits, and we integrated the GPS chip within the inertials for faster, independent alignment. The other services used more expensive, heavier, but more accurate stand-alone inertial platforms and would not accept any compromise to their derived mission requirements. So there was no commonality.

We all accepted ICNIA as satisfying the basic programmable radios and interrogation equipment functions, even if we didn't need a GPS module. Therefore, all three services planned for ICNIA commonality.

With regard to aircraft survivability equipment, the INEWS provided radar and laser warning and jamming functions across the threat spectrum and looked good for commonality.

All three services had different control and display requirements, based mainly on differing space constraints for the aircraft displays, so we had no commonality there.

A majority of the software was unique to the aircraft or unique to tactical missions that were different for each service. We agreed that basic navigation functions might use the same software. It was left to each service to reuse software where possible.

As we looked at INEWS, we realized that three quarters of the system (and weight) was dedicated to jammers. We felt our threat detection and signature reduction systems allowed us to evade the threats or eliminate them when we were outside their threat engagement envelope, so we didn't want to carry the extra space around in a rack. At that juncture, we deferred further efforts to integrate INEWS. Our initial plans were to simply install existing radar and laser warning systems.

That left ICNIA as the one "programmable for any service" system for commonality.

But ICNIA was originally developed for advanced tactical aircraft that had liquid cooling on the aircraft for all avionics. Also, fixed wing aircraft had a higher frequency vibration environment than helicopters, so environmental qualifications had to be reconciled. Therefore, we started to look at our rack environment for ICNIA modules.

As it turned out, we all had an air-conditioned cooling system, and the modules were designed for edge cooling between the module and rack, so the thermal profile remained the same as long as we could supply the necessary heat exchange. So we almost had SEM-E commonality; however, the Air Force had selected a different pin and socket connector at the end of its modules than we had selected. So module commonality could be achieved if we agreed on a single connector approach. But after months of technical discussions, the Air Force believed our connector would fail "within hours" of use in a fixed wing environment, and we could not accept their solution for helicopters.

Finally, an Air Force general declared we had done enough and declared "victory" (meaning, stop the effort). We kept our connector and the Air Force/ Navy kept theirs.

So the JIAWG quietly went away for Comanche.

THE MEP SYSTEMS

MISSION COMPUTER

Boeing retained overall MEP lead and responsibility for the mission computer hardware and software. The mission computer retained the parallel processing architecture favored by the team, and all modules were designed to fit a military-standardized SEM-E module size. The SEM-E size and host racks were used for all electronics with more than two modules.

The electronics within the mission computer included 32-bit processing cards, an 800-Mbps sensor bus, a 50-Mbps data bus, and dual 1553 data buses. The module types included data processing, signal processing, graphics processing, video generation, bulk memory, data bus interface, data network switch, and power supplies. The modules were projected to have a 95% level of automatic fault detection and a 98% level of automatic fault isolation to the line replaceable unit (LRU) level. The electronic racks were installed in the rear of the fuselage at waist-high level where they were easy to access for maintenance. Compartments on both sides of the aircraft provided separation for improved survivability. The racks and SEM-E modules were exposed in the nuclear, biological, chemical (NBC) environment, and modules had to be removed by maintainers in NBC gear. Only the cooling air that went through channels in the top and bottom of the racks was NBC protected.

The design included 30% spare processor capability and a 50% bus throughput margin in reserve. Fourteen open rack positions were included in the Dem/Val program stage for new modules and capabilities, such as the LONGBOW preplanned product improvement.

As a measure of how the Comanche processing compared to other helicopters at the time:

- The Apache computer operated at 160 kilo-operations per second. Comanche's parallel processing gave us 160 Mega-operations, a thousand times faster.
- The combined video signal and data processing rates of the mission computer were equal to almost 500 SUN workstations at the time.

The mission computer clusters consisted of parallel processing boards based in the Intel 380i chip set, which was the most modern chipset available during Dem/Val. It was an unpackaged Intel Pentium processor die, operating at 233 Mhz. It is believed that these boards were originally supplied by Westinghouse.

The high-speed data bus (HSDB) linked between the processor modules and video data distribution network (VDDN) modules to send data to "dumb" head-down displays in the cockpit. Fiber optic transmission lines became the required media for high-speed data transmissions. Two similar high-speed modules were developed: the HSDB was optimized for data; the sensor data distribution network (SDDN) was optimized for the targeting and designation system (TADS)/NVPS and other high-rate digital sensor feeds, such as LONGBOW. Each used a 1-GB, fiber channel, point-to-point fiber optic link.

As a cryptographic security measure, the entire computer system was "trusted." To meet this requirement, the development team could not use the available Intel operating systems. It was a requirement that the trusted operating system be developed (by Boeing Wichita) in a secure facility. All program code for the mission computers was also developed in semisecure environments, using the then-standard Ada programming language, which facilitated clear code reviews and eliminated the possibilities of Trojan horses or other software hidden threats.

The trusted operating system, and the software partitioning requirements that allowed multiple parallel programs to be running on the multiple separate 380i computer boards (see Fig. 3.2), led to the development of the operating system being a very complex task. The operating system capabilities were released in multiple stages to match the needs of the early limited software used on the prototype aircraft. The full features of the operating system were never flown, due to program cancellation.

Fig. 3.2 Modular electronics for ease of maintenance and growth.

The mission computers also included fiber optic HSDB boards, fiber optic VDDN boards, terrain map conversion boards, 270-V power supplies, and MIL-STD-1553 interfaces for lower data rate systems. The complementary HSDB and VDDN boards were also installed in the ICNIA, HIDSS, cockpit displays, NVPS, TADS, and LONGBOW radar. Harris supplied the HSDB, VDDN, and terrain mapping boards. It also supplied the large head-down LCD displays, which purposely contained little processing capability. Consistent with the goal of maximum data fusion, all processing was done in the mission computers.

The mission computer software controlled the formulation of all cockpit displays and provided menu selections on the MPD and MFD for pilot input. The displays were automatically changed with mission phase to minimize the need for pilot inputs.

The mission computer software also received and deciphered all information about the aircraft subsystems and presented simplified, coordinated information to the pilot to minimize his need for attention. For instance, multiple sensors monitored each subsystem, and if several showed that a subsystem had problems, the mission computer would reduce the symptoms reported to a simple status reading to the pilot.

The software also included embedded programs for in-cockpit training of the various systems. The software is discussed in more detail in Chapter 10.

The navigation system derived within the mission computer was fully coupled to the flight control system to provide automatic flight along selected waypoints. Navigation and tactical overlays were provided on a digital map for tactical awareness and easier, more thorough mission planning. Optimum route selection was updated in real time. Target data could be transferred from the aircraft to the tactical operation center.

A large portable memory device was needed to support the transfer of mission planning information (threat locations, mission navigation routes, terrain maps, secure communication information) developed at the headquarters to the aircraft. It is believed that this portable device was also the repository for the mission computer software programs that eventually were loaded on local memory on each 380i processor board. The individual software programs could be reloaded to other processors if a particular processor board, or mission computer, failed. In the architecture diagram, this was called the mass storage unit—video recorder (MSU-VR).

Surprisingly, this capability to run multiple separate programs in parallel was used early in the prototype flight testing. We originally planned that the basic aircraft management and cockpit display software would need, at most, two processor boards per cockpit; however, due to software growth and operating system speed limitations, we used many more initially.

The early software development was hindered by "redundancy zeal," suggesting that the software partitioning should allow a single remaining board to do the elemental tasks for aircraft operation (sensor monitoring, mode control). This "graceful degradation" was a neat talking point early in the program, but it was unrealistic to envision multiple hardware failures within one mission cluster during a two-hour mission, while the other computer cluster was destroyed due to battle damage. (The benefit of hindsight.)

But, as a result of this low level of software partitioning and a nonoptimized operating system, data transfers slowed the overall net speed of the parallel system, so we had to use more processor boards than originally planned for the initial software loads. Fortunately, the architecture could handle the extra parallel boards. We think that we ended up with five boards in each cluster—a tribute to the initial insistence on reconfiguration and growth. Eventually, we would have made the system more efficient and accommodated extra software tasks within the processing cluster.

Before we leave the mission computer, there are a few development notes to document. The first is the retention of the Intel 380i processor. The second is the MEP reset button, which caused some consternation during the flight test program. The third is electronics packaging innovations, which will be discussed in a separate subsection because it affected all systems.

We selected the Intel 380i processing chipset early as the most advanced chipset available. Sikorsky had suggested using a reduced instruction set chip (RISC) during the early development because it speeded processing of geometric calculations and was not hindered by requiring the capability to do little-used commands. The unpredictable future demands of processing, along with the corollary need to standardize on the Ada language and available compilers, dominated the decision to use the 380i. (Also, Westinghouse had the design in place.)

As discussed, the trusted operating system became a major development; there was a small team of Boeing Wichita engineers, all highly skilled, responsible for the operating system development. The development was initiated when we first won, but progressed slowly.

Commercially, Intel was releasing first the Pentium chip and then Pentium II during this timeframe. Either would have simplified the software development tasks due to larger on-board processing memory and speeds. However, a change would have necessitated a new trusted operating system development, with attendant delays, so we hung on to the 380i.

In EMD, Intel announced it would no longer produce the outmoded 380i chips. As we looked toward low rate initial production, we were convinced it was cheaper and would cause less delay to the program to buy a "lifetime" supply of 380i chipsets than to transition to Pentium, so the purchase was approved. The program was canceled before we needed them.

The MEP reset switch was effectively a computer reboot switch in the cockpit. It was necessitated by a quick but far-reaching decision. We noted that we had all sorts of validity checks on the data (Within the expected range? No abrupt changes? Consistent with other sensors?) coming into the mission computer. But what if an inconsistency was noted? How would the mission computer know what to use? Might we be displaying erroneous data to the pilot?

It was suggested that it would be dangerous for the mission computer software to display anything to the pilot that might be erroneous. It would be better in this phase to simply stop the processing if an error was detected. The flight controls and standby instruments were independent of the mission computer, and because the mission computer was doing multiple parallel tasks, it wasn't clear which processor to shut down, so the corollary decision was to shut them all down—no displays!

This decision gave the MEP a bad reputation during the flight test program. We experienced MEP (processor) shutdowns and were forced to either abort the aircraft test or use the MEP reset, which required a van to pull alongside the aircraft, open the electronics compartment door, and insert a cable to reboot the computers. It took almost a minute to reboot. Part of the problem was that the MEP (and software) was the long pole in the development tent. The aircraft, rotors, transmissions, flight controls, and engines quickly came together on the propulsion system test bed (PSTB) and first prototype aircraft. It was ready to fly. The MEP, meanwhile, was in initial software development, and although the necessary subelements of code for aircraft management were written, they had not been thoroughly stress tested in a multiprocessor environment. However, enough was working to use the MEP for ground runs on the PSTB and taxi runs on the aircraft.

MEP shutdowns, however, continued occasionally and randomly at West Palm Beach. We couldn't find a root cause. We stressed to the pilots that the MEP was stable if the "tested" procedures were followed. Meanwhile, we continued MEP development and testing at both the Stratford and Philadelphia development labs (Philadelphia was the final release lab) and at a mini-systems integration lab at West Palm Beach. The software releases were pacing aircraft test desires. And we went to mini-releases.

The first flight press day arrived, and we were ready with an "adequate" MEP capability, including display features, to show the press after first flight. As the pilots had the rotors at full speed and ready to taxi, people were on the tarmac waiting to see history. The aircraft pulled into a hover, landed quickly, then went into a second hover. Then the loudspeaker, which was relaying cockpit radio communications, emitted a soft curse . . . MEP shutdown.

The pilots had two options: shut down the aircraft or ask for the MEP reset while running blind without displays. To chief test pilot Rus Stiles, there were two reasons to continue: one, he wanted the world to know there was nothing wrong with the aircraft and wanted to prove it by finishing the test card, and two, we had only been given clearance for one flight because of a main transmission issue, which gave further impetus to finishing the planned test card to discover if we had any other issues. Initially there was some concern about opening the MEP bay while the aircraft was running, because it was under the engine exhaust. But we had tried this several times during ground runs, and the ground crew decided it wasn't an issue. So the pilots stabilized the aircraft for a minute while the MEP van rolled out, rebooted the MEP, and the displays resumed. They then successfully performed the first flight demo, and the MEP held together.

Shortly afterwards, one of the software engineers temporarily sent to West Palm Beach determined the cause of the first flight shutdown. The copilot was sitting in the back seat for the flight test and decided to input or adjust the barometric altitude using the MEP control panel. This wasn't part of the "tested" script. A divide by zero error occurred, and the entire MEP shut down. Yes, the MEP software was very finicky in those days.

Interestingly, one mysterious issue that accompanied an MEP reset was the SPU always failed. It restarted fine, and we later found out that this was caused by an overzealous adherence to the requirement to automatically respond to faults. When the MEP reset, it sent a test command to ancillary systems, one of which was the fire detection processor. When the fire detectors reset, they sent commands to illuminate the fire lights to ensure the circuits were good. The MEP had software that automatically shut down the SPU in event of a fire—which was activated by the fire light. So when the MEP reset, the fire light came on and the SPU shutdown. This was later corrected by a simple software change.

PACKAGING INNOVATIONS

In the 1970s, electronics were developed by function (UHF radio, HF radio, flight control computer, instrument displays, etc.) and packaged to minimize weight or to conform to commercial Aeronautical Radio INC (ARINC) box sizes for rack mounting or to cockpit dzus rail dimensions for cockpit installations. All these mounting arrangements facilitated standardized environments for environmental qualifications, allowed the use of shock mounts where necessary, and allowed for easy removal of components at the field level.

Racks of computer equipment were being used commercially in buildings, and a movement was generated to apply the rack concept to airborne military equipment. This was generally led by the more advanced electronic installations in the Navy and Air Force. The Army, concentrating on helicopter avionics, where weight was always critical, tended to still develop the lightest, smallest package for each function.

The LH requirement for increased reliability, lower maintenance costs, and the desire to eliminate the intermediate level of maintenance led to adoption of rack-mounted modules that could be replaced at the field level. The goal was to identify faults down to the module using onboard diagnostics and replace the modules in the field. Thus, we adopted a standardized military form factor for modules, called SEM-E, and required all multiboard equipment to be configured for rack mounting and module replacement.

There is now an Institute of Electrical and Electronics Engineers (IEEE) standard for SEM-E modules (1101.4-1993), but that standard (released in 1993 and later deactivated) was preceded by a military standard to support LH. The form remains the same. Interestingly, the standard only allowed two connector configurations—a 250-pin connector and a 396-pin connector. Both are very high density connectors, typically attached with ribbon wire to the printed circuit boards. Sikorsky was responsible for aircraft wiring and took the lead for the rack design also.

As seen in Fig. 3.3, most modules provided direct pin access to sensitive integrated circuits and contained a warning sticker on the outside, warning of static electricity damage. Part of the maintenance learning curve was how to handle module replacement in an Army field environment (mud, rain, Army arctic gloves). We had to develop grounding techniques between the maintainer and aircraft for safe replacements of modules. Eventually we developed module cases and nonstatic boots to cover the connectors. These were all successfully demonstrated on the prototype aircraft.

At the same time, fiber optics use was still in its infancy on aircraft. The first fiber optic cable installed by Sikorsky was on a simple control communications path between an HF radio transmitter and its control. Now we had multiple fiber optic lines carrying high-speed data and video all over the aircraft, and even distributed within the racks.

Sikorsky, as wiring lead, had to develop field-level repair kits for fiber optics—another first forced on the program by Army field environments. Our fiber optic engineer became a leader in the field and eventually went to NAVAIR to further fiber optic innovations.

Fig. 3.3 Perkin Elmer SEM-E form factor 270-V power supply module.

How difficult could a rack be? Sheet metal and a few connectors? Well, we found out a lot more was involved.

First we went to composite materials for weight, but then needed metal grooves to align the SEM-E module connectors to the back of the rack connectors. We also needed heat transfer from the modules to air-conditioning plenums within the racks. Heat dissipation was always a problem on Army helicopters, which rarely had air conditioning. We always used ambient air, occasionally supplemented by fans. But the new high-density modules required active heat dissipation to keep the integrated circuits from exceeding their thermal ratings. (As a side note, the reliability group enforced a further thermal cushion on integrated circuits, below the vendor spec rating, resulting in more heat transfer demands on the racks—but more reliable modules.) So the plenums were built into the top and bottom of the racks as winding aluminum channels to spread the heat transfer. The 270-V power conversion supplies, like the one pictured in Fig. 3.3, had the most heat generation and typically got a favored position in the racks, but other high-density modules also cried for favoritism.

Rack length was well established by SEM-E module form factors, but the depth was constrained by the electronics bay size, which was determined by the aircraft's outer mold lines. Naturally, there was no room left when we added the dimensions of the SEM-E module and the mating connector. We managed to bring rack flat ribbon wiring out to traditional aircraft connectors on the top of the rack. There we had barely enough room to allow the maintainers to connect the wiring.

The front of the rack was hinged downward and held with quick release fasteners to allow access to the modules. The modules were retained by top and bottom extraction claws that forced the module into the high-density (and high-resistance) rack connector and allowed for module release. This was another development project to meet the Army's field environment.

Lastly, the racks had to be installed and mounted to the aircraft. Typically this was done by bolting the racks to embedded nuts in the shelf; however, the rack length required rear bolts, and there was no space to get a bolt and wrench back there. So Sikorsky created a wedge clamp design for the back fasteners. A spring leaf wedge was bolted to the rear of the shelf. The rack contained a rear lip that was forced under the wedge. Then the front lip was bolted to embedded nuts. We applied this fastening method to the mission computer, ICNIA, HIDSS, and AVICS.

So the Comanche advanced the art of electronic installations.

COCKPIT HEAD-DOWN CONTROLS AND DISPLAYS

The Comanche cockpits were redundant, with each having identical displays and controls capable of providing the same functionality. During normal aircraft operation, each pilot had unique mission roles and associated cockpit functionality enabled. In the event of failure, lost functionality in one cockpit could be regained in the other cockpit, permitting the mission to continue with minimal degradation. The cockpit and its operation are described in detail in Chapter 9, The Cockpit.

From a hardware perspective, the dominant feature was two large (for that time) 8 × 10 multicolor (but night vision compatible) LCD displays built by Harris Corporation. They were called multifunction displays (MFDs) because the information changed from aircraft data to mission data depending on the pilot's needs. The major challenges were to get a decent yield out of the LCD array, to mechanically and electrically attach the miniature lead wires to the display edge, and to find a night vision goggle (NVG)-compatible back light for the LCD shutters.

Originally, one of the MFDs was monochrome, to achieve better resolution and not degrade the target video coming from the sensors. As LCD performance improved, Harris was able to make both MFDs color. Much effort was expended in matching the displays to the sensor to ensure the sensor information was not compromised by the display. All targeting performance requirements were based on the image presented on the crewstation displays. As with most aircraft programs, many issues were being worked on related to the display of symbology on both the helmet display and the panel displays. The Comanche program had pioneered "contact analog" symbology, which allowed symbology to be stabilized relative to the outside view vs a more conventional stabilization based on the display surface. This issue was to be resolved in flight test and could have resulted in a major change in how symbology was displayed.

Many of the controls were integrated with the flight control grips (multiple multipurpose buttons on the grips), with the additional ability for either pilot to manually activate the redundant control push buttons within the cockpit. One of the reasons for this redundancy was that we only wanted one pilot to have control of the active flight control grips, to avoid accidental feedback of control inputs to the sidearm controllers.

Additionally, there were dedicated label buttons for major functionality requests across top and bottom and unlabeled, mission-task-sensitive menu buttons down the side of the display. The side labels were shown as text on the display and changed with the phase of the mission or the sublevel of information selected. (This sounds self-evident and like typical smart menu structures today, but it was a major departure from the traditional fixed legend pushbuttons of that era.) It was a flying Apple Macintosh computer display, and the buttons could be selected (highlighted and activated) by a cursor control built into the pilot's flight grip, similar to a mouse control.

While Harris supplied the MFDs, Sikorsky separately supplied the MFD bezel frames with the unlabeled buttons. There were both political and technical arguments for doing this. Politically, Boeing and Sikorsky were planning on downstream production splits, each bidding to provide sublots of yearly procurements and support of equipment changes. Boeing had primary control of the mission computer and software. Sikorsky needed to stay in control of all interfaces in the cockpit and aircraft to ensure knowledge of and participation in all changes. There was also a reliability-driven technical reason to keep all failure-prone back-lit switches (a big maintenance driver) in an easily removable/repairable assembly. We did not want to replace, or stock at field level, an expensive MFD for a burned out bulb.

But we learned that trying to specify what were typically internal equipment interfaces was a difficult effort for system-level engineers. It took a lot of effort to achieve an easy mounting bezel whose wiring didn't interfere with the LCD side attachments. We would have been better served by giving both the display and the removable bezel to Harris for optimum interface control, and let them relay the switch information to the mission computer.

We also had a side display (MPD, multipurpose display), about 4×6 in., to use as a primary aircraft mode selector for menuing down to subsystems. It allowed control of radios and other subsystems. It had a 2×10 pushbutton array with front legends that were programmable by the mission computer depending on the state of the cockpit operation at the time. This display was originally monochromatic. We then made it a color LCD display, called it an MPD, and had similar grip-mounted highlight and designate functionality as the main MFD displays.

There were many other cockpit buttons and annunciators that had to be both NVG compatible and easily repairable. Bulb and switch failures dominated the fault tree, due to sheer numbers of components, so we expended considerable effort to auto-detect and auto-isolate failures. This led to circuit designs that added more wires and parallel resistors to each switch and light to isolate whether a bulb burned out or a wire had broken. The differentiator was voltage drop sensed within the electronics in the HIDSS, EDEU, and AVICS. Although it saved aircraft maintenance turnaround time, it certainly added to weight and cost.

Electro-Optical Sensor System (EOSS)

The "see" part of "see without being seen" came from the target acquisition system (TAS), along with the night vision pilotage system (NVPS). These were mounted on the nose of the aircraft to provide maximum field of view, unobstructed by the rotor, and to reserve the mast-mounted position for future inclusion of the LONGBOW system. Figure 3.4 compares the LH capabilities with the AH-64.

The electro-optical sensor system (EOSS) combined the TAS and NVPS into a single unit. The TAS provided forward-looking infrared (FLIR) and solid state TV (SSTV) imagery with multiple zoom lenses for targeting and an

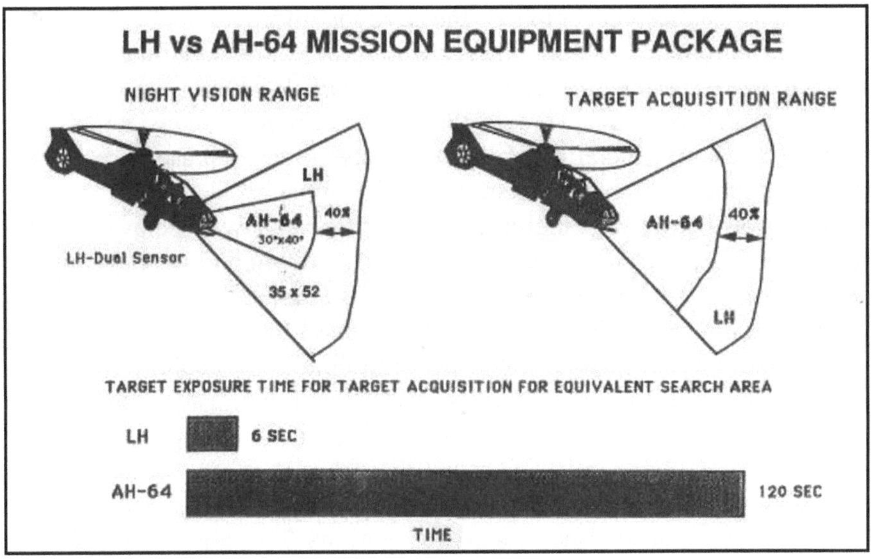

Fig. 3.4 LH vs AH-64 Mission Equipment Package.

additional wide field of view as a backup pilotage system (BUPS), a laser for ranging and designation (for missile guidance), and a laser spot tracker. The NVPS was the primary pilotage FLIR imaging system, to be displayed on the pilot's HIDSS.

The EOSS was a complex unit consisting of independent FLIRs, stabilization platforms, turrets, processing, and data/video distribution networks. The Comanche EOSS with its second-generation FLIR and its modern advanced electronics was designed to achieve an order of magnitude reliability improvement over predecessor systems.

TARGET ACQUISITION SYSTEM

The electro-optical targeting and designation system (EOTADS), which contained a second-generation forward-looking infrared, television, and laser RF/D, was perhaps the most advanced electro-optical targeting system yet developed for a tactical aircraft. The EOTADS provided the capability for rapid, wide area target searches in an automated mode in which the system would quickly detect and classify potential targets. These targets were presented to the crew as cropped images of the potential target with an automated assignment of target classification according to a crew-selected prioritization. This allowed the crew to quickly confirm the classifications and select a course of action to pursue the target or assign it to another weapon system. Several additional modes of operation were available from the system. One was an automated search and storage of the search area, which could be replayed to the crew from a masked position, allowing the crew either to confirm the targets and develop an attack plan or to pass the targeting information to other systems. The use of automated detection/classification algorithms to find targets was thought to be essential with a second-generation FLIR targeting system. Studies have shown that manual review of target scenes requires considerably more time with a second-generation FLIR sensor due to the increase in scene detail and content inherent in second-generation FLIR imagery.

A key contributor to the increase in detection range was the availability of linear detector arrays, which came to Comanche from parallel developments in Army research labs. First-generation FLIRs used small (1×4) thermal energy detector arrays. These had to be scanned through complex rotating prisms to paint a multipixel, 3×4 aspect ratio picture. Resolution was determined by the number of pixels in the picture. Typical raster picture refresh rates were 20 times per second to avoid flicker. So the array had only a momentary view of any point in space (its field of view) to gather its thermal energy and output an analog signal of thermal strength. The second-generation FLIRs used larger arrays and less prism rotations. This allowed longer dwell times at each pixel point in the picture and resulted in greater sensitivity.

The second-generation FLIR detector used in the Comanche electro-optical systems was a 480×4 scanned focal plane array operating in the 8- to 12-micron band, which the Comanche program has always believed provided the best worldwide environmental performance capability. It provided a 40% increase in standoff and detection range and the ability to work in adverse weather conditions and with battlefield obscurants. Large-format staring focal plane arrays were becoming available, but most of these operate in the 3- to 5-micron band. It was recognized that the 3- to 5-micron systems do provide greater focal lengths and support longer range detection and recognition when operated in the manual targeting mode, but the 8- to 12-micron band is more compatible with automated targeting. Little work has been done in adapting staring focal plane arrays for rapid scanning of target areas to support automatic target searches. Comanche was planning to incorporate the longer range capability of the 3- to 5-micron system through a future upgrade to the TV side of the EOTADS.

The target acquisition system also had an aided target detection and classification (ATD/C) function with files of known enemy vehicles that could be compared with the view being received. This included a search-on-the-move function; moving target detection; and automatic multitarget prioritization, tracking, weapons selection, and fire control. Hover search times were less than 6 seconds (pop up, scan target area, and pop down/move away); scene review was conducted on the MFDs after the aircraft had remasked. When a match was found, this information was displayed to the pilots, and a weapons cueing/rangefinding function initiated. This resulted in a 95% reduction in target acquisition time and three times more target location accuracy compared to systems then in use. False alarm rates were projected to be five times lower than the Army's requirement.

Fig. 3.5 "The eyes of the Comanche".

In early 1993, Martin Marietta completed tests of the EOSS at Camp Grayling Figure 3.5 illustrates one of these tests. The tests showed that the system delivered at least 30% better resolution and 50% better identification than the Apache's existing target acquisition systems. The first complete EOSS was delivered in June 2003.

The laser rangefinder designator in the targeting system was a solid state, diode-pumped, dual frequency laser in a very compact package. The 1.54-μm "eye safe" mode was to be used for range finding and training, and the 1.06-μm mode would be used primarily for designation with the Hellfire missiles. The laser development had been completed for some time, but it had not been tested as a part of the overall targeting system.

NIGHT VISION PILOTAGE SYSTEM

The NVPS provided the imagery projected on to the helmet-mounted display (HMD), to allow nap-of-the-earth and terrain avoidance flying at night and in limited adverse weather conditions. The imagery provided navigation and reconnaissance capabilities and gave a clearer definition of the external scene than previous systems of the horizon.

The NVPS at that time was second generation and could detect smaller temperature differences than earlier FLIRs. (The actual temperature numbers were classified.) It also included an image intensifier function (as in night vision goggles) to see in the visual rather than thermal spectrum. It had a 30-deg by 60-deg sensor field of view, but due to HIDSS weight/processing limitations, only a 30-deg by 52-deg field of view was displayed on the pilots' HMD. This was still a significant increase over what was currently being used on the AH-64 Apache, which used only a monocular display system of approximately 30 deg. The visual range of the system was a 40% increase over the Apache's first-generation FLIR.

The original First Team proposal used a Northrop FLIR operating in the 3- to 5-micron region. It provided pictures similar to a monochromatic TV camera and was smaller and lighter than the existing Apache pilot night vision system operating in the 8- to 12-micron region (thermal imagery). The Northrop FLIR had applications with the U.S. Air Force and U.S. Navy. But like all comparisons between competing technologies, there were positive and negative attributes to both the 3- to 5-micron medium wavelength and the 8- to 12-micron long wavelength IR technologies.

Simplistically, the 8- to 12-micron technology was a thermal imager and could detect smaller differences in temperature than the 3- to 5-micron technology. This was due to both the radiation characteristics of the targets and backgrounds in the two regions and the sensitivity of the detectors in both regions. Atmospheric absorption was an additional factor, and the 8- to 12-micron version lost less signal under battlefield smoke conditions. The Army night vision labs (NVLs) led the push to change the FLIR technology. NVL, through the Defense Advanced Research Projects Agency (DARPA), was also funding advancements from the second-generation scanning arrays to 8- to 12-micron starring arrays that had more dwell time on target and therefore more sensitivity. These arrays could be backfitted to the second-generation FLIRs with little change to optics and processing.

So shortly after we won the program, we were asked to change the NVPS technology to 8- to 12-micron. This meant dropping the Northrop design and going with a Martin Marietta design. Sikorsky, who had lead at that time for the NVPS, resisted for a while, but the NVL technical pressure was continuous. Finally, we agreed to the change, and for contracting simplicity transferred responsibility to Boeing and Martin Marietta.

Martin Marietta and NVL worked closely on the TAS and NVPS designs for Comanche, and it was another case of the technology growth from scanning arrays to starring arrays running parallel with the program. The yield on these new detectors was poor, and we were always in line (and waiting) for the first or second array out of DARPA to the TADS/PNVS.

At the beginning the detectors were noisy, and software coding was necessary to filter out the noise. We installed one prototype system on the Shadow aircraft, and the picture was unusable. But overnight, the Martin software engineer recoded the integration times, and the picture was beautiful.

Later in the program, we had another detector array noise problem, but due to program delays, that software engineer had left Martin Marietta. This time it took a lot longer to get the noise problem corrected. It was not unusual with the long program delays and lack of money to lose all the experts on staff.

LONGBOW RADAR

The LONGBOW system was planned to be installed on one of every three Comanches and would complement the electro-optical vision system on certain specific missions requiring standoff range in adverse-weather operations, and autonomous weapons firing under conditions of heavy battlefield obscurants. This was another example of technology development running in parallel with Comanche. Extra processing capacity was reserved within the mission computer to accommodate the additional video and control/display menus.

HELMET-INTEGRATED DISPLAY AND SIGHTING SYSTEM (HIDSS)

The HIDSS was a critical technology for the Comanche, providing the two-eye imagery overlaid with symbology for heads-up and eyes-out operation of the helicopter. The symbology had to have daytime brightness for use with the see-through visor and be overlaid in the NVPS imagery and NVGs at night. Symbology consisted of flight graphics and alphanumerics, which were overlaid over the imagery. The HIDSS was capable of displaying symbology from the mission computer either alone or with imagery from any of the image sources.

Each HIDSS consisted of the following:

- The high-cost aircraft-retained unit (ARU), consisting of a frame that mounts left and right imaging system optics containing the display source, a helmet tracking sensor, and an aircraft connecting cable with a quick-disconnect connector.
- The low-cost pilot-retained unit (PRU), consisting of the helmet shell, energy-absorbing liner, custom fit liner, and communications system. It had provisions for mounting an image intensifier sensor. The combination of ARU and PRU was designated as helmet display unit (HDU).
- The expanded display electronics unit (EDEU), consisting of the display electronics, helmet tracking electronics, and a control and display interface to collect and encode most of the cockpit switches and annunciators for transmission to the mission computer.
- The helmet tracking source assembly.
- Three helmet visors, for day, night, and laser protection use.

The major HIDSS items are shown in Fig. 3.6.

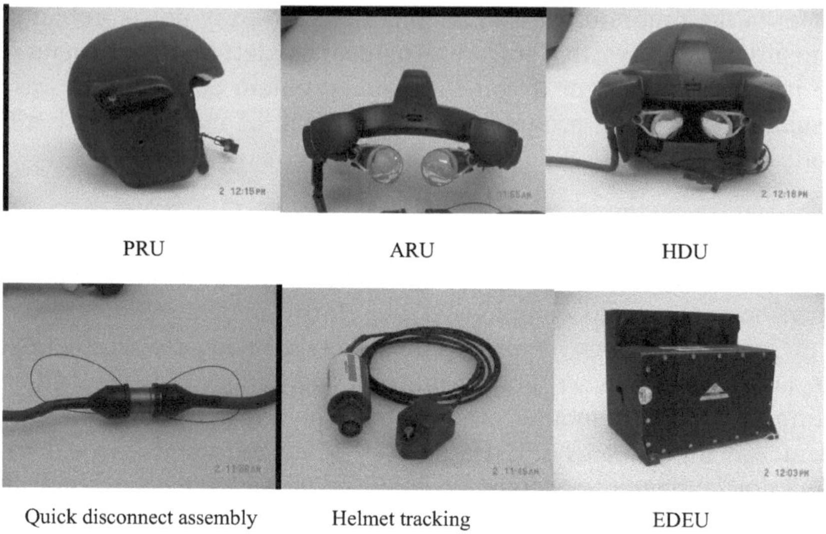

Fig. 3.6 Components of HIDSS.

HELMET DISPLAY UNIT. The HDU had many, sometimes conflicting, design requirements:

- Minimize vignetting (obscuration of the eye's FOV).
- Fit personnel in the 5–95% range of the pilot population.
- Provide for removal of any HMD-related visual obstruction in the event of system failure.
- Provide for biocular/monocular viewing of the video imagery or the image intensifier sensor imagery with a 40-deg × 52-deg field of view, partially overlapped between the two eyes.
- Minimize heat buildup in the helmet shell and liner.
- Maximize safety during emergency egress in an explosive atmosphere and provide for rapid cable disconnect from the aircraft in emergency egress.
- Maintain the crash protection characteristics of the Army pilot's HGU-56P helmet during modification of the HMD, and consider the same protection for the full HMD.
- Provide laser eye protection.
- Maintain noise attenuation performance of the HGU-56P for the full HMD.
- Accommodate NBC protective gear
- Provide a monochrome Aviation Night Vision Imaging System (ANVIS) compatible display
- Meet the weight and center of gravity (CG) requirements set forth by the Army for safety and long duration wear comfort.

Meeting all these requirements was equivalent to putting 10 lb of displays in a 5-lb bag, and it finally took the use of early LED technology to achieve all requirements.

AIRCRAFT-RETAINED UNIT (ARU). The ARU was detachable from the PRU for stowage in the aircraft cockpit when not in use. The ARU was too expensive to be individually issued to every pilot. The ARU consisted of a frame on which left and right imaging system optics were mounted. Each contained relay optics, display combiner, and display image source. The ARU incorporated adjustment mechanisms that allowed independent interpupillary distance (IPD) adjustment of the imaging optics, in addition to providing vertical and fore-aft accommodation of the imaging optics for the range of crewmembers. Also included were the aircraft interconnect cable assembly, including the crewmember's side of the quick-disconnect connector, aircraft side of the quick-disconnect connector, and helmet tracking sensor. Protective visors mounted to the ARU included day (shaded), night (clear), and other special purpose (eg, laser eye protection).

PILOT-RETAINED UNIT (PRU). The PRU included a protective helmet shell, energy-absorbing liner, custom-fit liner, communication hardware (consisting of a microphone and earphones), and helmet retention system. The custom-fit thermal-protection liner, in conjunction with the proper helmet shell size and energy liner, accommodated the nominal helmet positioning for the head of a particular pilot. The PRU maintained the ARU exit pupil relationship throughout the vibrational environment of the aircraft as transmitted through the pilot's body. It also had provisions for mounting image intensifier sensors.

EXPANDED DISPLAY ELECTRONICS UNIT (EDEU). The EDEU included the display electronics, helmet tracking electronics, and controls and displays interface. Each EDEU accepted video (scene imagery, scene imagery overlaid with symbology, or symbology only) from either of the two mission computers, and it outputted the waveforms and drive signals needed for the display sources in the ARU. The EDEU accepted video from, and also supplied power for, the helmet-mounted image intensifier sensor.

HELMET TRACKING ELECTRONICS

A magnetic helmet tracking system was used to calculate the pilot's line of sight for pointing the NVPS sensor. This was a big improvement over the infrared sensor used on the Apache, because it was lighter, more accurate, and provided all-axis helmet position detection. This allowed the imagery to be stabilized to the horizon even if the pilot rolled his head, a capability lacking in the earlier generation infrared system. This also enabled the incorporation of a new approach to symbology, which kept it referenced to the outside scene regardless of pilot head position. This dramatically different technology resulted in a surprising level of controversy within the Army pilot community. Although it was a theoretically better approach (and mirrored the approach used in fixed wing head-up displays), it was different than what the pilots were used to. Their resistance to the concept gave the program a hint of things to come when new technologies directly affecting pilot operations were proposed.

Helmet Tracking Filters The EDEU incorporated adaptive filters that could be selectively used to smooth the helmet tracker's calculated position and orientation data output to reduce residual noise in the data. The filters were designed to not introduce latency and automatically adjusted algorithms such that lag was minimized under highly dynamic conditions. Filter variables were adjustable through an external data interface during the development program. As the pilots noted, these filters never worked very well. If they were tuned to reduce the jitter, latency was introduced. If you increased the response times, jitter was reintroduced. What was fundamentally needed was an increase in processing speeds. But we were constrained by central processing speeds, sensor resolution and data rates, and the inability to convert to new processors by the trusted operating system. Our continuous bane was our inability to transfer what we could create on Apple products and simulation to an airborne environment.

HTS Compensation Map Configuration Control A cockpit compensation map was required for each cockpit for use with the helmet tracking system; the data for that map were stored in aircraft software. The map defined any magnetic field distortions within the cockpit. There were some concerns that pilot-introduced metallic items would interfere with the helmet tracking system, but this did not appear to be the case in operational use.

Controls and Displays Interface (CDI) As explained in the Cockpit Heads Down Controls and Displays section, Sikorsky was responsible for converting all passive and low data rate signals for the cockpit and aircraft and sending them via the 1553 bus to the mission computer. The AVIC handled the noncockpit signals; the HIDSS display electronics unit was "enhanced" by the addition of this function. While Kaiser handled the display areas, Hamilton Standard handled the CDI function and its parallel design in the AVICS. The CDI also provided power for annunciators and switches in the cockpit

The CDI interfaced with the following:

- Left multifunction display
- Right multifunction display
- Cockpit-integrated keyboard (CIK)
- Dedicated cockpit switches and annunciators

- Cyclic grip switches
- Collective grip switches
- Left wing panel (TAS)
- Armrest switches

Some interfaces were cross-connected from the other cockpit to provide ballistic redundancy.

CHALLENGES IN DEVELOPING THE HIDSS TECHNOLOGY

There were a lot of challenges in developing the HIDSS technology. These were driven by the Army requirements, human–machine interfaces (light weight, large exit pupil, ease of adjustment, ease of gas mask donning, and large variability of the head sizes), and safety concerns. Safety of the pilot and copilot in the event of an unforeseen crash event was given utmost consideration. In cooperation with DARPA, we initiated development of the miniature LCD technology with Kopin Inc. (the manufacturer of small high performance displays) for the HIDSS. This provided additional safety for the crewmembers compared to the original high voltage CRT close to their heads.

During preliminary design studies for LHX, prior to ARTI, the crew station design team proposed a 40-deg × 80-deg display field of view (FOV) as the goal for the HIDSS. We decided to conduct "out the window" simulator tests to determine whether this was optimal for flying.

Hamilton Standard, a contributing technical advisor, made a helmet mockup based on a motorcycle helmet for technical evaluation in the simulator. It was very insightful. We learned that the optics were huge and made the helmet bulky and heavy. Further, writing speeds, FOV, and brightness requirements for the CRT displays exceeded existing technology. The weight became an issue for pilot fatigue, and a 20-*g* crash would likely exceed the neck's ability to retain the helmet. The simulator installation incorporated a counterweighted sling device that effectively made the helmet weightless. This permitted pilots to evaluate the helmet FOV without focusing on the discomfort of weight. Some engineers joked about the need for pilots with 20-in. necks, like NFL linemen.

The HIDSS symbology was partially overlaid on the real world. This "contact analog" approach was shown in simulations to allow the pilot to fly "heads out of the cockpit" while retaining necessary elements of navigation and targeting. The NVPS was also three-axis stabilized to the pilot's head motion, so the symbology overlay was the same in night operations.

Because the contact analog symbology philosophy differed dramatically from what the Army had fielded in its only other helmet display–equipped aircraft (the Apache), it became another contentious issue with the operational pilots assigned to the Training and Doctrine Command (TRADOC) office.

This was due in part to the helmet tracker not providing the accuracy or response needed for ideal symbol stabilization. This resulted in perceptible lag when the head was moved rapidly, and jitter when the head was held steady. Common complaints centered around attitude and heading data, which were designed to appear as a "virtual" HUD, anchored to the aircraft centerline. Operational pilots were used to seeing a screen-referenced horizon line even when it was not overlaid with the horizon, even though the old screen-referenced approach was acknowledged by the training community to require a difficult mental disassociation from the imagery. (The Apache IHADSS was not commonly used in the daytime.) Other aspects of the contact analog approach were accepted as true improvements, such as weapons sighting, sensor and laser line-of-sight position, visualization of the real-time flight path, and navigation reference.

When flying at the extreme low altitudes required for the Comanche mission, maintaining navigation awareness is as critical as it is difficult. The navigation symbology consisted of a series of flags denoting the location of navigation waypoints. Using contact analog techniques, the flags appeared to be grounded at the location, in the real world, as shown in Fig. 3.7.

One advantage to the contact analog design approach was that the symbology need not change in any way for daylight, night, or adverse weather operations. Because the sensor imagery, which created a synthetic depiction of the outside world at night, was stabilized in all three axes—pitch, heading, and head roll—it accurately overlaid the real world. (Note that the first-generation imaging system found on the Apache was not head roll stabilized, so it rarely overlaid the outside world unless the pilot held his head in a perfectly level position.)

The display was see-through, so the entire FOV was overlaid with the outside world. There were many evolutions of the display FOV. Initially, the

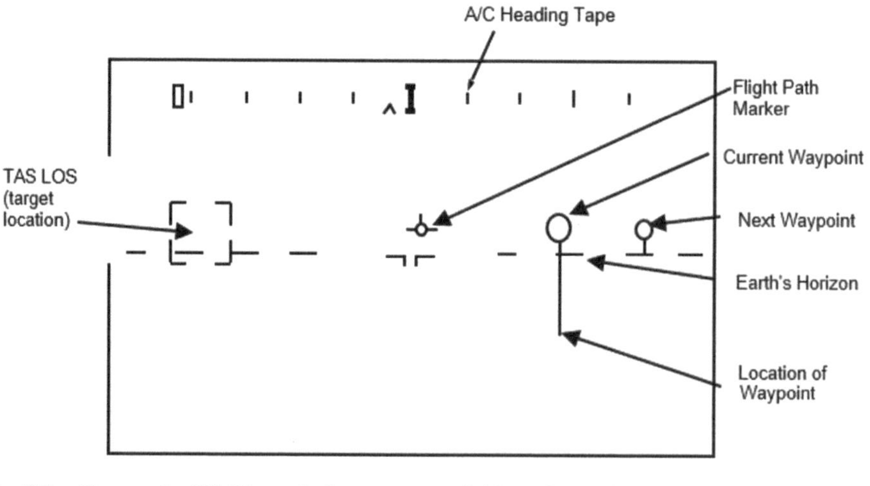

Fig. 3.7 Comanche HMD symbology was overlaid on the real world.

objective was for a fully overlapped system, where both eyes see exactly the same imagery and symbology. But it was apparent this couldn't be achieved with anything much above about 40- to 45-deg FOV, and this meant not meeting weight and CG requirements; nor was it compatible with the NVPS sensor, which provided only a little over 30 deg of vertical FOV. Ultimately, the HIDSS provided a partially overlapped biocular FOV of about 35 × 52 deg, with 30 deg of the vertical portion used for imagery and symbology filling the remainder. This meant that each eye saw about an 18-deg central section of fully overlapped imagery (the same in each eye), and the 17-deg outer section of the FOV was unique to each eye.

Hamilton Standard originally engaged a Honeywell division that made the existing Army helmet sighting and tracking system used on Apache. Its existing system measured the helmet's three-dimensional position and pointing angle in the cockpit and also displayed it on monocular optics that clamped to the pilot's helmet. It was an infrared system. Additionally, Honeywell had a division in England that supplied small, lightweight CRTs.

At some juncture, we added image intensifiers to the HIDSS. It is thought that the reason was that the copilot needed a way to see outside. With the TAS devoted to autonomously scanning for targets and the NVPS being used by the pilot, the copilot was effectively blind at night. There was also a lower level requirement for the pilots to be able to deploy out of the aircraft with night vision capability (for escape and evasion). Image intensifier technology was improving rapidly and was quickly becoming equivalent—in resolution and performance—to the FLIR. There was a strong desire to use Comanche as the launch platform for helmet-mounted image intensifiers. Mounting it in this way resulted in a much larger FOV capability than standard night vision goggles (NVGs), which provided only about 30 deg maximum. In reality, this was another example of the Army seeing a gadget it couldn't refuse. Naturally, this aggravated head weight and moments of inertia concerns and eliminated the infrared tracking system that would saturate the NVGs.

Kaiser Electronics came on board with a magnetic tracking system compatible with NVGs. Its additional optics and display capability soon made it the lead for the helmet portion of the design, while Hamilton and Kaiser designed the modules for the EDEU.

The HIDSS design evolved into two separate but symmetrical display paths for symbology and imagery, one for each eye. The electronic imagery was reflected off coated optics in front of the eye (coated to maximize reflection at the video frequency). The optics also provided "see-through" capability to the outside world. Each path allowed for insertion of an NVG tube. The two optical trains were mounted on a rigid frame with adjustments for interpupular eye variations.

The pilot could see the outside world through the optics during the day and overlay the symbology on the NVGs at night. Additionally, the NVPS was selectable to one of the pilots as a single FOV thermal video source, split by the mission computer into two overlapping videos for each eye.

The optics design became very complex and expensive because of the needs for visibility through the optics to the real world, crisp night videos over a wide field of view, and readable symbology over a wide ambient light range. The expense and the fear of accidental damage drove the decision to make the helmet system detachable from the pilot's personal helmet. Gentex, a supplier of Army crashworthy helmets, came on board at this juncture to provide the pilot's portion of the HIDSS.

Naturally, now that the pilot could leave with a skeleton helmet, and because scenarios could be envisioned where the aircraft crashed at night behind enemy lines, an additional requirement evolved to carry the NVGs off the aircraft attached to the helmet. This capability was not available during Dem/Val, but the design was solidified and would not have been difficult to incorporate. The main challenge was where to put the battery. It ended up in the survival vest and interfaced the aircraft connector. The design shown Figure 3.6 at the beginning of this section shows NVGs clamped over the ARU, rather than inserted in the optical train.

The HIDSS field of view always was the primary driver for optics weight and complexity. As time went on, it was ultimately about 35 deg vertical, but with partial overlap, 52 deg wide. Every degree helped.

Two bumps in the HIDSS development road occurred post-ARTI. First was hyperstereopsis and the second was CRT obsolescence.

Hyperstereopsis was a term promulgated by our human factors/crew station designers who were intimately involved in the HMD design. Everything was neatly packaged and balanced in the original HMD design, and then came the addition of image intensifiers. At that time, the Army flew Apaches and Blackhawks with dual night vision goggles, which were image intensifiers with integrated displays screens on a simple eye lens in front of each pilot's eye. But there was no way to put the NVGs spatially in front of the pilot's eye through all the wide field of view optics.

So we created a removable lens element that could either complete the required optical path from the CRTs or add the image intensifier video in parallel with the CRT video. Every additional optical element degrades overall transmissibility, but we had some limited ability to tolerate it and still meet the visual specs. But the image intensifiers needed to be parallel to the pilot's sight line—and ideally spaced with the same interpupillary distance as the (adjustable) final eye lens.

The optimum location was above the helmet. It provided one of the shortest optical paths and could be fed close to the final eye. Unfortunately, it provided a high moment of inertia and added weight beyond what we believed was tolerable. Additionally, it interfered with the interface with the pilot-retained portion of the HIDSS.

A simpler solution was to side-mount the image intensifiers on the HDU and interrupt the optical path closer to the side-mounted CRTs. This minimum path put the image intensifiers at cheek level and resulted in about a 10-in. separation between the two NVGs. The increased moments of inertia were not that significant given the existing moments of inertia of the system.

We iterated the optics design and proceeded to go forward with new optics and image intensifier mounts.

Then a senior crew station designer, who was a wizard with making cardboard mockups, walked into the design area wearing two periscopes. What a strange sight. Especially when he continued to bump into desks and chairs that were below his vision.

The point of his mockup was to show what the effect would be if the two eyes had visual sight points that were separated by more than the normal eye distance (i.e., the image intensifiers). He used simple tubes with angled mirrors like a periscope. The effect caused by the separated lines of sight was an exaggerated depth effect; things looked closer than they actually were. This was hyperstereopsis.

It certainly caused us to have problems with near-proximity tasks using the mockup. We spent enough time with the phenomenon to show it to the Army and broadened the base of concerned people. In hindsight, it might have been used as a good excuse to drop the NVGs, but we didn't drop the requirement.

Instead, we compromised the design a little more and mounted the NVGs closer together. Eventually, we received permission to clamp the existing Army night vision goggles outside the ARU (another design iteration) to eliminate the hyperstereopsis effect and simplify the logistics of protecting the classified/costly NVGs separately from the ARU. The Army night vision lab conducted an experiment to assess hyperstereopsis, because this would inevitably be an issue regardless of what kind of helmet was developed. Results were somewhat inconclusive. There was little performance degradation encountered with the hyperstereopsis displays, but the pilots reported they didn't like it. The hyperstereopsis effect was not noticeable beyond a few feet, so objects outside the cockpit were viewed without any effect.

The second bump was our lightweight CRTs. We needed extra brightness out of the existing CRTs used on Apache because we were painting more daytime stroke symbology on the CRT, and faster writing to stay within the refresh rate lessens brightness. The higher daytime outside ambient light also lowered the resulting contrast. Our English supplier from Honeywell thought that it could provide new CRTs to the new requirements. The supplier built a limited production run that met our needs. The CRTs required higher beam voltages (on the order of 9000 V).

But the high voltage became a safety concern. (Old CRTs were at 6000 V and heavily protected.) To satisfy that concern, we added cable weight to double insulate the wires as they traveled past the pilot. But this made the cable bulky and stiff and added to helmet weight, so we did "draping" studies to find a means to offload the cable weight from the helmet. We tried cockpit ceiling hooks and cables attached to the pilot's shoulder. It is believed that the final design came over the shoulder.

We successfully integrated the new CRTs and the NVGs into our HDU. But the HIDSS solution was a lot like reducing aircraft vibrations. You dampen it at one location and it pops up somewhere else.

As part of the HIDSS/cockpit integration, there was a (somewhat secondary) requirement for the pilot to egress in emergencies with the HDU still attached to the pilot-retained unit (his helmet) to minimize egress times. We accepted the requirement, and Sikorsky's wiring experts found a quick disconnect for the HDU wiring that would separate with an active pull.

But some intelligent safety engineer questioned whether a spark might be produced if the quick disconnect released in a crash environment. So back to the quick-disconnect drawing board. We had a supplier create a quick disconnect that sealed the connector until the high voltage pin was disengaged, thus isolating the spark from the air. Another small step for mankind.

Somewhere around the transition from Dem/Val to EMD, we tried to order additional CRTs. The supplier couldn't remember how it created the old ones and failed every time it tried to produce them. There was no other viable source for lightweight, high voltage CRTs, so the HIDSS team turned to liquid crystal display (LCD) shutters similar to the MEP's large flat panel displays and light-emitting diode (LED) sources. Of course, this led to additional design iterations. Thank goodness, LED technology had emerged around the same time with brighter sources evolving each year, so the final HIDSS design was within the weight and CG requirements. Another example of technology catching up with the delayed Comanche program.

INTEGRATED COMMUNICATIONS, NAVIGATION, AND IDENTIFICATION AVIONICS (ICNIA) SUITE

The integrated communications, navigation, and identification avionics (ICNIA) included an integrated suite of communications equipment providing radio frequency (RF), digital, and secure communications. The equipment provided reliable voice and data communications interoperable with Army aviation command and control nets, ground force nets, interservice nets, and allied force net. Identification, Friend or Foe (IFF) capability and an audio distribution system (ADS) were also provided. The ADS functioned as the aircraft intercom system. As an integrated suite, ICNIA enhanced functional reliability by providing select redundancy and graceful degradation in the event of equipment failure that allowed the mission to continue.

The integrated communication/navigation/identification system incorporated Global Positioning System (GPS) navigation with accurate backup provided by both Doppler and inertial inputs to ensure mission completion. The navigation system was fully coupled to the flight control system to provide automatic flight along selected waypoints. Navigation and tactical overlays were provided on a digital map for tactical awareness and easier, more thorough mission planning. Optimum route selection was updated in real time. Target data could be transferred from the aircraft to the tactical operation center. The system was designed to be highly fault tolerant, reliable, and survivable. The system was installed in separate installation racks in the MEP rack system. It was designed for straightforward future growth.

The ICNIA system was initially funded under the USAF PAVE PILLAR program and included programmable VHSIC modules (SEM-E size) to generate those unique waveforms needed to replicate the existing dedicated military communication radios and identification transponders. ICNIA was initially focused towards the USAF F-22 Advanced Tactical Fighter and the Navy's Advanced Tactical Aircraft programs. There were specialized RF generator modules for specific frequency bands, but the modulating waveforms were generated in common waveform-generating modules that could be reprogrammed in flight. Because of the fighter aircraft environment, the radio modules were originally focused toward SATCOM and UHF (HAVE QUICK) radio waveforms, IFF Interrogator and Transponder functions, and GPS.

The Army, operating in a nap-of-the-earth environment, needed VHF (SINCARS) and HF radio functions for Army interoperability with ground troops. We were assured that both the RF generators and the waveforms could be developed, so we developed the necessary antennas while minimizing the impact on the aircraft signatures. We designed the basic rack structures, similar to the mission computer racks, and reserved shelf space for the racks, partitioned into two racks for ballistic tolerance.

But as time went on, we found that ICNIA had basic funding problems. PAVE PILLAR was basically a demonstration program for advanced systems and there was no plan to develop a fully fielded ICNIA system. So the services needed to add funding. The Navy and Air Force had identical needs and jointly funded their efforts. The Comanche team had not planned initially for the added cost of Army-unique VHF and HF modules, and the development of these modules was delayed as funding restrictions emerged. The Army eventually supported the Joint Tactical Radio System (JTRS). The UHF/VHF waveforms from JTRS were compatible with ICNIA hardware and eventually would be used.

As the prototype aircraft was being built, we reverted to off-the-shelf radios for the test program. By EMD, we had dropped the HF radio requirement and showed dual SATCOM antennas in the configuration. A small side step, if we fast forward to when the Comanche program was terminated, the Comanche

Integrated Communications Navigation Identification Friend or Foe (IFF) Avionics (ICNIA) developed by team member TRW included:

- (Single-Channel Ground-Arborne Radio System – Enhanced System Improvement Program (SINCGARS ESIP) clear and secure voice
- VHF-AM/UHF-AM clear and secure voice
- HAVEQUICK I/II
- Link 16 (with combat net radio message translation)
- Enhanced Position Location Reporting System (EPLRS)
- Digital battlefield connectivity (improved data modem data compatibility)
- Full Joint Variable Message Formats (Multiservice) (JVMF) message set
- JVMF to Tactical Data Instructions Link – J (TADIL-J) message translation and reverse message translation
- Embedded high frequency (HF) with clear/secure voice/data
- Intelligence downlinks (TRIX, TADIX, TIBS, TRAP)
- UHF Satellite Communications (SATCOM), Damand-assigned-multipurpose-access (DAMA) transmit/receive (voice and data)
- Imagery transfer over U/V and Link 16 links
- MK XII IFF transponder with modes 1, 2, 3A/C, and 4
- Auto frequency scans
- Simultaneous transmit—Vertical Message Formats (VMF) messages out on multiple radios
- Communications Security (COMSEC) fill via JMPS
- All hosted in two 40-slot racks with 2,132,482 Source Lines of Code (SLOC)

Five shipsets of the Comanche ICNIA were complete at termination. It is flying on the F-22. Over 42% of the Comanche ICNIA software was common to the F-22 ICNIA system. Messages were automatically filled/formatted for transmission by the crewmember with a single button push. ICNIA was and remains the only truly software-programmable radio that provides simultaneous multithread operations. The Army went on to develop the Joint Tactical Radio System (JTRS) utilizing wideband networking waveform (WNW) and soldier networking waveform (SNW) for interconnectivity and network data links. The JTRS today has less capability than the ICNIA system.

Aircraft Survivability Equipment (ASE)

Comanche aircraft survivability equipment (ASE) provided radar, laser, and chemical agent detection along with crew notification of detected threats. Continued processing of ASE sensor data was supported by the redundant MCCs.

We originally planned to incorporate the integrated electronic warfare system (INEWS) into Comanche, but because of the Army's operating environment and the low-signature approach of Comanche, the jammer portion of INEWS, which was a significant portion of its weight and cost, was considered to be low in effectiveness. It was more cost effective to carry laser and radar warning systems and take out threats before Comanche was in their range.

So we went into EMD with off-the-shelf radar and laser warning systems. Naturally, we had to integrate their signals into the overall mission computer database and cockpit display systems, but this was a minor iteration because we planned to get the same information from INEWS.

The original LHX requirement was for a look-ahead NBC environmental detector, for early detection and possible avoidance. The technology for this was found to be too immature and heavy. The requirement evolved to a point chemical detector, which sniffed the outside and would tell the pilots that the aircraft had entered a contaminated area and needed to be cleaned before pilot egress. The pilots were kept in an air-conditioned cockpit. Refreshed, sanitized air came from the environmental control unit using the pressure swing adsorber to clean the air.

ANTENNAS

A major success within the program was the development of conformal communications antennas that not only provided adequate performance without compromising the signature control aspects of the aircraft, but also were able to be integrated into a very limited amount of space. The primary locations for the communications antennas were the horizontal stabilizer and the vertical "endplates," which attached to the stabilizer. The UHF, VHF AM/FM, IFF, Link 16, and SATCOM antennas were located in these components. In testing, the SATCOM antenna, which consisted of separate 10-in. transmit and receive elements, performed as well as the standard external SATCOM antenna.

AIR VEHICLE INTERFACE CONTROL SYSTEM (AVICS)

The AVICS was the interface to all simple aircraft sensors (pressure switches, temperature sensors, etc.). It was built by Hamilton Sundstrand and included diagnostics to isolate failures between the sensor and wires. There were several interface cards in each of the two AVICS units.

The goal, which was achieved as we recall, was to make all the interface modules identical. This took some balancing among the various aircraft subsystem designs to stay within the total available interfaces. The diagnostic community was an excellent arbiter of necessary sensors.

Additionally, we also incorporated the same interface modules in the HIDSS EDEU. The EDEU covered sensors in the forward aircraft, and AVICS covered the rest. So we achieved a high degree of commonality.

ARMAMENT

Armament was a separate segment of Comanche, and the MEP engineers didn't get involved other than to write the ballistic equations for rocket and gunfire launches. It was planned that Comanche would fire the RF and Semiactive Laser (SAL) Hellfile missiles as well as the family of 2.75-in. Hydra 70 rockets, but these capabilities were still in the early stages of development at termination. We had an unbelievable requirement for single shot gun accuracy (one bullet on target) derived from some Army lab experiments that were impossible to realistically meet in a moving aircraft at long distances without knowing the wind velocity and direction along the path. A new rotary air data system was an essential invention to provide wind vectors at the aircraft.

An additional invention to attempt to meet the single shot requirement was a small inertial system mounted to the armament bay doors, to check the last-best alignment of the weapons launch rail to the targeting system after the weapons bay door was opened. It was flown and tested, but never fully integrated with the TAS.

The armament segment was always delayed by funding cuts. GE Burlington built the prototype 20-mm gun and developed lightweight ammunition for it, but it was never integrated into the aircraft.

ROTARY AIR DATA SYSTEM (RADS)

The other innovation to support the armament and flight control requirements was the rotary air data system (RADS), which provided accurate airspeed and altitude information. RADS was part of the MEP. RADS was a unique method of calculating airspeed in low airspeed environments.

Normally, helicopter airspeed and barometric altitude were measured by redundant pitot-static tubes that were placed at spaced (for fault-tolerant reasons) locations, generally in front of the rotor hub. The difference between the ram-air pitot pressure and the static pressure was converted to "forward" airspeed. The locations of pitot and static ports were chosen to provide the "least inaccurate" measurements over various airspeeds and aircraft attitudes. In forward flight, these conventional pitot/static probes provided acceptable accuracy for piloting, but measurements below about 30 kt were affected by rotor downwash and were generally unusable. This traditional system did not meet the needs of weapons for ballistic adjustments due to wind and air density.

Subtracting groundspeed/direction (from onboard inertial navigation systems) from airspeed/compass heading in forward flight yielded an approximation of instantaneous wind speed and direction. Wind parameters were essential inputs into ballistic equations used for accurate targeting of rockets and guns. Near hover conditions, where much of the weapons delivery took place, the airspeed/groundspeed derivation of wind did not work. A method of measuring the impact of wind on the airframe from all directions was required.

Thus, armed aircraft like Apache employed separate (but nonredundant) low airspeed systems to derive more accurate wind calculations in low speed and hover.

Forward airspeed and barometric altitude were basic inputs into the flight control stabilization system and needed to be redundantly available under all weather conditions. Hence, they included anti-icing capability.

There were three competing systems in the marketplace for measuring airspeed while at low airspeeds. One, which used radar or laser air particle motion detection measurements several lengths in front of the aircraft, was heavy and was not in production. A second used a gimbaled weather vane sensor looking at the combination of rotor downward thrust and forward airspeed impacts on the thrust, and had a large radar cross-section.

The third was mounted above the rotor hub on a standpipe and featured a set of spinning arm with pitot sensors at the tips of the arms. This system had a constant pitot pressure generated by the speed of the spinning arms, but additionally, provided a sinusoidal output created by the sum and difference pressures due to the airspeed impact on the sensors. The rotational point of maximum pressure provided the direction of the airspeed. This system was called low range airspeed system (LORAS).

Years before LHX, Sikorsky, Hamilton Standard, and the United Technologies Research Center evolved an idea similar to LORAS, but using a sensor mounted on the tip of the main rotor blade. The main challenges were to design a sensor that could withstand the very high *g* forces at the tip of the rotor blade and survive the dynamics of the environment. Additionally, it had to send its signals through the rotating hub. Although slip rings were common helicopter instrumentation devices, they were difficult to use with low-level signals due to brush contact noise.

The uniqueness of the RADS is the method by which the system periodically samples pressure data from a main rotor blade–mounted pitot scoop integrated with the high accuracy pressure sensor and computation of velocity vector that is resolvable into the aircraft's coordinate system. Mathematical techniques are employed that provide accurate airspeed computations along with inherent noise immunity. For example, the implementation technique provides the capability for special processing of pressure data sampled as the main rotor blade passes the turbulent flow created by the fuselage or tail wake.

Although the three previously mentioned low airspeed systems were not redundant, and therefore couldn't be used for flight-critical functions, the RADS was inherently as redundant as the number of blades. So the plan was to eventually replace the pitot-static probes with RADS. The RADS performed the combined functions of a conventionally redundant air-data system and a low airspeed sensing system in an integrated, seamless manner. Airspeed information was sensed and computed from zero to airspeed up to the maximum flight speed of the aircraft. In addition to the normal computation of the airspeed, the RADS computed, with the aid of the inertial system, the direction of the relative wind. This feature was most useful when the pilot had to execute hover or low speed, low altitude maneuvers in restricted visibility. The availability of relative wind velocity information coupled with navigation data enhanced the ability of rotary aircraft to perform accurate low altitude hover, fire control, and other autopilot maneuvers. The complete RADS as installed (three blades) on an aircraft consisted of three separate channels that independently provide airspeed information to each of the three flight control computers.

The RADS had the following components:

- The blade-mounted pressure sensor (BMPS) for measuring dynamic pressure when the blade is rotating—three per shipset
- The rotary interface unit (RIU) for transferring the BMPS power and signal from rotating to stationary fuselage—three separate channels
- The digital air data module (DADM) for computing airspeed—three per shipset
- A blade-mounted counterweight (BMCW) for balancing blades without BMPS—two per shipset
- Five pitot scoops for dynamic air pressure collection on a rotating blade

In order to compute the air-data information, the RADS required additional input data from the following redundant sensors:

- Ambient temperature sensor (T)
- Rotor speed (Nr)
- Static pressure sensors (Ps)
- One per rotor revolution signal

Among these sensors, the location, accuracy, and resolution of the static pressure sensor were very important. In the low air speed situations, the static pressure could get corrupted due to main rotor downwash and tail rotor effects. The cost of sensor, installation cost, weight, reliability, and maintainability were important factors in overall system consideration. However, the RADS IPDT developed an alternate algorithm that eliminated the static pressure sensor completely. This would have been a boon to reducing system complexity and cost.

The sensor was mounted on the trailing edge for signature and icing reasons. It was field removable and did not require blade rebalancing when replaced, although the field-level maintenance did allow blade balance checks in flight.

We used flat ribbon wire, normally just found within electronic boxes as backplane wiring, within the blade and out the root end to the rotating portion of the rotary slip ring. We could not, as I recall, pass the wire through the flat end of the blade root. Instead, it came out along a fiberglass section and was "glued" along the external root end. Again, this ribbon wire installation was a development project itself due to the severe vibration environment of the blade.

To reduce costs, we planned to install the sensor in three blades and use dummy sensors in the other two. That also allowed us to reduce the rotary transformer (slip ring) channels to three.

Originally, the drawback to this plan was a longer loose wire cable to allow it to reach adjacent blades. This created additional fatigue problems on the wires. The IPDT had several options:

- Increase the number of blade sensors and slip ring channels to five.
- Carry five sensors (identical blades), but only three channels on the rotary transformer, and swap blade positions if a sensor or wiring failed, thus eliminating the extra cable length and damage potential.
- Retain two dummy sensors, but keep the cables optimized for only three blades. This involved at least two maintenance actions if any blade failed (replace the sensor depending on location and the blade).

The third option was the leading candidate, but the final decision was deferred, because program cutbacks deferred armament development.

We tested the system on the whirl test stand, early in the program. All responses were positive.

Without the need for accurate wind calculations, until guns and rockets could be tested, the RADS development was also halted. The first prototype Comanche instead used an instrumentation boom to accurately measure airspeed and direction.

After years of delays due to other priority flight testing, we did spend a considerable amount of flight test time developing this system. The main challenges were in correcting errors due to angle of attack. The system was very sensitive to changing sensors inflow angles, due to changes in blade angle. We found, however, that the system could be calibrated using inertial vertical velocity (derived from flight control sensors, which had the requisite redundancy). The accuracy during high-*g* maneuvering remained to be developed, but the system looked promising. Late in the flight test program, the MEP display of airspeed was derived exclusively from the RADS.

Figure 3.8 shows only the forward-facing pitot opening of the sensor mounted on the blade. The sensor itself was within the blade. This shows the sensor inboard from the tip of the blade. The replaceable blade tip section was rather long.

The RADS would have been a tremendously sensitive and accurate input into the flight controls, as well as assisting in solving the "first shot" accuracy requirement of LHX. It would have allowed quick response to wind gusts in both flight controls and ballistic solutions.

Fig. 3.8 Forward-facing pitot opening.

CONCLUSION

The Comanche electronics were clearly the most advanced helicopter avionics in the world and were ideally suited for the advanced missions that Comanche would tackle and survive. From the early days, the MEP team relied on technology insertions to meet its stringent requirements. Although the crew station designers sometimes outstripped available airborne hardware and could only demonstrate the viability of their approach through the massive capabilities of simulators, airborne technology usually found a way to catch up (unfortunately, thanks to program delays).

If we look at today's computers, Comanche might seem somewhat simple, but we were always about 10 years ahead of other airborne equipment. We think we drove the Army into the present day.

Chapter 4

Low Observables

Bill Harper and Bruce Kay

When the LHX was originally conceptualized, there were no specific requirements for low observables; however, concurrent with early LHX development, others were studying the feasibility of making aircraft less detectible by an enemy and analyzing the effects that capability would have on the modern battlefield. As the decade progressed and dramatically lower aircraft detectability appeared to be feasible (and a real game changer on the battlefield), the LHX requirements were modified with specific detectability requirements. These were so low that aircraft then currently in production could not meet them with any type of feasible modifications.

To survive in combat, a helicopter has to avoid being detected; if detected, it must maximize its survivability. The longer a helicopter can remain undetected, the better its chance for survival.

Helicopters can be detected at long range by radar and electronic emissions. As they get closer, visual and aural detection come into play. Heat can also be a source of detection by infrared sensors.

Electronic emissions will not be discussed here, other than to say that power management and other low probability of intercept techniques were used to manage electronic emissions.

The Advanced Composite Airframe Program (ACAP) in 1980 (see Chapter 1) was intended to determine the feasibility of composite fuselages. It also included airframe shaping as might be included in a low radar cross-section (RCS) design.

In 1983, the companies received LHX preliminary design contracts from the Army that included signature reduction for the first time. In 1985, an Army letter of agreement was the initial requirements document with low observability (LO) requirements. In 1987, an Army Operational Requirements Document (ORD) and Required Operational Capability (ROC) document were released that defined improved survivability through reduced visual, aural, radio frequency (RF), and infrared (IR) signatures.

By then Sikorsky was teamed with Boeing. We were introduced to Boeing's low observable group in Seattle and continued as a team to develop our approaches.

Radar Cross Section

There are a number of parameters that influence the detection of a helicopter with a radar system—aircraft signature, altitude, speed, line of sight between the radar and the helicopter, and background clutter. The Boeing Sikorsky team evolved specific RCS design values through detectability modeling and simulation, exposure statistics, and radar processing modeling.

The original Army RCS requirement was one number—one very low number. We called it the "bug" criteria. The Army had based its requirements on similar studies of fixed wing aircraft. We felt that because helicopters flew in a different operational environment—low altitude with terrain clutter—the helicopter requirement need not be so severe. We ended up recommending LO requirements less severe than the Army's original requirements.

Our early analysis showed that helicopters were being detected primarily due to the dynamic time-varying returns of the rotor hub and potentially other time-varying returns. We modified our computer detectability models accordingly.

The next step was to understand clutter. Ground clutter is a complex function of ground parameters (conductivity, dielectric constant), radar frequency, polarization, grazing angle, surface slope, surface roughness, vegetation, and environmental factors such as rain, snow, and ice cover. Establishing a clutter value was important in the development of the RCS specifications. This was analyzed by both the Army labs and our team. Bill Harper was the technical lead at the Joint Program Office (JPO) and worked extensively with the Army on this. He was able to generate enough supporting information to convince the Army and the Rand Corporation that a single clutter number should be used to establish the Comanche RCS requirements. This was an important step towards concurrence that the Comanche would operate in a clutter-influenced environment.

Bob Hubbard of the Army program office arranged a meeting at Army Research Laboratory (ARL) with John Shipley. Bill was given the opportunity to present the Boeing Sikorsky position on the detectability analysis approach and the resulting signature level design requirements. Our position was that the aircraft radar cross section should be the size of a duck not a bee. The design requirements should vary with radar frequency classes and be specific to static and dynamic returns as they vary around the aircraft azimuth angle. Bill was not allowed to attend the session where Shipley presented the ARL position that had set the original bee requirements, but at the end of the day, Bob Hubbard concluded the meeting by telling John Shipley and Boeing Sikorsky that the Army Comanche Program Office was going to accept the Boeing Sikorsky recommendations.

We later learned that the special operations team wanted to fly at 300 ft above the terrain at high speed at night, and John Shipley didn't think the signature levels suggested for Comanche supported that type of operation.

This was a breakthrough moment in the LHX design. The lower frequency class threats required less signature reduction, and the higher frequency class threats, although requiring more signature reduction, were easier to reduce and control with treatments. Fuselage returns became less of a design issue, and rotor blade returns became less of a driving design problem. This was a major step toward establishing RCS levels that were achievable and effective for Army combat helicopters.

The next issue was whether a radar warning receiver was necessary. Army transport helicopters typically operated between 100- and 200-ft altitudes, above local terrain features, to stay out of small arms threat engagement range. If the crew has a warning of an operating radar threat in the area and reduces the aircraft altitude to 25 ft, the probability of the radar having line of sight with the aircraft is greatly reduced. Therefore, a radar warning receiver (RWR) is a critical subsystem needed on those helicopters to improve their survivability. The need increases with increasing conflict intensity and complexity. The RWR function was initially recommended as a necessary subsystem for the Comanche.

Dave Unsworth, Boeing Sikorsky weights engineer, published a technical paper in 1985 titled "Weight Impact of Advanced Technologies on VTOL Weight Prediction," which showed the weight impact of helicopter RCS reduction. This analysis showed that a jammer solution was not realistic, was extremely heavy, and would be costly if designed without signature reduction. With the Comanche level signature reduction, the jammer would still weigh over 200 lb and have a unit cost around $1 million. The bottom line was signature reduction alone was the best solution. Comanche did not use electronic jamming techniques or chaff.

At this time it was believed that RCS reduction increased the aircraft weight by about 500 lb but provided a survivable, mission, and cost-effective solution without the jammer. The signature reduction did offer a survivable solution option of adding a jammer when operational conditions required the platform to operate at higher altitudes for extended periods of time without the ability to mask the helicopter when painted by an active radar.

Not only was Comanche able to get very close to enemies without being detected, the helicopter could hide behind terrain, and in conjunction with its

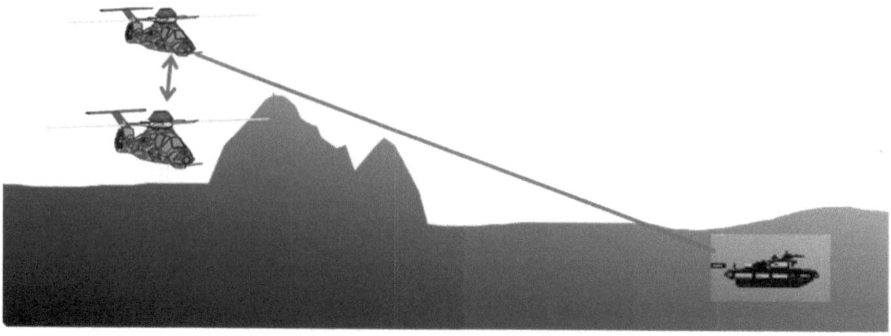

Fig. 4.1 Pop-up maneuver from cover.

Electro-Optical Target Acquisition and Designation System (EOTADS) was able to effectively use a pop-up maneuver to detect and destroy enemy forces with minimum exposure. The helicopter would quickly pop up, the EOTADS would scan the scene and put in memory any enemy targets that were detected, and then return to concealment. The crew would then analyze the scenario, select appropriate weapons, pop up again, fire the weapons, and destroy the targets before they ever knew what hit them (Fig. 4.1).

However, in another interesting analysis, it was shown that in some instances a helicopter behind and below the ridgeline was more detectable than a helicopter exposed to the threat radar with a background of terrain. In one case, operational testing confirmed that the probability of being detected dropped by 35% when the aircraft was in front of the terrain, decreasing signal to noise (S/N) or basically introducing more clutter return back to the radar. This was an interesting discussion with the user when Boeing Sikorsky attempted to explain that, at night, positioning the aircraft in front of the ridge to conduct search or engagement tasks was better than popping up from behind the ridge line.

Because the Comanche would be directly imbedded in general Army aviation units, it was considered impractical to maintain classification for its LO design features. The Army program manager made the decision that this was the best option, was supportable in the field, and was consistent with the Department of Defense (DoD) security classification guide. Furthermore, the design techniques used general signature control approaches that could be achieved with techniques and materials described in the public domain.

The specific levels for Comanche low observables were classified.

Impact on the Aircraft Design

Low observable factors played a profound role in establishing the helicopter configuration. Their consideration dominated the detail design of the fuselage, propulsion, and rotor systems. Analyses were performed by a dedicated RCS group who did their calculations in a secure facility. These analysts were also part of the integrated product development teams (IPDTs), and interacted with them on a daily basis. Key members of the IPDTs also attended a training course on electromagnetic theory and principles in order to foster better communication.

During the preproposal stage, the Army provided threat information and draft specifications for low observables. In addition, it gave the design teams the option of including some radar signature reduction hardware as a kit. Boeing Sikorsky chose that option to better meet the mandatory 7500-lb empty weight limit. The kit included sharp, pointed, treated chines to cover the rounded chines on the baseline configurations.

Kits would not be procured for all aircraft, reducing program costs. The use of kits also eased the issue of handling some radar signature–reducing materials that tended to be somewhat fragile in the field. Additionally, the Army felt that the use of the kits only for stealth missions would help keep those design features classified [1].

Tailoring of the specifications continued through detail design, though it became more difficult to change requirements as time went on. In order to do this effectively, it was necessary to understand the threats and translate them into meaningful effectiveness metrics. These metrics had to be boiled down into survivability/kill ratios or mission capability for various scenarios. To do this, simulations were used to lay down typical threats and assess how potential configurations would perform against those threats.

Signature reduction was achieved by shaping and judicious use of radar-absorbing materials. Fuselage shaping to reflect incoming radar waves away from search radars was the first means of reducing radar signatures. Conductive surfaces will reflect radar waves similar to optical reflectance laws. Any small protuberances can reflect radar, so it was necessary to provide smooth external surfaces. Landing gears and weapons had to be retractable, and other traditional external features such as antennas had to be made flush with the outside mold lines. Transparencies had to be treated with very thin conductive materials to prevent radar from entering the cockpit. The most important aspect was the nose-on signature, because that would reveal an approaching helicopter. Velocity is also what allows threat radars to separate aircraft returns from clutter returns. It was not necessary to reduce lateral signatures very much, and the fuselage sides incorporated only moderate shaping. Signature-level specifications were provided as a function of azimuth position and radar frequency. Pitch attitude was also considered. Specification levels were azimuthally grouped in bands. Actual signatures were anything but simple and consisted of a series of peak and valley spikes (Fig. 4.2).

Static signature returns, shaping, and all the dynamic returns were analyzed with special RCS computer codes and evaluated with scale and full-scale model tests. Back then everyone except Sikorsky had to buy time on a Cray

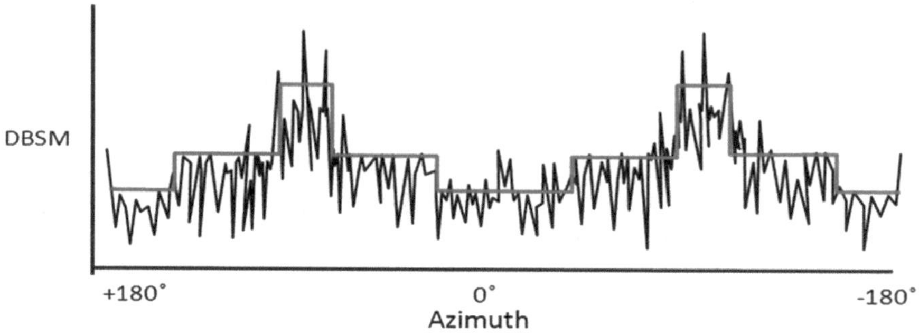

Fig. 4.2 Azimuthal radar cross section.

computer to do RCS analysis. Sikorsky was the only company that could run RCS on a PC.

The more difficult and insidious challenge was to understand and mitigate the source of dynamic signature return spikes. There were numerous sources of these signature spikes, mostly attributed to the main rotor, fan, and rotor/body interference (RBI) returns, and all of them had to be addressed. The main rotor blades and antitorque system by themselves required special treatments that are discussed in their design description chapters. Main rotor blade leading edges were of particular concern along with additional challenges associated with dynamic rotor body interactions (RBIs).

RBI occurs when incoming radar beams reflect off the fuselage onto the blades, or vice versa, and back to the radar receiver (Fig. 4.3). Local radar-absorbing material on the main rotor pylon, underside, and leading edge of the blades was used to mitigate this phenomenon. Rotor hub reflections were prevented by special-shaped hub fairings.

The Fantail did meet the low observable requirements; however, there was a concern because of the Doppler requirement for the Fantail blades. Also

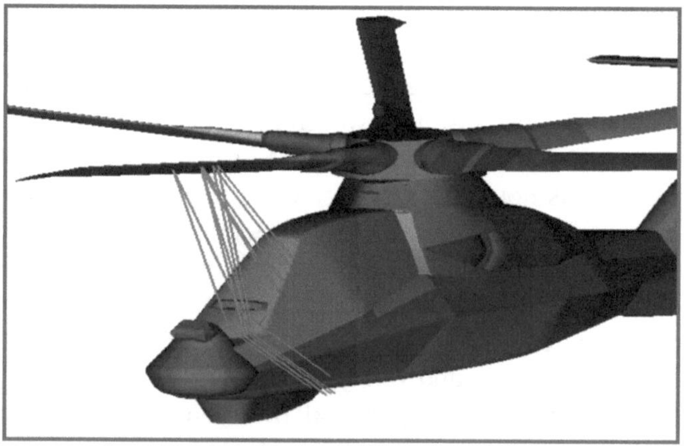

Fig. 4.3 Dynamic rotor body interference.

considered was how an improvement in the overall RF characteristics could be made in order to achieve better than the specification levels. It was initially decided to try a screen kit for the Fantail. It was lightweight and provided acceptable performance for missions on which it would be employed; however, we did tests and found that the openings had to be large to prevent airflow blockage, and that made them ineffective as an RF cover. The screens were eliminated and were never part of the LO kit.

A lot of attention and careful integration were also necessary to prevent big returns from large cavities, like the engine inlets, exhausts, and air management apertures. Unique, flush The National Advisory Committee for Aeronautics (NACA) engine inlets were designed to defeat frontal radar threats. Tuned screens covering a convoluted inlet duct prevented radar waves from reflecting off the front face of the engines. The engine exhausts were fairly concealed in the tailcone structure and shielded from most aspect angles. Where the ducts were exposed, they were treated with radar-absorbing material. Other smaller air inlet and outlet apertures used open cell treated honeycomb to minimize returns. The honeycomb formed a grid that allowed air to pass through, and the cell walls were treated to attenuate radar signals.

In keeping with the smooth exterior surface philosophy, we wanted to eliminate anything that disrupted a perfectly smooth exterior surface. Hinges, latches, skin laps, protruding head fasteners, and so forth were banned outright and replaced by flush alternative designs. Stringent step-and-gap criteria were imposed on all doors and removable panels. Considerable effort was expended to develop the design, tooling, and manufacturing processes required to achieve the very tight step-and-gap tolerances. There was a significant benefit in assembly, because the parts fit together without the need for any tedious, labor-intensive adjustments.

It's interesting to review a few examples of some out-of-the-ordinary design solutions for common applications. It is commonplace in the helicopter industry to use latch mechanisms to secure engine cowlings and other frequently operated access panels. In order to minimize local radar hot spots emanating from latches, they were replaced with screws on panels, which did compromise maintainability to a small extent. Screw heads in radar-absorbent panels were covered with radar-absorbent material caps.

The quest for smooth external surfaces went even further than just eliminating obvious feature disruptions. It was felt that by virtue of their large quantity, even countersunk fasteners could adversely contribute to radar detectability. This resulted in a major change in the way the external skins were joined to the substructure. T-blades were co-cured as part of the skins, and the substructure was then fastened to the blades instead of through the common skin flanges (Fig. 4.4). To avoid any skin waviness, honeycomb sandwich skins were employed throughout the airframe. In earlier helicopters, skins were allowed to buckle under load. A small weight penalty was paid for this accommodation.

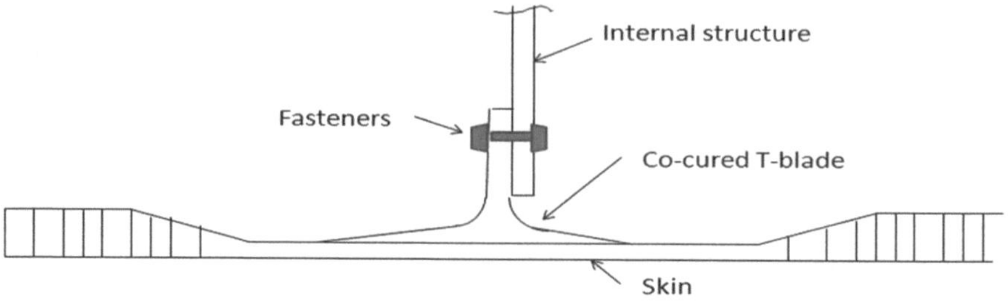

Fig. 4.4 External skin construction.

Where it was not practical to use shaping to reduce signatures, radar-absorbing materials were utilized. These conditions were applied primarily to the main rotor pylon and the tail. The treated surfaces were constructed with honeycomb sandwich material (Fig. 4.5). The outer surfaces were either Kevlar or fiberglass, materials that were transparent to incoming radar waves. The inner surfaces were conductive graphite, and the honeycomb cells were treated to attenuate radar waves as they entered and reflected off the inner surfaces. By tuning the sandwich depth to match radar wavelengths, reflected signals could be reduced by phase cancellation. Using this sandwich construction technique, there were only slight weight penalties for treatment. This is in stark contrast to fixed wing fighter planes and bombers that use magnetic radar-absorbing materials that can weigh several pounds per square foot, which is more than helicopter skin structures weigh by themselves.

The next level of treatment was associated with traveling radar waves that could cause local emanations from any kind of surface discontinuities. Even when reflected, some residual radar energy could travel along conductive

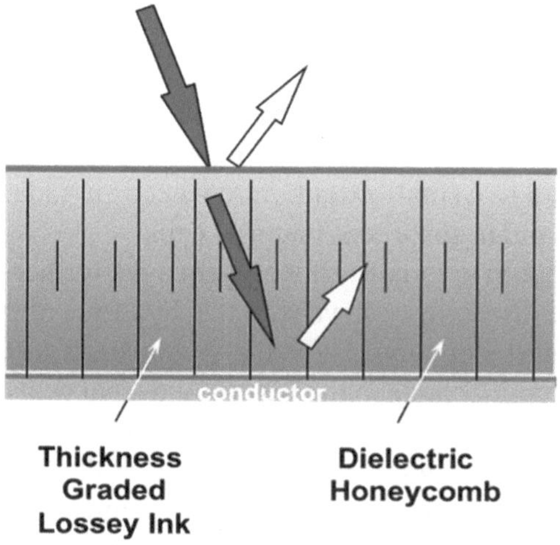

Fig. 4.5 Radar-absorbing honeycomb sandwich.

surfaces and reradiate at discontinuities. Suppressing this condition was a lot more difficult to implement. Discontinuities included contour changes, skin joints, and local design features like access covers, antennas, and fasteners. These were the features that caused local radar signature spikes. In addition to tight step-and-gap tolerance control between panels, the geometry and construction of the joints had to be fine tuned to reduce radar signature. Elastomeric seals between movable reflective panels were loaded with conductive material to provide electrical continuity. Edge treatments for both fixed and movable sandwich panels were tediously designed to achieve desired electrical properties. To do this, R-cards were incorporated in the composite layups. The R-cards were basically plies of special material with graded electrical resistance. Gaps between radar-absorbent panels did not have seals, but the geometry and materials used in the panel closeouts were rigorously controlled to minimize adverse signals.

Many of the special RCS materials were procured from very small businesses that in common parlance might be called "garage shops." This caused the procurement and material engineering departments to have difficulties with standard procedure compliance.

There were several other significant problems associated with incorporating this level of RCS control. Although it was theoretically understood that these features would cause adverse effects, quantification was another matter entirely. There were no pragmatic analytical prediction methodologies available, so element tests were used to measure returns. The element tests were performed in Sikorsky's secured anechoic chamber facilities. All of the features were catalogued, but assimilating the cumulative effects into the overall aircraft signatures was not possible. There was just not enough accurate signature characterization from the element tests, or computer capability to geometrically overlay the local data onto the overall signature models. So without solid data, the RCS engineers were faced with the difficult task of trying to convince hardnosed, veteran designers to incorporate unusual features in their designs. This task was made even more difficult because most of the RCS engineers were relatively young and had limited experience dealing with the old timers. Furthermore, the designers had enough RCS training to enable formidable, if not always correct, push-backs. It was a contentious relationship between ideologues and pragmatists.

The RCS engineers were also handicapped because of security compartmentalization regulations. Even if they had successful experience with a particular approach from another classified program, they were unable to bring that information to the debate. Classified data from other programs could not be used, and if needed, would have to be regenerated anew. Nevertheless, persistence paid off, and palatable compromises were usually reached.

The effectiveness of treatments was eventually evaluated with a full-scale RCS model (Fig. 4.6). Different radar frequency signatures could be accu-

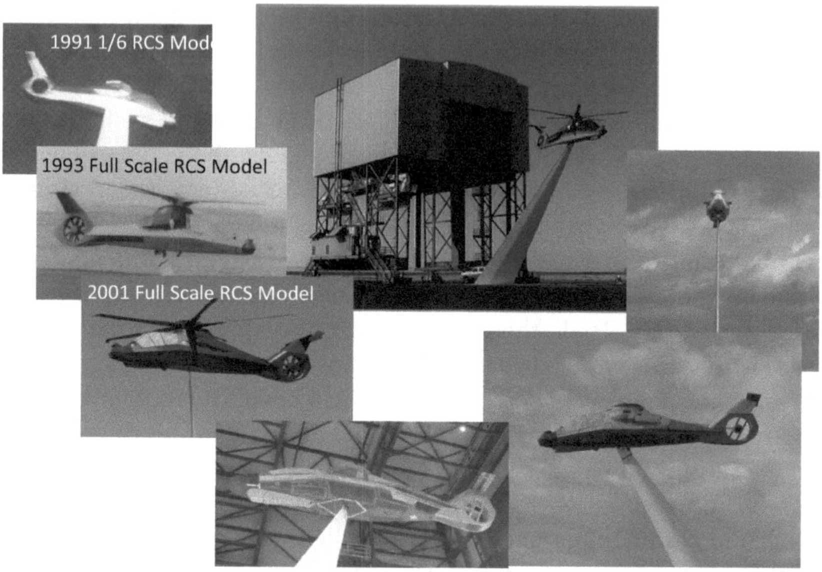

Fig. 4.6 RCS test models (Boeing Photo).

rately evaluated, which was not possible with an earlier 1/6 scale model. Rotor body interference effects could also be measured by clocking the rotor blades.

For security reasons, it is interesting to note that these tests, performed outdoors, had to be conducted on moonless nights to prevent any observations by snooping foreign satellites (or so we were told).

In all, over 5000 hours of RCS testing was conducted.

Infrared Signature Reduction

In order to defeat heat-seeking missiles, control of infrared (IR) signatures was a basic defensive necessity for Comanche. The principle sources of IR radiation from an aircraft are its emitting and reflecting surfaces (Fig. 4.7). Reduction of these sources is required to degrade or deny detection and tracking. This can be accomplished by:

- Reduction of the temperatures
- Shielding or masking radiating areas
- Reduction of surface reflectivity
- Reduction of surface emissivity

The goal was to reduce the apparent IR contrasts between the helicopter components and the environmental background. For Comanche, this was based on mission analyses that considered flight conditions, operating environment (daytime, night, summer, winter), and expected threats. A number of different computer programs were used to model IR signatures and guide the design.

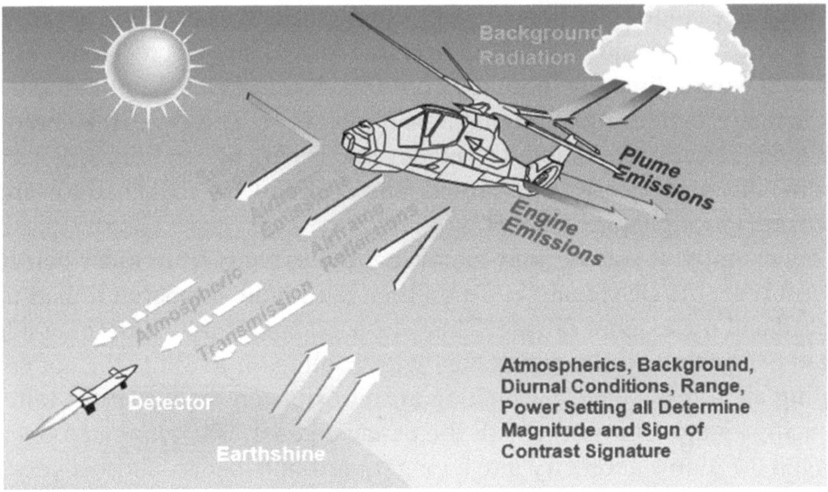

Fig. 4.7 IR signature contributors.

During a meeting held at Boeing in April 1992, it was discovered that there were two schools of thought on the IR missile lock-on calculation. Aviation Applied Technology Directorate (AATD) used one method, and the Boeing Sikorsky team was using a method recommended by Army Material Systems Analysis Activity (AMSAA) of the Defense Intelligence Agency (DIA). A DIA document review revealed a third technique was also used that only used 50% of the missile detector waveband. A literature survey showed that all the techniques had been and were being used by the Army, Navy, Air Force, RAND, and the Institute for Defense Analysis (IDA) to calculate infrared missile lock-on ranges over the past 30 years. In order to determine a single approach to be used by the Army and the Boeing Sikorsky team that best represented a missile's capability, we contacted the Missile and Space Intelligence Center (MSIC) for its recommendation. It indicated that the method recommended by AMSAA was the appropriate method to calculate lock-on ranges. This was important to ensure that the Army and the Boeing Sikorsky team were getting consistent results in their simulations. But there were other agencies like the RAND Corporation and the IDA that were using their own methods. How RF and IR threats were modeled was extremely important to get realistic results and to ensure the results represented the capabilities offered with Comanche.

The most attention was given toward reducing hot engine exhaust signatures. Engine exhaust IR signature suppression was not new for Sikorsky. Production UH-60s were delivered with IR suppressors. These units were mounted on the engine exhaust and were optimized for hover conditions. However, this approach was considered a nonstarter for Comanche. The suppressors were too big, had large radar signatures, and did not provide the necessary amount of suppression to survive Comanche missions.

Early trade studies had considered locating the exhausts on top of the fuselage, where they would be shielded from the lower hemisphere. However, this approach would direct the hot gases onto the rotor blades, heating them and thus actually increasing IR signatures. There were also concerns about the high temperatures weakening the blades under parked conditions. For these reasons, this approach was discarded, and efforts were dedicated toward an improved, low signature concept.

Conceptually, it was desired to conceal the exhaust from radar detection and minimize its IR signature. To do this, the plume temperature had to be reduced and made more diffuse. This was accomplished by designing a long slotted exhaust manifold, augmented with cooling air to dilute the hot gases. Cooling air was obtained by sucking air into slots on the top of the tailcone and using ejectors to mix it with the exhaust gases. The elongated exhaust changed the plume geometry into a long thin ribbon that produced a very low IR signature. Extensive lab scale tests were conducted at the United Technology Research Center to optimize geometries, gas/air mixing ratios, and ejector efficiencies. Exposed exhaust nozzles were corrugated in order to maximize their contact areas with cooling air.

An internal UTRC newsletter dated May–June 1991 discussed the development of the IR exhaust: "Senior Army officers took a close look at the technology during a visit to the Research Center. The team made this possible by building a full-scale prototype exhaust system and also by producing a video to simulate its in-flight performance using scientific visualization technologies developed at the Center."

Bob Frawley, who led the development of this novel approach, was awarded UTC's prestigious George Mead medal for his accomplishment. The Mead medal is given out annually by the corporation to honor the most significant technical achievements of the year.

The environmental control system and the transmission and hydraulic systems heat exchangers also produced heat that had to be vented overboard. The outlets for these systems were located on the top of the main rotor pylon where they would not be visible to lower hemisphere IR threats (Fig. 4.8).

Low emissivity paint was used to complete the Comanche IR signature reduction suite. The paint also had to have a matte finish to minimize light reflection and had to be resistant to chemical warfare gas agents and their associated decontaminants.

Lockheed Sanders conduct a hardware-in-the-loop (HWIL) test to establish whether a missile could track and intercept the Comanche aircraft at the short lock-on ranges identified by the physics equations, as well as at long range. The HWIL use of the Stinger missile seeker and lock-on before launch was not a criterion. The missile was allowed to acquire the target during flight. There were more than 50,000 simulation runs at various aspects, altitudes, and speeds. For long-range shots at the low altitudes, there was insufficient signal

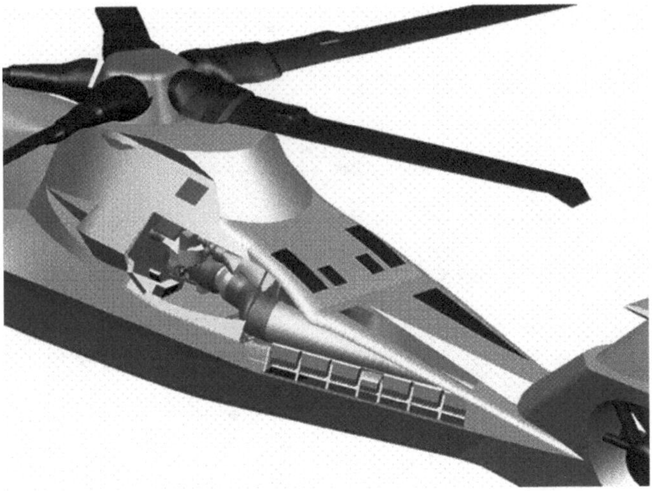

Fig. 4.8 Low IR signature engine exhaust and hot air outlets.

in the seeker electronics for tracking of the missiles. This ended in ground collisions. For the short-range launches, the issue was no well-defined track point associated with the plume or hot spot on the aircraft. This caused ambiguity in the seeker track point, which manifested into a "wandering" track point along the aircraft boundary. Statistical analysis of the simulation data showed that at short range with an aerodynamically limited capability due to missile speed, the probability of intercept of the Comanche platform ranged from 1.0% at low speed to 7.0% at high speed. The probability of intercepting a current technology signature control helicopter in this interkinematic range was 78%. Improvement in survivability over the current Army helicopters with engine IR suppressors ranged from 93% to 99% with the Comanche signature reduction approach.

NOISE REDUCTION

Most people associate helicopters with noise. Ubiquitous Huey helicopters in Vietnam made their incoming presence known long before they became visible. Even modern helicopters are fairly loud and could not be considered stealthy in any way. So, it was obvious that noise abatement was necessary for Comanche.

The main sources of noise for a helicopter are the main and tail rotors. Most of the familiar "wop-wop" noise emanates from the main rotor blade tips. We were able to optimize tip volume to mitigate most of that sound; however, additional means were required, and this was accomplished by incorporating a low-speed quiet mode in the helicopter flight control system. Rotor speed was reduced 10% for that condition.

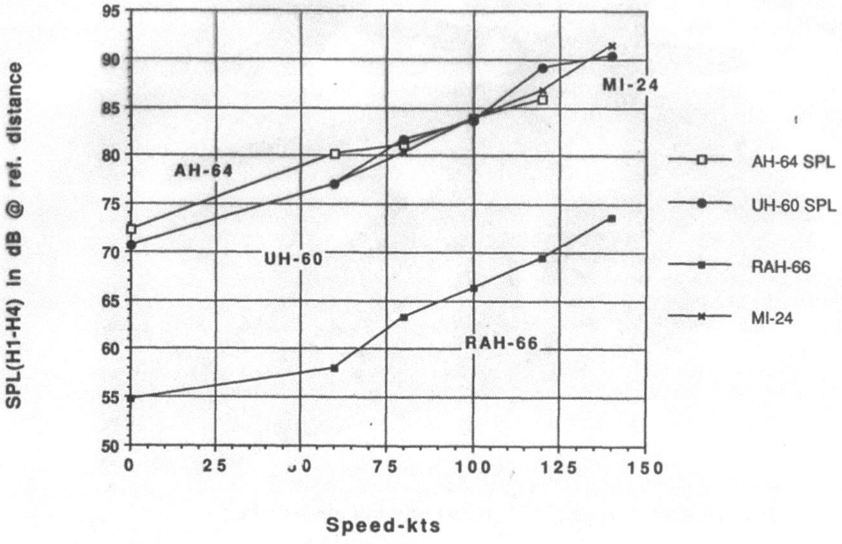

Fig. 4.9 Helicopter noise levels.

Eliminating whine from the Fantail entailed tuning the blades and stators. The Fantail was also specifically designed to eliminate the "siren" effects of then-current tail fans, again by using geometries specifically designed to reduce noise.

Acoustic characteristics were measured during wind tunnel and whirl stand testing. Sound arrays also were used to collect noise levels on the ground from various military and commercial helicopters to gather data on noise level vs detection range. Sound measurements taken during Comanche flight testing validated the low noise levels. Although the helicopter was not silent, it certainly did not announce its approach by any means. Chapter 26, Marketing, explains the dramatic way we demonstrated this during flight demonstration to VIPs. Figure 4.9 compares Comanche to then–current generation helicopters. The 15-dB difference is dramatic. Comanche noise levels were comparable to standing on the sidewalk and hearing an automobile passing by.

VISUAL DETECTION

Visual detectability is primarily affected by the vehicle size, contrast with the background environment, and sunlight reflections. Size was pretty much a given, and not much could be done in that regard. Blade flicker is also a factor and was minimized by using five blades instead of four. Comanche followed the common attack helicopter practice of using flat-sided transparencies to restrict glint to small angles of reflection.

During the proposal stage, the Army requested antireflective coatings, similar to that used on eyeglasses, to be applied to the canopy. This was the only

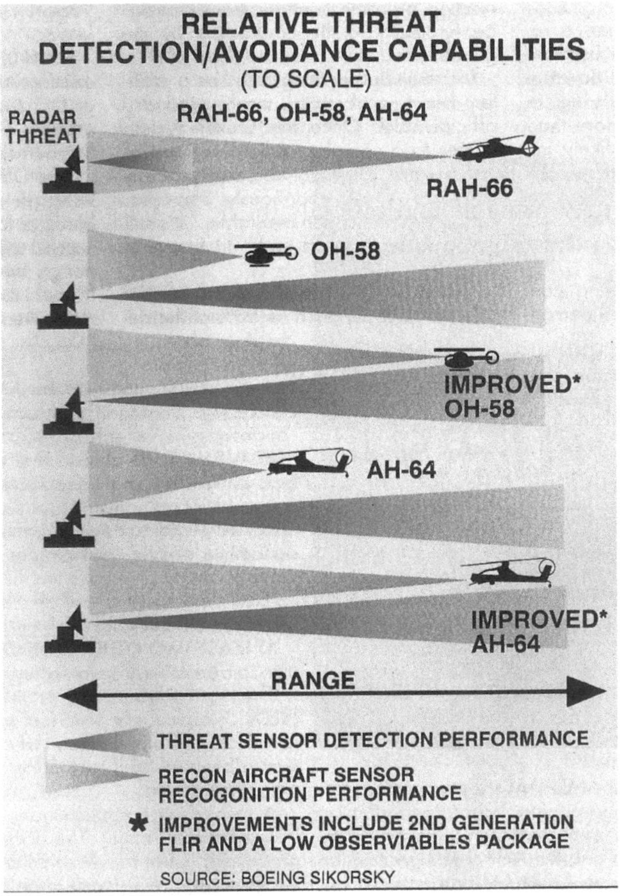

Fig. 4.10 Relative Threat Detection/Avoidance Capabilities.

specification that Boeing Sikorsky could not comply with, because durable large-scale antireflective coatings did not exist and were beyond the state of the art.

Further details on all these designs can be found in the appropriate aircraft design chapters.

Press Status Reports

Boeing Sikorsky's progress on developing the low observable characteristics of the aircraft was periodically reported in the aviation press. The following are some of these reports:

> With the armed reconnaissance weapon load stowed internally, the Boeing Sikorsky LH is "very close" to the Army's goal for low radar cross section, which is less than 1% of those for existing helicopters,

> [Army LH Program Manager Major General Ronald K.] Andreson said. Boeing Sikorsky "beat the heck out of the IR (infrared) requirement, which was much more stringent that anything that exists in the field today," he added. [2]

From *Aviation Week & Space Technology*, 19 July 1993 [3]: "According to Boeing Sikorsky officials and officials in the Army's Comanche program office, radar, infrared, acoustic and visual detection tests and simulations performed over the past two years indicate that the current configuration of the RAH-66 will offer significant stealth improvements over existing aircraft."

The article quotes then–Boeing Sikorsky Program Director Rick Hellyar as saying that the LONGBOW Apache had a frontal radar signature about 630 times larger than Comanche, and the OH-58D was 250 times larger.

By 21 October 1996, *Aviation Week & Space Technology*, in an article titled "Tests Confirm Low-Observability of Comanche Scout Helicopter" [4], said that "Both Army and the Boeing/Sikorsky team program officials said that preliminary tests indicate the Comanche will have a radar cross section (RCS) that is 1/400th of the [non-LONGBOW] AH-64 Apache and infrared cross section with 1/4th of that attack helicopter's suppressed IR signature…".

Also:

> program officials said they have long-term plans to make these emissions even smaller to meet expected improvements in sensors that may be developed or brought on the world market by potential foes after 2006.
>
> Follow-on programs associated with the RAH-66, including a number of classified projects, involve work on reducing radar reflection, or "flash" from the rotor blades with coatings of radar absorbent material (RAM). Other initiatives are to hide sensor openings from radar with transparent coatings and, possibly, to adapt the use of bi-static radar. The latter is not foreseen as being operational until 2015 or later. [4]

References

[1] David F. Brown, "Teams Competing for LH Contract Differ on Need for Low-Observables Kits," *Aviation Week & Space Technology*, 22 October 1990, p. 16.

[2] David F. Brown, "Boeing Sikorsky LH Operating Cost Targets 40% Below Army Estimate," *Aviation Week & Space Technology*, 22 April 1991, p. 27.

[3] Stanley W, Kandebo, "Boeing Sikorsky Findings Underscore RAH-66 Stealth," *Aviation Week & Space Technology*, 19 July 1993, p. 22.

[4] David A. Fulghum, "Tests Confirm Low-Observability of Comanche Scout Helicopter," *Aviation Week & Space Technology*, 21 October 1996.

Chapter 5

ARMAMENT

BRUCE KAY

For the basic armed reconnaissance mission, Comanche carried 4 Hellfire and 2 Stinger missiles in internal compartments, and 320 ammunition rounds for the nose-mounted gun. For the air combat mission, 2 Hellfires, 2 Stingers, and 500 rounds were employed. A wide range of other weapon load-outs was possible, as shown in Fig. 5.1.

The Hellfire missile is primarily an antitank weapon. It contains a seeker head that locks on to a target designated by the electro-optical target acquisition and designation system (EOTADS). Once the target is handed off to the seeker, the helicopter does not have to maintain line of sight and can return to a concealed position. Stingers are heat-seeking missiles, primarily used for air-to-air combat. Rockets are used for ground attack and are not very accurate, but fired in salvos they can create a lot of havoc.

As part of the first preliminary design studies, the missile launchers were concealed inside a sponson on the side of the fuselage. Aerodynamic drag and radar signature concerns led to the retractable configuration, featuring a sliding rack stored inside the fuselage. The retraction mechanism and cutouts in the fuselage shell were especially complex with this arrangement. In the retracted position, the launcher racks also compromised fuel cell packaging. The third and final iteration, evolved along with a simple, continuous box beam core structure, was to hinge the launchers. This was a much simpler and cleaner design solution. To facilitate loading the missiles, a hinged door was mounted to the launcher structure. In the upright position, this provided unconstrained access for armorers (Fig. 5.2). The door also served as a work platform to facilitate main rotor pylon maintenance activities.

Fig. 5.1 Comanche weapon load-outs.

For a full Hellfire load-out, an external stores rack was installed that could hold eight additional Hellfires (Fig. 5.3). Or, additional Stingers could be carried. The external rack was also used to carry two range extension tanks for self-deployment ferry missions up to 1260 n mile range. The external store rack was a bolt-on kit.

During the proposal stage, there was a considerable amount of negotiation with the Army user community about LHX gun requirements. Based on stringent weight restrictions, the First Team was recommending a 50-caliber machine gun, which was capable of suppressing enemy troops on the ground; however, the user community wanted to be able to defeat light armored vehicles. The customer prevailed, and a stowable, externally powered,

Fig. 5.2 Retractable missile launcher.

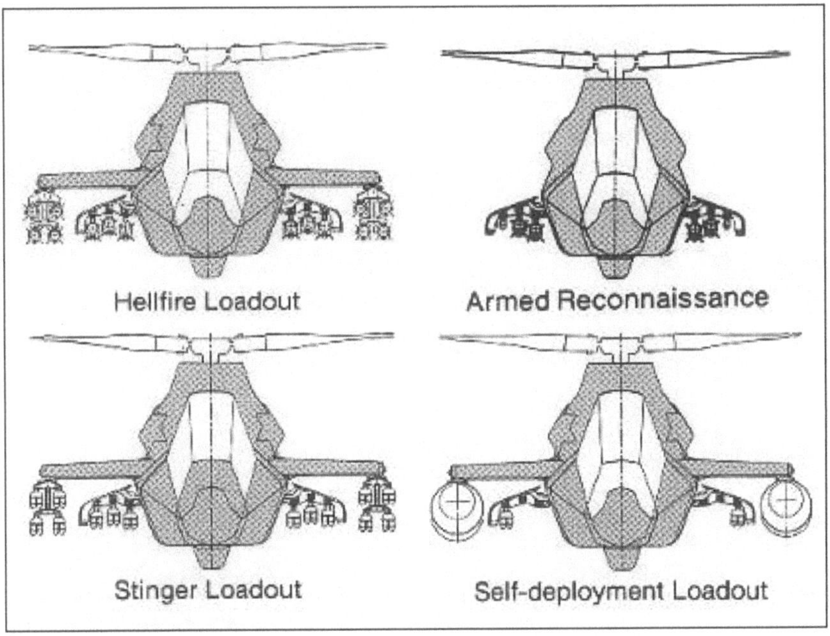

Fig. 5.3 Armament load-out options.

turreted XM301 three-barrel, 20-mm Gatling gun produced by General Electric in Burlington, Vermont, was selected (Fig. 5.4). (GE was later sold and became General Dynamics Armament and Technical Products.) The gun was mounted on a GIAT composite turret (weighing 127 kg) under the nose of the helicopter. This gun was optimized for Comanche and was the lightest

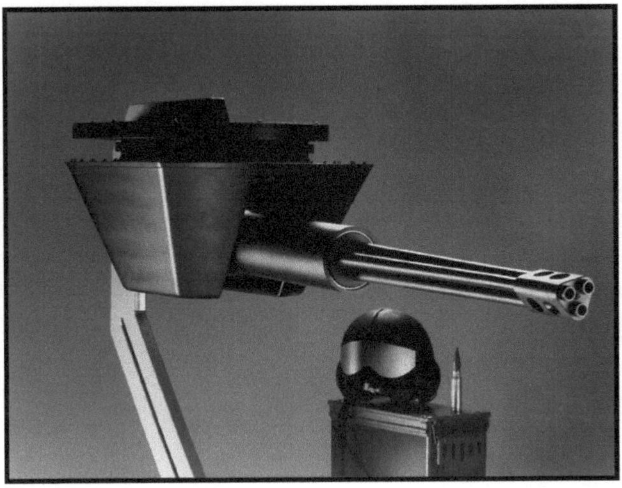

Fig. 5.4 20-mm turreted Gatling gun.

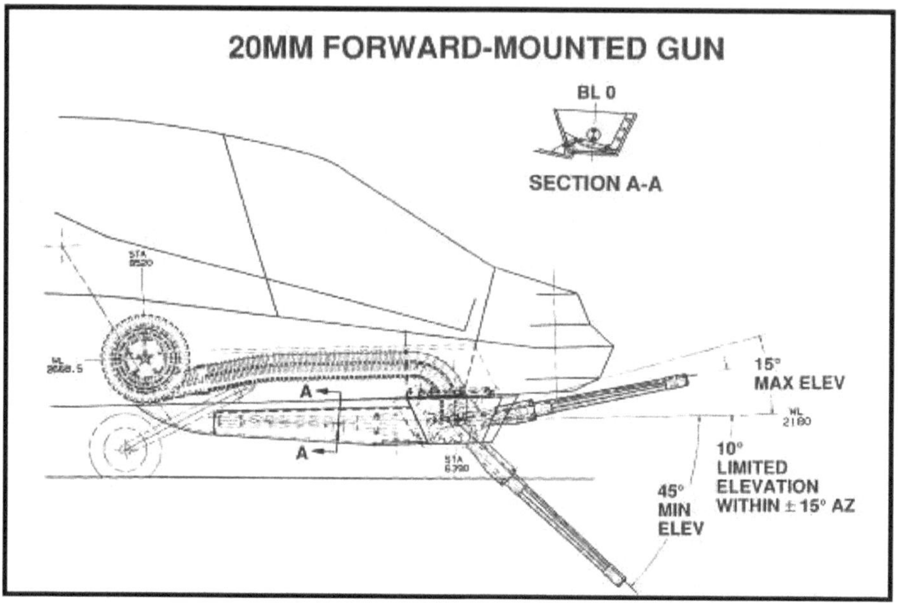

Fig. 5.5 Nose-mounted 20-mm Gatling gun.

20-mm Gatling gun in the world. It featured firing rates of 750 and 1500 shots per minute. Its round dispersion pattern was a tight 2.2 mrad. The turret had a side-to-side slew range of ±120 deg, an up elevation of 15 deg, and a down elevation of 45 deg (Fig. 5.5). Slew rates were 80 deg/s azimuthally and 60 deg/s in elevation. It was controlled by a helmet-mounted sighting

Fig. 5.6 Faired gun turret and ammunition reel.

reticle. Dynamic bore sighting was accomplished in flight electronically, eliminating the need for mechanical alignment tools and targets. All of these features made it an ideal weapon for air-to-air and air-to-ground lethality, in a lightweight package.

The gun turret was housed in a fairing under the fuselage nose section. In the retracted position, the gun rotated 180 deg aft to become fully concealed inside the fairing. In addition to aerodynamics, the fairing also was designed to minimize radar cross section.

Ammunition was stored on a reel mounted inside the main fuselage box beam and was fed to the gun via a flexible chute (Fig. 5.6). The ammunition path was carefully designed to avoid jams and for easy loading. The 500-round ammunition supply system could be reloaded in less than eight minutes by two crew members. Reliability was expected to be at least 100,000 mean rounds between failures (MRBF), more than four times higher than single-barrel weapons.

Chapter 6

HANDLING QUALITIES

JOHN OCCHIATO
HANDLING QUALITIES PROGRAM LEAD

From the beginning, we knew that LHX had to be able to outmaneuver and precisely and quickly target any airborne or ground-based adversary. Additionally, it had to have superior advanced scout capability, so it was important that the pilot could confidently operate aggressively and with precision close to the ground in dense terrain. This meant that it had to have almost fixed wing fighter–type maneuvering performance and precise handling characteristics.

In the design phase, every design feature that impacted these attributes was carefully scrutinized. The handling qualities team was passionate about delivering these attributes. The air combat requirement alone drove engineers to a single-rotor design over a tilt-rotor design for which there were advocates. After winning the LHX competition, the Boeing Sikorsky response to air combat requirements resulted in the Comanche becoming the most maneuverable, agile, and best-handling helicopter ever built to that point in time, and indeed, well beyond. The RAH-66 represented a significant leap forward in the world of helicopter handling qualities, flight control systems, and control law development, and, even though it was never put into production, it changed the culture and approach used ever since in these areas of helicopter design.

There were several contributing factors that gave the Comanche superior maneuvering/handling performance. Certainly, the bearingless main rotor (BMR) and fly-by-wire (FBW) flight controls with advanced control laws were the chief contributors; however, there were other factors as well. For example, a significant amount of attention was paid to engine integration, antitorque fan performance and control, and the empennage design. What was critically understood from the beginning was that, with such enormous attention being paid to air combat capability, precise control and pointing accuracy were important, and that would mean that any weakness in handling performance would show up dramatically. Boeing Sikorsky engineers knew that they had to pay significantly more attention to the details than had been the case in previous designs if they wanted to succeed.

It is appropriate to start by offering some basic definitions. Having excellent maneuvering performance meant that it had to have high levels of both maneuverability and agility, and these attributes drove the design in different ways:

- *Maneuverability:* The ability to change the energy state of the system measured in acceleration and/or change in direction
- *Agility:* The ability to change the maneuver state measured in quickness and accuracy

The maneuverability of an aircraft is a measure of its "muscle," whereas its agility is a measure of its "finesse and repositioning precision." Maneuvering capability is largely a measure of an aircraft's aerodynamic performance capability, with power and thrust margins being the critical determinants. Agility is directly related to an aircraft's handling characteristics. Control power, damping, and response to control are the critical agility determinants. Whereas a system's maneuverability is solely a product of the basic air vehicle design, agility can be significantly augmented through clever automated control to enhance basic air vehicle fundamentals.

Maneuverability Requirement

The central maneuvering requirement for both the Black Hawk and the Apache helicopters was an obstacle avoidance maneuver requiring a pull-up to $Nz = 1.75\ g$ followed by a push-over to $0\ g$ Nz (normal load factor, with $1\ g$ being balanced nonaccelerated flight) under 4000 ft/95°F ambient conditions. That specific callout determined the main rotor solidity requirements and affected the hinge offset selection, to some degree, for those designs.

For LHX, the high speed obstacle avoidance maneuver, again under 4000 ft/95°F ambient conditions, was different, demanding a pull-up to 300 ft followed by push-over recovery to within 15 ft of terrain masking level, all within 6.5 seconds. That combination of obstacle height, timing, and terminal accuracy forced an increased Nz range over the Black Hawk and Apache to about $2\ g$ pull-up to $-0.5\ g$ push-over. This was a very influential driving requirement for the main rotor design decision.

Handling Qualities/Agility Requirements

The most significant set of LHX requirements, however, came from the new handling qualities specifications. The Army crafted a significantly different specification, one that virtually required not only FBW flight controls, but also a high level of attention and detail in the design phase. A summary of the more significant changes that were made moving from the old MIL-H-8501A spec over to new ADS-33 spec includes:

- ADS-33 added several mission task elements (MTEs) composed of all of the maneuver elements that are required as the helicopter is operating at low speed near terrain. It called for the handling to be designed to meet Level 1 for these elements. (Level 1 means simply suitable for the task with acceptable workload, whereas Level 2 would mean slightly degraded and requires more piloting work to meet the task performance goals. If rated Level 3, significant pilot workload is required to perform the task, and improvements are required.) Specific levels of time and precision were called out for each element. This meant addressing specific control techniques to optimize the maneuver performance.
- ADS-33 required consideration of a useable cue environment that meant the design had to be equally adequate for combat operations at various levels of visual cueing. The control laws needed to allow for mode changes to accommodate all conditions. Simulation testing had to adapt to consider several levels of usable cueing.
- The new specification suggested that Level 1 be the goal for all flight operations at all envelope conditions and not just at the more critical flight maneuvers.
- The new specification also called for different response characteristics in near-earth, low-speed maneuvering vs up-and-away flight maneuvering; modes such as low-speed turn coordination and velocity stabilization were expected.

Needless to say, these changes, and others, demanded a lot of attention in design and a lot of added simulation to demonstrate that they were being met. It was fundamentally these additional requirements that forced FBW on the LHX design along with a comprehensive set of control laws in order to meet these demands.

The attention paid to Comanche's handling qualities design, as well as some of the more interesting design process details and stories, are discussed in the remaining sections of this chapter and in Chapter 8, Flight Controls.

To a large extent, the new LHX requirements for maneuvering and flight path control revolutionized the ways and means in which handling qualities and flight controls designers had to respond. The new handling performance requirements alone essentially demanded that the flight control system be a full-authority FBW system. They also added specific complex features to the flight control architecture. Once it was realized that the flight control system would be an FBW design, it was apparent that the cockpit would not be equipped with conventional full-throw sticks but would use some side-arm controller scheme for most if not all control axes.

Starting a program like LHX with all these major system changes and with new maneuvering and extended handling expectations forced a dramatic change to the flight control and handling qualities design culture that was the norm at Sikorsky at the time. It also forced a new level of work intensity and coordination between the piloting community and the handling qualities/flight controls folks, which in turn demanded an order of magnitude increase in piloted simulation throughout the entire program. This was absolutely required in order to adequately reduce risk in the design prior to and during test.

The effort started in the design period, went all the way through the Dem/Val phase, and continued until the end of control law development flight testing. Simulation needed to be even more aggressively used during the test phase to help manage development costs. To a large extent, the HQ analysts were able to meet the new demands through FBW control law design, but it also often meant battling with the other design communities more aggressively than we had ever had before to ensure that there was sufficient power and thrust margins out of the main rotor and antitorque systems to meet the maneuvering goals. That battle continued through the development test phase, when and if rotor or antitorque system performance levels or the empennage didn't meet projections from the design estimates.

As a result of the combined effort put forth through the prototype risk reduction testing, the design and test phases, and the attention paid to detail, the final handling qualities product achieved the goals set by the U.S. Army and the Boeing Sikorsky team. The combined effect of the BMR rotor and the full-authority FBW control system provided the quickness, predictability of response, and high level of agility that allowed for maneuvering to and stopping at any desired pointing position, making maneuvering and handling easier with lower workload than any helicopter to date.

It was strongly believed that at the time of cancellation the Comanche had met the goals for improved survivability that the Army had envisioned for LHX. And in the words of the project pilot, Rus Stiles, "Comanche was the best helicopter I have ever flown." He was speaking from the point of view of its handling and maneuvering performance perspective, of course.

The Comanche teaming arrangement had Sikorsky owning the handling qualities lead and Boeing owning the flight controls lead. Both worked together to form the final excellent results. Both companies came at these technologies quite differently based on their years of experience and their product lines, and it took a bit of an effort to get to a good working culture, but in the end we succeeded.

THE SHADOW PROGRAM

The Shadow program was Sikorsky's first real venture into an attempt to get some experience with and knowledge about designing and adapting the pilot's station using FBW with a side-arm controller, both of which were inevitably critical for LHX. It was also used to test a night vision system that was also critical for LHX. The primary objective was to demonstrate how the overall piloting workload could be reduced with such a concept. The Shadow's experimental pilot station was equipped with a full authority FBW flight control system. Safety pilots in the aft seats monitored the experimental pilot's control activity and aircraft response. The details of the Shadow program are described in Chapter 8, Flight Controls.

S-76 Fantail

Even prior to the Dem/Val program, we had decided to use a fan-in-fin antitorque system. The S-76 Fantail demonstrator allowed us to get some LHX-specific performance/acoustics/handling data and more experience with the build and integration concerns. The Fantail design and testing is described in more detail in Chapter 12.

Comanche Design: Major Air-Vehicle Configuration Decisions

Having gained the knowledge from the technology programs and the Shadow and Fantail prototype demo programs, we now had a reasonably good command of the key features that were needed in the Comanche design to provide superior handling performance.

The handling qualities group generally assumes the responsibility of all flight response and handling characteristics, so they would be the owners, keepers, watchers, and spokespeople for LHX's maneuvering performance. And with maneuvering performance considered critical to LHX success, it was very important that they got involved early and influenced the major air-vehicle design decisions and sizing where possible. That meant looking after and helping to select those parameters that would affect LHX thrust and power margins as well as agility. HQ is never involved in the engine selection and is traditionally consulted only marginally when it comes to sizing helicopter main and tail rotors. For LHX, however, it needed to be more involved, and we were. The T800 engine was selected early on and would deliver 1200 shaft horsepower—we could make no impact there. However, we did get heavily involved in the fuel control design and used our adaptive fuel control program experience to influence some key control techniques. For the fan design, the HQ group was only marginally consulted. But in the case of the main rotor sizing and hub configuration, they got much more involved.

Main Rotor Design

The main rotor hub configuration for LHX was far and away one of the most difficult decisions to make. It took place after the Boeing Sikorsky teaming agreement was reached. Sikorsky had the main rotor lead, so the responsibility and the decision was ours. The fact that the signature maneuver for the 300-ft obstacle avoidance timing and accuracy requirement brought operation to negative Nz into the LHX envelope was a big influencing factor. Yes, LHX was to operate under extended negative Nz conditions, and it was clear that the new attack and recon missions required aggressive high agility for air combat success. Did these factors necessitate a specific hub configuration? As the lead HQ person on the Sikorsky side of the program at that time, John Occhiato tried his best to frame what was required from a rotor response perspective. This was framed around this obstacle avoidance maneuver and having to maintain adequate HQ down to negative Nz.

The most basic question was: Should we use a low-risk articulated rotor or a high-agility bearingless rotor?

The stiffness of articulated rotors is indicated by their hinge offset, the location of the flapping hinge as a percentage of the rotor radius. For bearingless rotors the same term is used, with “effective hinge offset” used to relate them to hinged rotors. John had determined that as a minimum, we would need an 8% equivalent hinge offset to get the high level of agility required.

Could a high-offset articulated rotor get the job done? Perhaps it could offer marginal capability, but would that be close enough? That design would certainly be a lower risk solution. The rotor design group conducted a very thorough trade study between the two concepts that they were considering—a conventional high (6%) offset articulated design and a bearingless main rotor hub design. Their BMR design had an equivalent offset of about 11% and for sure it would offer very high agility.

So here we were, knowing that the BMR would allow exceptional agility and ensure sufficient control at negative Nz conditions, but we also knew there could be potential risks. Were we prepared to face those risks—risks of potentially higher and, perhaps, excessive rotor stresses and vibrations that we may not be able to restrain? What about the risk of ground resonance or an untenable regressive lag mode response? These potential risks made the hub decision very difficult.

There was no time to build a prototype and conduct a risk reduction program to resolve these potential risks. The rotors group presented its case for either a BMR or a competitive 6% articulated rotor hub as choices. The tradeoff study resulted in a 250-page report. After the risks were discussed and the case was made for the significance of the agility and added control that the BMR rotor would bring to LHX, the conclusion was made to go with the BMR hub design. The development program proved that the BMR decision was the correct decision.

The case for the BMR was based on several handling factors that can best be explained with some illustrations and general data. The fact that LHX would be called upon to hug the terrain for masking purposes and perhaps

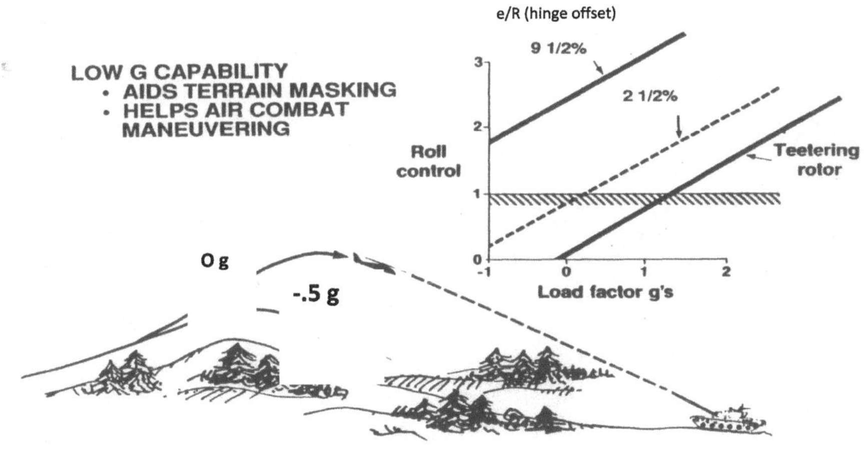

Fig. 6.1 Stiff (high hinge offset) rotors provide control at negative load factors.

have to fly aggressively to negative Nz levels at times helped make the case for that rotor system. Figure 6.1, although not specific to the two rotors compared in the design process, shows the marked difference in the response across rotor effective hinge offset. The higher-offset rotor allows tighter terrain following and more roll agility as well. These same factors carried over into an air combat scenario and could have a profound effect of the outcome.

Another case can be made for the BMR over an articulated rotor based on the increase in system bandwidth. The higher bandwidth of the bearingless rotor can be translated into a crisper and smoother response to higher frequency of input.

The effect on agility is another factor in rotor offset selection. The higher offset rotor produces more hub moment per stick deflection and that translates into a quicker, better-damped rotor response. This translates into generally quicker response across all maneuvers; for example, the difference that hinge offset can have on the time through an 80-kt, 90-deg, 2-*g* steady turn just by the quickness of rolling in and out of the turn can be seen in Fig. 6.2. An 8% hinge offset is almost 1 second quicker than a 2% offset, for example.

When these improved turn times (and higher bandwidth) are incorporated into a real battlefield scenario, they translate into a difference in survivability. The quicker the turn rate, the better chance of getting to the ground or air target before it gets to you. Now instead of just talking about turn times, the argument for quickness is converted into "real" meaningful significance—survival factors. A 1-second difference in turn time could change the probability of survival by 20%.

The tradeoff study results were close, with mixed pluses and minuses. Within the team, there were strong advocates on both sides of the trade. Eventually we concluded that the BMR would be required to give excellent agility and ensure sufficient control at negative thrust conditions.

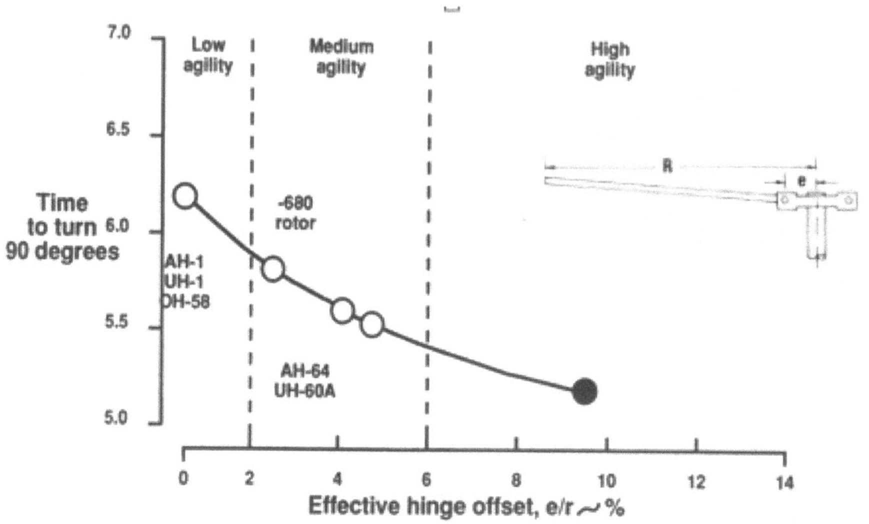

Fig. 6.2 Stiff (high hinge offset) rotors reduce the time to turn.

In addition to its control power advantage, the simplicity and technical elegance might have been the strongest winning points for the flexbeam design. It was also felt that Boeing Sikorsky would compromise its offering if it proposed a "low tech" articulated rotor solution. So, after the risks were discussed and the case was made for the significance of the agility and added control that the BMR rotor would bring, the decision was made to go with the BMR hub design.

Antitorque System Decision

Long before the main rotor design options were considered for LHX, Sikorsky's antitorque system choice was already in place—the fan-in-fin, or Fantail. The Army knew that in order for LHX to be more survivable, the helicopter would be required to have reduced acoustic and/or radar signature levels and that it had to be better able to operate in an nap-of-the-earth (NOE) environment when compared to its current fleet of light helicopters. This decision was made very early on, prior to LHX teaming days and before the Dem/Val phase in the mid-1980s. The fan-in-fin design was based on a similar design used on Aerospatiale/Eurocopter/Airbus helicopters for many years. The French designs were very noisy and gave the fan-in-fin concept a bad reputation. We concluded that with basic geometry changes, the fan-in-fin could be very quiet, and designed ours accordingly. It was known that the fan would require more power to achieve the same effectiveness as a tail rotor, but Sikorsky engineers decided that this was not a large enough factor in choosing the fan concept.

The role of the handling qualities team was to ensure that the yaw axis handling met the maneuver and handling requirements through comprehensive automatic flight control system design where necessary.

EMPENNAGE DESIGN

Designing the horizontal tail for the Comanche was fairly straightforward using conventional practices. Typically we size the horizontal tail to recover about 80% of the unstable moment that the main rotor and fuselage create with changing angle of attack ($M\alpha$). All rotors and fuselages have unstable angle of attack stability, as depicted in Fig. 6.3, and the horizontal tail is then required to reduce that level somewhat.

We did not try to neutralize that instability completely, simply because, if we did, it would result in an impractically large tail. We basically followed that same design practice for Comanche. It is a simple method that historically satisfied the need.

We typically use the following relationship to resolve the horizontal tail size:

$$M\alpha_{\text{tail}}/[M\alpha_{\text{rot}} + M\alpha_{\text{fus}}] = -0.8$$

We first determined the value of the rotor instability, estimated using GenHel simulation analysis, and then determine the fuselage instability, generally based on wind tunnel data. If those data are not available, they can be reasonably estimated using standard techniques. Knowing both the rotor and fuselage $M\alpha$, we can then determine the tail size. We assume a dynamic pressure loss at the tail and then, knowing the airfoil lift curve slope, we determine the angle of attack stability $M\alpha$ that the tail produces as a function of tail size; that is,

$$M\alpha_{\text{tail}} = Cl\alpha_{\text{tail}} \times \Delta\alpha \times b \times qt/q \times S$$

where

b = tail moment arm to aft CG

qt/q = dynamic pressure at tail relative to free stream

S = tail area

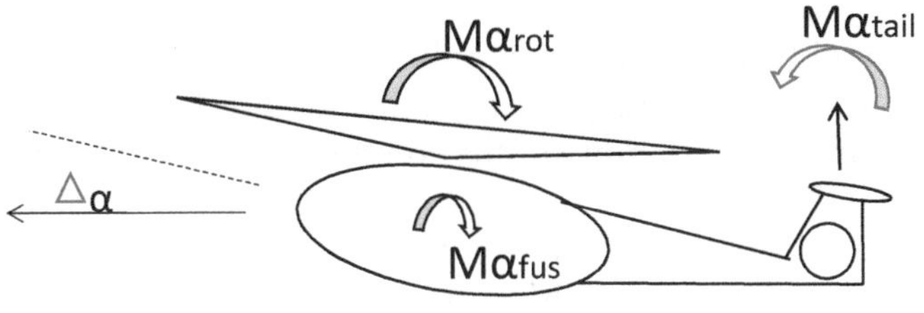

Fig. 6.3 Illustrating the terms in the above equation.

The tail size then falls out of satisfying the previous closed-form calculation.

The BMR system creates a larger unstable $M\alpha$ than an articulated rotor, so it seems that it would demand an even larger horizontal tail. But the BMR also provides more pitch damping than an articulated rotor, which in turn creates a higher bandwidth, "crisper" response that makes the handling easier for the pilot; therefore, there is less need to have the same net level of $M\alpha$ from the tail. The bottom line was that the Comanche tail was still sized to this traditional criterion, and that resulted in a tail area of 23 ft^2. Other factors are involved.

The LHX configuration of choice from the start was that the tail be placed high for several reasons. It was important to keep the tail away from any possible fan and/or main rotor wake interference. These wake impingements often create either trimming or dynamic response issues. Keeping the tail high also meant that it would be in the best free stream flow field, and in that position could offer an end-plating factor for the vertical tail, making it more effective and allowing for a smaller surface area.

A similar method was used to size Comanche's vertical tail. In this case, the method of sizing attempts to provide a modest level of static stability with the sideslip (i.e., $N\beta$, yaw moment with sideslip). The practice at Sikorsky has been to allow for the vertical tail to neutralize the static yaw axis instability created by the fuselage and then get the positive stability from what the tail rotor provides. Unlike the pitch axis, we are able to obtain yaw axis static stability with a reasonably small vertical tail.

But with the fan-in-fin antitorque system, because the fan is embedded in the tail shroud, it does not sense sideslip as a conventional tail rotor would and therefore offers no countering stabilizing forces or moments. The vertical tail alone must provide the desired level of static yaw (sideslip) stability, similar to a fixed wing design. For the Comanche, we chose to provide for a 10% greater positive static stability ($+N\beta$) than the fuselage value of negative stability ($-N\beta$), $N\beta_{vt}/N\beta_{fus} = -1.1$.

Following a process similar to what was done for the horizontal tail sizing, we again estimated the local dynamic pressure and the lift curve slope of the vertical tail airfoil and sized it accordingly. Credit was given to a slightly more effective lift curve slope due to the end-plating effect created by the top-mounted horizontal tail. The fan and fan shroud were canted 20 deg (top of fan tilted to port side), and it was decided that the vertical tail would be canted in the reverse direction to keep its center of pressure over the aircraft centerline as much as possible to minimize any roll to slip coupling. The only area that was considered effective for the vertical tail was above the fan, as depicted in Fig. 6.4. The directional stability of the combined fan/fan shroud was considered to be negligible.

The vertical tail incidence was next set by determining the setting that would balance the fan thrust at 120-kt airspeed. This was all determined using the Comanche GenHel simulation program.

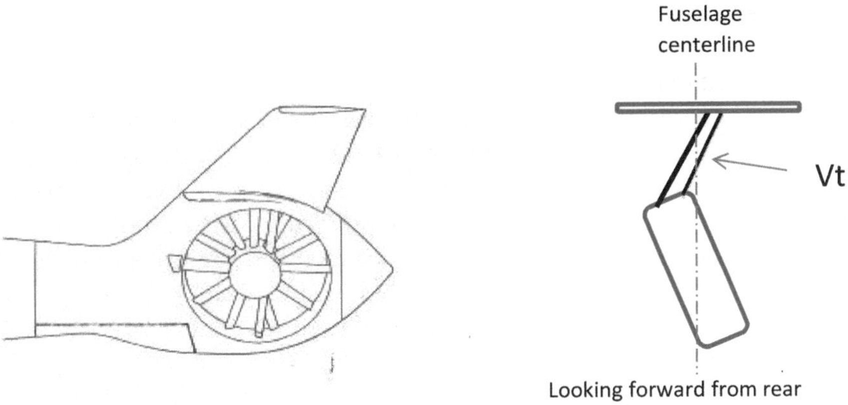

Fig. 6.4 The Comanche tail design showing the canted Fantail rotor.

With the total empennage design set, the effectiveness and adequacy of the design, both vertical and horizontal tails, would have to be substantiated in test.

FLIGHT DEVELOPMENT PROGRAM

There were concerns about the BMR among some in both the handling qualities and dynamics communities as the flight development program approached, mainly because Sikorsky was entering a development program with an untested rotor system. Although we had flown rigid rotors (the advancing blade concept) before, the characteristics of this particular rotor were quite different from those of anything we had ever tested. More often than not, at Sikorsky we had entered development testing of a new helicopter design with an articulated rotor system that in many aspects represented a growth or variant of a previous design. The hinge offset might be slightly changed or the number of blades might not be the same as a previous design, but it was still an articulated rotor. We had familiarity with typical loads/stress/dynamics issues that may result and knew typical methods of dealing with them.

We simply had no experience flying the Comanche rotor, and there was some apprehension about potential issues like the lag regressive mode. At the time, our analytical tools were indicating that that mode could have very light damping, which could create disturbing roll coupling issues at some point of the envelope. If these issues did arise, could we deal with them or would they be showstoppers?

A second concern was the total effectiveness of the Fantail in supporting both the pointing precision and aggressiveness that was demanded of LHX. We knew of the lack of sensitivity and control dead-band near zero thrust and were concerned about the potential low-bandwidth response that would entail.

The piloted simulation sessions were indicating that the directional control would be adequate, but until we witnessed the real fan response we could not be totally confident. Would the weak response in and about zero fan thrust be the cause of a serious control response issue that would not work for the aggressiveness and precision required of LHX?

Tail aerodynamic effectiveness potential issues were of less concern mainly because we had plenty of experience modifying tail feathers on previous helicopters and always came up with an acceptable design. We were also less concerned about flight control/control law issues because, once again, our experience told us that with adequate flight time we could always deliver an acceptable, production-ready system.

The aircraft flight program was initiated with the core primary flight control system (PFCS) mode operational, which was similar to non-FBW programs starting with the automatic flight control system (AFCS) turned off. (The various control law modes are described in the follow-on section.) The aircraft demonstrated reasonably good handling in core PFCS mode, which contained only yaw rate damping. Envelope expansion took place in this mode up to a point. Full envelope expansion took place following the maturity of the core AFCS mode. In that mode, the aircraft had demonstrated superior handling qualities. Through the course of the development flight testing, all of the specified maneuvers were flown as well as the specific mission task elements (MTEs) that were to be used to rate Comanche handling performance. Flight test results met or exceeded expectations based on what was simulated. Although some of the more focused flight control law features were still to be completely developed, the risk of doing so unsuccessfully was considered to be very low because it simply meant testing and tweaking gains that were already working successfully in piloted simulation. The simulation results had been carrying over to the development program almost one-to-one.

Additional handling qualities issues came up that had to be dealt with in test.

REGRESSIVE LAG MODE

As mentioned previously, bearingless rotors are desirable for their design simplicity as well as their ability to produce large moments quickly, thus improving aircraft response and maneuverability. Although the stiffness that is characteristic of these rotors results in typically higher frequency modes, there often exists one mode low enough in fixed frame frequencies to couple with the dynamics of the rigid body below. Typically, the regressing lag mode (RLM) is the lowest frequency rotor mode and hence the one that most strongly couples with the body modes.

If the RLM is lightly damped and starts to oscillate, it can couple into the rotor roll mode. This potential roll mode coupling was noticed somewhat early on in the development program, but when we routinely started flying higher load factors, it became clear that we had a problem that needed our attention. In a high-speed, 2-*g*, sustained turn maneuver, the flexbeam in-plane bending starts taking off into an undamped oscillation at about the 2-*g* point and couples into the aircraft's roll response.

At the time, there was no immediate fix for the problem. It was hoped that the problem might be treatable with added roll-rate stabilization. Several modest changes in the AFCS command model and feed-forward control path reduced the propensity to excite the RLM mode oscillations for what might seem to be more normal operational maneuvers where load factor wasn't sustained for more than a second or two. However, when in a sustained high-*g* descending turn, the RLM became very noticeable and intrusive. Like other bearingless rotors, the Comanche main rotor employed a snubber-damper between the root end of the torque tube and the flexbeam to provide auxiliary damping to the first lag mode. An improved snubber-damper design was expected to contribute 50% more damping to the first lag mode than was present in the demonstrator aircraft.

Meanwhile, a lead Sikorsky handling qualities engineer was experimenting with a potential controller scheme using the Comanche GenHel model. The idea of a cross-feed controller scheme had been promoted by Rotea and Weller in 1996 for improving aeromechanical stability for a bearingless rotor. The same two gentlemen tested the controller with some success in a United Technologies wind tunnel test program, and Sikorsky's lead engineer proposed using such a controller technique after the RLM issue started to become more intrusive as Comanche was flown to more aggressive load factors.

Sufficient data were taken during simulation analyses to demonstrate the impact that the increased Nz had on the RLM damping and the impact that the cross-feed controller scheme would have on the mode damping when turned on.

The controller scheme that Sikorsky was exploring simply fed a filtered pitch rate into both the longitudinal and lateral cyclic pitch and a filtered roll rate into both the longitudinal and lateral cyclic pitch. This fix was eventually tested on the aircraft and proved to resolve the RLM issue.

Directional Stability and Vertical Tail Development

Comanche's directional stability was noticeably weaker than what is typical for a single-rotor helicopter with a tail rotor. That characteristic was somewhat obvious in the simulator, but it was expected, and we hoped that with adequate augmentation from the FCS it would not be an issue in the flight test article. Unfortunately, that did not turn out to be the case. Even though the FCS added adequate rate damping, there was a consistent, noticeable tendency for the aircraft to drift in and out of small sideslip angles, creating a low frequency oscillation and needing constant pilot attention to keep the nose on a steady track.

Figure 6.5 Final EMD configuration.

Part of the problem was that the vertical tail was equipped with a gurney flap to keep the fan at or near zero pitch and power in cruise flight. This improved high-speed performance but unfortunately, the fan generated very little thrust per pitch change at this trim position, and that made the rate feedback very ineffective. That characteristic, plus the fact that there was probably insufficient vertical tail area for the fan design, meant that something needed to be done to get an acceptable production solution.

We decided to test endplates on the horizontal tail, and that eventually resolved the issue. Two sets of endplates were tested—5 ft^2 and 8 ft^2—mounted on both ends of the horizontal tail. The endplates made a huge difference in the sideslip stability, and although the 8-ft^2 plates were preferred by the pilots, consideration had to be made regarding potential vibratory issues with the larger plates. The final resolution allowed the 5-ft^2 plates to be incorporated into the configuration. Adding the endplates gave us a bit more pitch stability as well because the plates made the horizontal surface more aerodynamically effective. Figure 6.5 shows the aircraft in its final EMD configuration.

ATTITUDE QUICKNESS

The BMR rotor allowed us to exceed the Level 1 requirement in pitch and roll, whereas the fan fell short of meeting the Level 1 requirement for yaw attitude quickness, even with help from the control laws. This was not considered to be an issue for eventual operational effectiveness, however, because the fan still provided a high Level 2 capability, which allowed other advantages such as performing "snap" turns at 100 kt.

CONCLUDING COMMENTS

Designing and developing the Comanche scout/attack helicopter posed significant challenges to the Boeing Sikorsky engineering community, and it can be argued that the handling qualities engineering challenge was by far one of the most demanding for several reasons. We were asked to design to a brand new specification that was an order of magnitude more demanding than the previous spec, calling for Level 1 handling over the entire envelope. We had to do that with a starting configuration that was not a derivative Sikorsky design, one having a first-generation Sikorsky bearingless rotor, a fan-in-fin antitorque system instead of a tail rotor, a first-generation Sikorsky fly-by-wire flight control system, and a new nonderivative cockpit with sidestick controllers. We were literally starting from scratch, designing to meet a new design specification and demanding maneuvering requirements.

With all this in front of us, we first needed to conduct and learn from risk reduction technology programs before we started the actual design. We also had to develop and change the culture of how we approached flight control system design and also how we used piloted simulation to develop and risk-reduce the design. We made these changes, and did so with a partner that had its own engineering mindset and design culture. It was at times very trying, but the combined talent and wisdom of two major helicopter companies working together resulted in the impressive Comanche helicopter.

Chapter 7

SUPPORTABILITY

ARTHUR W. LINDEN

The Army had a strong interest in improving the supportability of its aircraft due to a history of high operational and support costs and poor aircraft availability. It wrote specific requirements into the LH specification to meet these goals. This was new to the industry; in the past, the myriad of characteristics that make up supportability were usually simply fall-outs of the design the preliminary design engineers came up with.

Supportability is composed principally of three organizations: reliability, availability, and maintainability (RAM); diagnostics; and logistics. RAM at Sikorsky was an existing organization within the engineering department, logistics was an existing organization within product support, and diagnostics was a new organization within Comanche engineering. All three disciplines were brought together under the supportability umbrella. As such, each organization had unique and shared responsibilities. The most widely shared responsibility was maintainability. Traditionally, RAM had the upfront maintainability design interface responsibility, with logistics providing the more downstream aspect related to maintainability fielding.

Due to the sophistication of Comanche avionics, built-in diagnostics were essential to meet the combat sustainability and affordability goals. As a result, the new discipline of diagnostics engineering emerged under supportability.

Comanche's extensive maintainability features were a collaborate effort within the integrated product development teams (IPDTs) with supportability providing guidance and evaluation of alternate designs and contributing to tradeoffs in order to provide the design feedback necessary to achieve specification compliance.

Top supportability/maintainability goals were:

- On-board integrated prognostic/diagnostics
- Elimination of Aviation Intermediate Maintenance (AVIM) maintenance level
- Automated log books
- Bar code inventory tacking
- Easy access
- Common tools
- Elimination/reduction of common and peculiar support equipment
- Automated fault detection and fault isolation
- Elimination/reduction of scheduled maintenance
- Equipment identification and modularization
- Preventing escalation of maintainer skill levels
- Design for male and female maintainers
- Design for maintenance wearing mission-oriented protective posture (MOPP) and Arctic gear
- Automated boresite alignment
- Automated rotor track and balance
- Simplified/Murphy-proofed maintenance
- Portable maintenance aid (PMA)
- Aviation Unit Maintenance (AVUM) battle damage repair
- AVUM composite material repair

We took the Army's supportabillty requirement very seriously. Engineering drawings could not be released without a sign-off by the appropriate supportability member on the team to ensure that maintainability and reliability allocations were being met. Many of the initial design decisions were the result of supportability considerations. Technology allowed simpler designs that inherently improved supportability, but other decisions were made in favor of supportability even if they had an adverse impact on other aircraft attributes like weight. Figure 7.1 illustrates some of the supportability features of the design.

At the time of the downselect we were ecstatic that we had won. A few days later at the debrief, we learned that it was our supportability solutions that had won the day, much more than our aircraft design approaches.

From a 15 April 1991 *Aviation Week & Space Technology* article [1]:

> Maj. Gen. Ronald K. Andresen, the Army's LH program manager, said the two teams were judged to be "very close" to each other in their technical and producability proposals, two major factors in the evaluation. He said there were "significant differences" in favor of Boeing Sikorsky in the other categories—cost, training and man-machine interfaces. The latter group accounted for 55% of the Army's scoring of the development proposals. [Our total score was 230.4 in these categories vs 172.4 for the SuperTeam.]

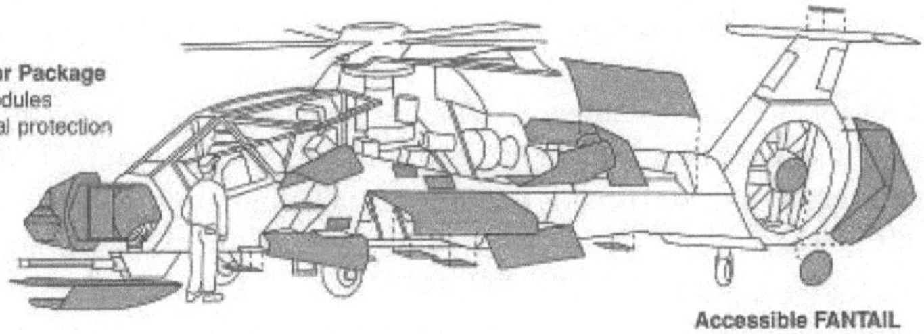

Fig. 7.1 Supportability considerations in the aircraft design.

> The winning team also was rated better in the unscored, pass-or-fail category of "operational suitability," an assessment of how well the overall proposal meets Army needs for an armed reconnaissance helicopter.

Some of our solutions favoring supportability included the unique fuselage design that permitted almost indiscriminate locations for doors and access panels, easing the maintainer's job; highly reliable, redundant modular electronics; a sealed and pressurized cockpit; battle damage–tolerant and repairable composite airframe; and the bearingless rotor system.

Two maintenance levels were planned compared to three for the then-current fleet. A 40% reduction in maintenance burden was projected compared to the then-current fleet. Skill levels [defined as military occupational specialties (MOS)], which were 15 on the AH-64, were reduced to 4.

From an April 1996 Boeing Sikorsky fact sheet [2]:

> The results of this approach was a revolutionary design which:
>
> Reduced manpower needs by 32%
>
> Reduced the number of maintenance specialties required from 9 to 4
>
> Reduced tools and support equipment by 83%
>
> Reduced the number of parts by 34%
>
> Reduced training costs by ½
>
> Reduced maintenance levels to only 2 compared with 3 for the current fleet

The results were significant: RAH-66 Comanche operating costs were more than 60% less than the AH-64A and almost 50% less than the OH-58D; an RAH-66 battalion would deploy with eight less C-5 sorties than the AH-64 and 1/6 the C-5 sorties for a tank battalion. Overall, the RAH-66 Comanche was projected to save $20 billion in operating and support costs over 20 years compared to today's fleet.

From *Aviation Week & Space Technology*, 15 April 1991 [1]:

> At a forward area arming and refueling point [FARP], a three-man team in arctic or chemical-biological gear can rearm an LH with four Hellfires, two Stingers and 500 rounds of 20-mm gun ammunition, and refuel as LH with 269 gal. of fuel in about 12.5 min.... This quick turnaround time, measured during tests on an LH mockup, is a direct result of the forethought that went into the design. Some of the features that made this possible are a single-point pressure (or gravity) refueling, powered loading of the linked 20-mm ammunition, and clear access to the missile launch rails which were at waist height.
>
> There was no requirement for separate automatic test equipment; digital diagnostics were integrated in to the design. Boresighting of the gun was automatic. Electronics included multilevel redundancy at all levels (sensors, buses, processors, modules, and displays). Line replaceable avionics modules were installed one module deep for easy access.
>
> Only 12 unique user-level tools were required. Front-line battle damage repair was minimized with damage-tolerant composites in the airframe and rotors. Composite repair was possible using cold bond adhesives.

In 1996, an update of Comanche's supportability features was printed in the May issue of *Army* magazine, authored by Frank Colucci [3]:

> In theater, the Comanche needs far less support equipment than the Cobra, Apache and other Army helicopters. Today's aircraft are maintained with some 312 standard hand tools, while the RAH-66 deploys with just 49 tools in one box. Apache battalions are followed by an electronic equipment test facility in two big trailers, and even the simple Cobra needs a heavy bore-sighting rig.
>
> The Comanche has integrated diagnostics and prognostics, and it boresights its sensors and weapons in flight. The RAH-66 was designed around a two-level maintenance scheme—field and depot—and needs no aviation intermediate maintenance facility.
>
> The notebook-sized portable intelligent maintenance aid stowed in the aircraft is made up of an electronic test set, a logbook, and a maintenance manual to inspect and maintain the Comanche air vehicle and MEP. Redundant, fault-tolerant avionics keep the RAH-66 flying despite malfunction or battle damage.
>
> Three cooled and pressurized MEP bays are filled with standard electronic module (size E) packs, which can be readily identified and replaced by maintenance personnel in the field. Built-in fault isolation speeds up troubleshooting and repairs. The caution/warning/advisory system

worked flawlessly on the first flight, and like the other Comanche technologies, it has important wartime implications for Force XXI.

Reduced maintenance requirements support a smaller force and high operational tempo. While today's Army helicopters are typically available to fly just 3.75 hours a day, the Comanche promises to sustain a tempo of 11 to 12 hours a day. Operational tempo is defined by the maintenance man-hours per flight hour (mmh/fhr), the turn-around time at the forward arming and refueling point (FARP) and the survivability of an aircraft. Comanche requirements call for 2.8 mmh/fhr. Projections are now around 2.3 mmh/fhr. By comparison, the AH-1 needs up to 7 mmh/fhr and the AH-64A from 9.7 to 12 mmh/fhr.

The Comanche returning from a mission will speed its own turnaround by transmitting fuel, armament, and battle damage or maintenance needs en route to the FARP in an automated spot report.

MAINTENANCE REQUIREMENTS

The overall aircraft design promised a 40% reduction in the maintenance burden relative to the then-current fleet (see Fig. 7.2 and 7.3).

Note: A major part of the overall supportability story is what was done to improve reliability and maintainability. This is covered separately in Chapter 18.

A full-scale aircraft mockup was built to demonstrate manpower and personnel integration (MANPRINT) and operational suitability. Figures 7.4 through 7.6 are some of the demonstration photographs used by Boeing Sikorsky to illustrate supportability features early in the program.

From the *Boeing news*, 15 April 1991 [4]: "The Portable Intelligent Maintenance Aid (PIMA) [shown in Fig. 7.7] points maintenance personnel to impending problems, details smart fault-isolation procedures, keeps records and displays paperless manuals."

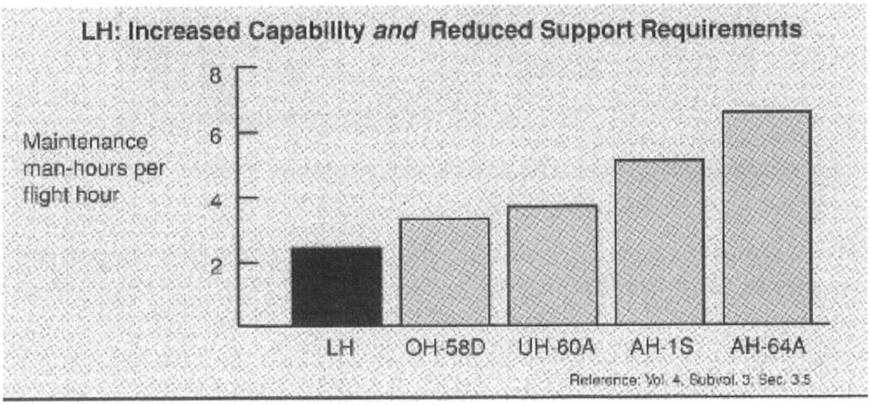

Fig. 7.2 LH maintenance requirements.

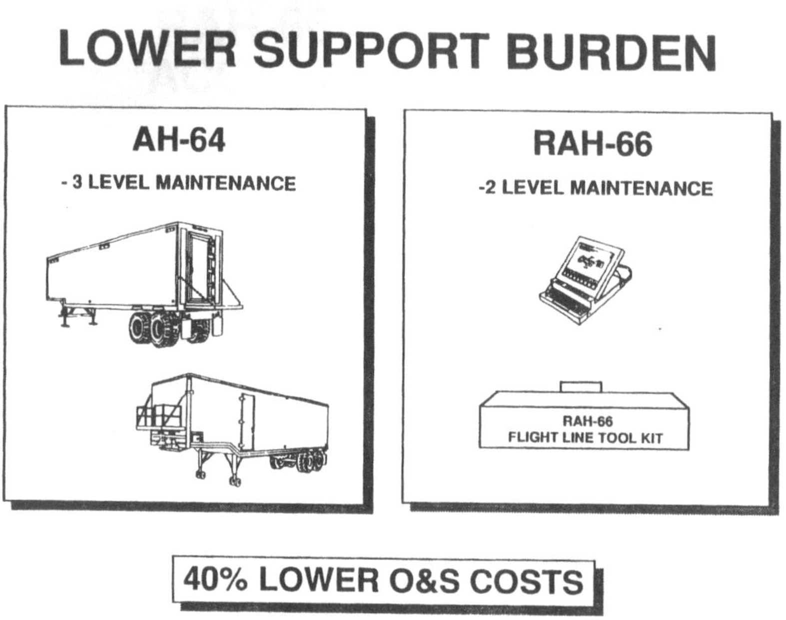

Fig. 7.3 The RAH-66 support burden was considerably smaller than the AH-64.

Fig. 7.4 Weapons bay doors can be used as work platforms.

Maintaining the aircraft with a simple toolkit was another supportability requirement. This toolkit was to be used for all line maintenance tasks ranging from opening panels to removing and disassembling parts. During the course of the Comanche program, adding any new tools to the kit was met with extreme resistance. The resulting Comanche toolkit contained only 49

Fig. 7.5 Access to the engines and transmission is provided by large moveable cowlings.

Fig. 7.6 A crewman can easily service the transmission even in chemical protective gear.

Fig. 7.7 The PIMA.

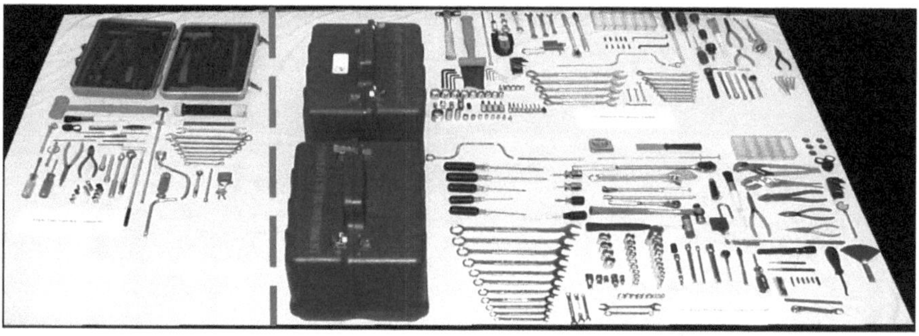

Fig. 7.8 Comanche toolkit compared to general aviation/armament mechanic's toolkit.

tools compared to 275 tools in a general aviation/armament mechanic's toolkit (Fig. 7.8).

INTEGRATED TRAINING

An integrated training system (ITS) was developed using fiber optics integrated into the helmet-mounted display. An innovative Field Assistance technology and Support Team (FAST) was developed to ensure early RAM maturity.

OPERATING AND SUPPORT COSTS

The U.S. Department of Defense (DoD) must secure appropriations from Congress before any weapon system procurement can be initiated. Estimates of total cost must be developed as part of the process. This includes operating and support (O&S) costs. O&S costs are maintenance driven. A larger fleet of unreliable aircraft that are difficult to repair costs more to own. Poor RAM characteristics not only drive up the cost of ownership, but also increase the size of the fleet's logistical tail, making the weapon system less mobile on the battlefield. Under the O&S umbrella is a significant reoccurring ownership cost with many contributing cost elements including:

- Maintenance personnel
- Spare parts and material
- Maintenance manuals
- Support equipment
- Training equipment
- School houses

Higher maintenance demand requires more maintainers, more spare parts, and more material. As a result, more manuals and support equipment are needed. Maintainers need to be trained, requiring an increase in training equipment and classrooms. In addition, a mobile weapon system demands the support system also be mobile. Mobile facilities, typically housed in trailers, are needed to support equipment, spare parts, and material. Quarters are needed for maintenance personnel.

Again quoting the 15 April 1991 issue of *Aviation Week & Space Technology* [1]:

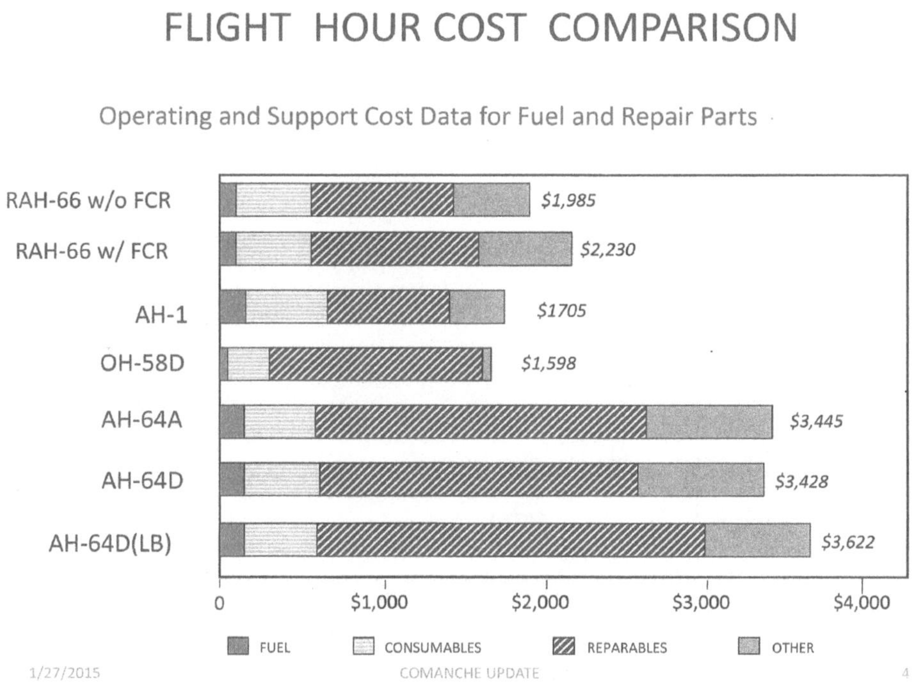

Fig. 7.9 Helicopter flight hour cost comparison.

> Operating cost targets proposed by Boeing Sikorsky for the LH helicopter are about 40% less tan those estimated by the U.S. Army when the service solicited final offers from competing development teams…
>
> In their cost proposals, the companies exceeded expectations most dramatically in their operations and support estimate. Last Spring, when the Army solicited development proposals, the target was $500 per flying hour for parts and Army maintenance personnel at the depot and field unit levels. The target was increased to $702 per flying hour during the summer, when Defense Secretary Richard B. Cheney cut back on the planned acquisition total and production rate. The Boeing Sikorsky offer came in at $442 per flying hour.

TRANSPORTABILITY

From *Aviation Week & Space Technology*, 15 April 1991 [1]:

> For rapid transport, an eight man crew will be able to disassemble and load one LH onto a C-130 in less than 20 min. and reassemble the aircraft, once delivered, in under 22 min, see figure 7.10. To accomplish this task. the First Team has developed a transportability kit that includes a blade stand, tow bar, jack pads, caster wheels and a hydraulic cart, and a kit with 12 basic tools.

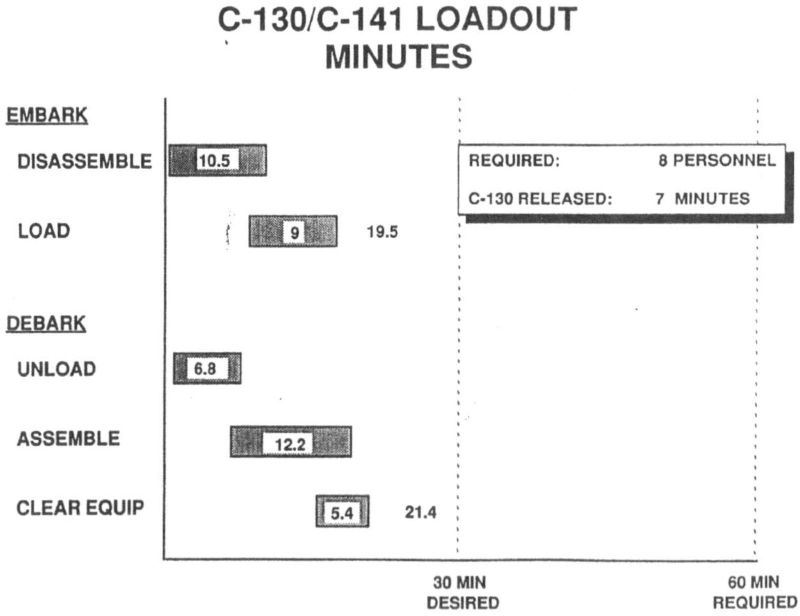

Fig. 7.10 Required: eight personnel; First Team: five personnel.

> When carried in larger transport aircraft, the crew size shrinks and the embarkation and debarkation times improve.... preparation includes "kneeling" the helicopter on its retractable main landing gear, manually folding down the aircraft's normally fixed T-tail, removing the main rotor blades and attaching them to the tailbone using a cradle, and mechanically lowering the rotor head by using the main rotor control servos. Boeing Sikorsky has designed the helicopter so that its main rotor blades can be removed without disconnecting the control system, a feature that eliminates the need for a check flight after the blades are reattached.
>
> In instances where self-deployment is required, the aircraft will be fitted with two stub wings or external fuel armament management system (EFAMS) pylons and external 450-gal. tanks, providing a 1,260-naut.-mi unrefueled range. For self-protection in this configuration, the helicopter will carry two Stinger missiles and 150 rounds for its 20-mm gun. A crew of five will be able to prepare the aircraft for departure in this configuration in about 15.5 min. and a crew of the same size at the helicopter's destination will be able to remove this equipment, rearm and refuel in under 19 min.

DIAGNOSTICS/PROGNOSTICS

Extensive diagnostics were built into the systems to report status, health, failure detection, and fault isolation to the flight and/or maintenance crews.

Although integrated diagnostic technology is extensively used today in automobiles, trains, and aircraft, it was still being developed in university labs when the LHX/Comanche program started, and probably was not really understood until the late 1990s and early 2000s. But it was a key to Comanche's approach of reducing maintenance requirements and cost.

> The [United Technology Research Center]...contributed artificial intelligence technology for the LH's integrated diagnostic system...[It] developed the qualitative reasoning basis for the Intelligent Fault Isolation Procedure (IFIP) troubleshooting system, which uses a qualitative model of the system, analyzes current aircraft information and tells the maintenance crew about possible component failures. The system also predicts unmeasured parameters and suggests the next test in order to obtain the greatest degree of fault isolation at the lowest cost. [5]

Comanche's integrated avionics architecture introduced the potential of fault propagation, which increased the complexity of diagnostics. In the federated architecture, when a function is lost it is tied to specific hardware (i.e., single computer chip) and communication network. In the integrated architecture, loss of a function can be caused by failure(s) in various subsystems that could be responsible for elements of functions associated to a number of operations/jobs/functions. In an integrated architecture, subsystems interact to achieve functions. A fault error can propagate to the next subsystem and/or network within the chain needed to provide the function and result in an erroneous fault message or location, making it impossible to isolate. A physical wire can host many different networks. It is typically impossible to identify a single function subsystem that caused the problem.

This was a difficult problem that required an independent fault detection/tolerant communication network or guardian network. The hypothesis was to partition functions/applications into specific messages from specific modules/hardware and monitor the event intervals over the networks. An error was established based on bit value loss in the time domain. The error message is not generated until the failure accrues within a specified number of assigned periodic intervals of the message. This shows how the loss of a message could help isolate the problem. The fault-tolerant communication network would not determine whether the message was correct, just that the message was in the correct time interval and the correct bit length (timeliness, specified length, and replication). Diagnostics was achieved through a voting structure that looks at function elements assigned to hardware to isolate to the specific failure item.

Figure 7.11 is from the 1990 proposal executive summary.

MARKETING BROCHURE

Boeing Sikorsky stressed the importance of supportability in a marketing brochure, shown in Fig. 7.12.

Integrated Diagnostics/Prognostics System

Demonstrated:

- Module fault detection/isolation
- Rotor track and balance
- Mechanical systems fault detection/isolation
- Engine electrostatic emission monitoring system (EEMS)
- End-to-end system performance (fault signal; processor; data bus; caution, warning, advisory; PIMA)
- Testability analysis tool (TAT)

Developed:

- Hierarchical approach and architecture

Fig. 7.11 Integrated Diagnostics/Prognostics System status.

Comanche

Shoots Down Operating and Support Costs by 40%

Design for reliability. And, "Make the system fit the soldier," according to the Army's MANPRINT doctrine.

Success in meeting these goals is the main reason why Comanche will incur up to 40% lower life-cycle costs than the current fleet.

Pilots and maintenance personnel achieve maximum performance with a low-cost integrated training system. It features electronic classroom presentations, automated training using the aircraft on the ground, and safe, realistic simulator-based training and hands-on maintenance training using hardware but not aircraft.

For reliability, the composite, hingeless rotor hub eliminates many components, bearings and seals. Redundant electronics are built in rugged modular form and mounted in temperature-controlled racks.

For faster, lower-cost maintenance, these modules are hand-sized and positioned one-deep for easy removal and replacement. In addition, integrated diagnostics and built-in test capability quickly pinpoint failed components and eliminate the cost of separate test equipment.

Weapons-bay doors double as work platforms, allowing all parts to be serviced without the use of ladders. Only 34 tools are required for flight line maintenance.

In even the harshest field conditions, three soldiers can completely refuel and rearm a Comanche in fewer than 15 minutes.

All these factors produce extremely high mission availability... a critical contributor to cost-effectiveness.

Comanche's T800 engines, modular for ease of repair, require only 6 tools for all flight line maintenance. Below: Maintainers use a unique Portable Intelligent Maintenance Aid (PIMA) that helps diagnose and predict problems in electronic and mechanical systems. This "paperless" system is also used to display maintenance procedures and record logbook data.

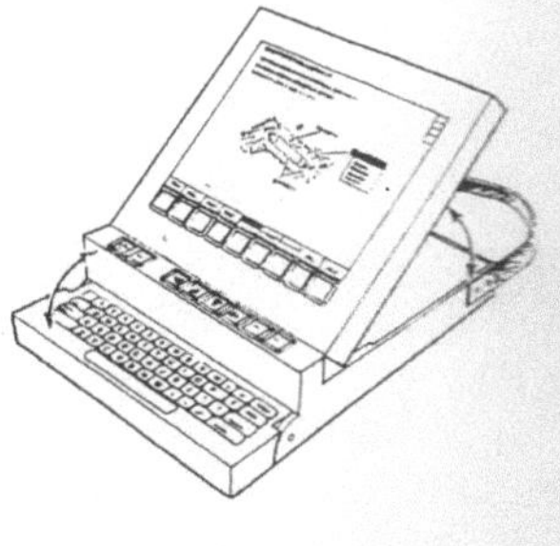

Fig. 7.12 A marketing brochure discussing supportability.

ONE IDEA THAT DIDN'T WORK

One idea that didn't make it through the preliminary design phase was the swing-away nose. It was detailed in the 1990 Dem/Val proposal (Fig. 7.13).

Initial design work showed that swinging the nose, which weighed over 400 lb, could be made to work on level ground; however, if the aircraft was on anything other than level ground, the heavy nose would be difficult to open

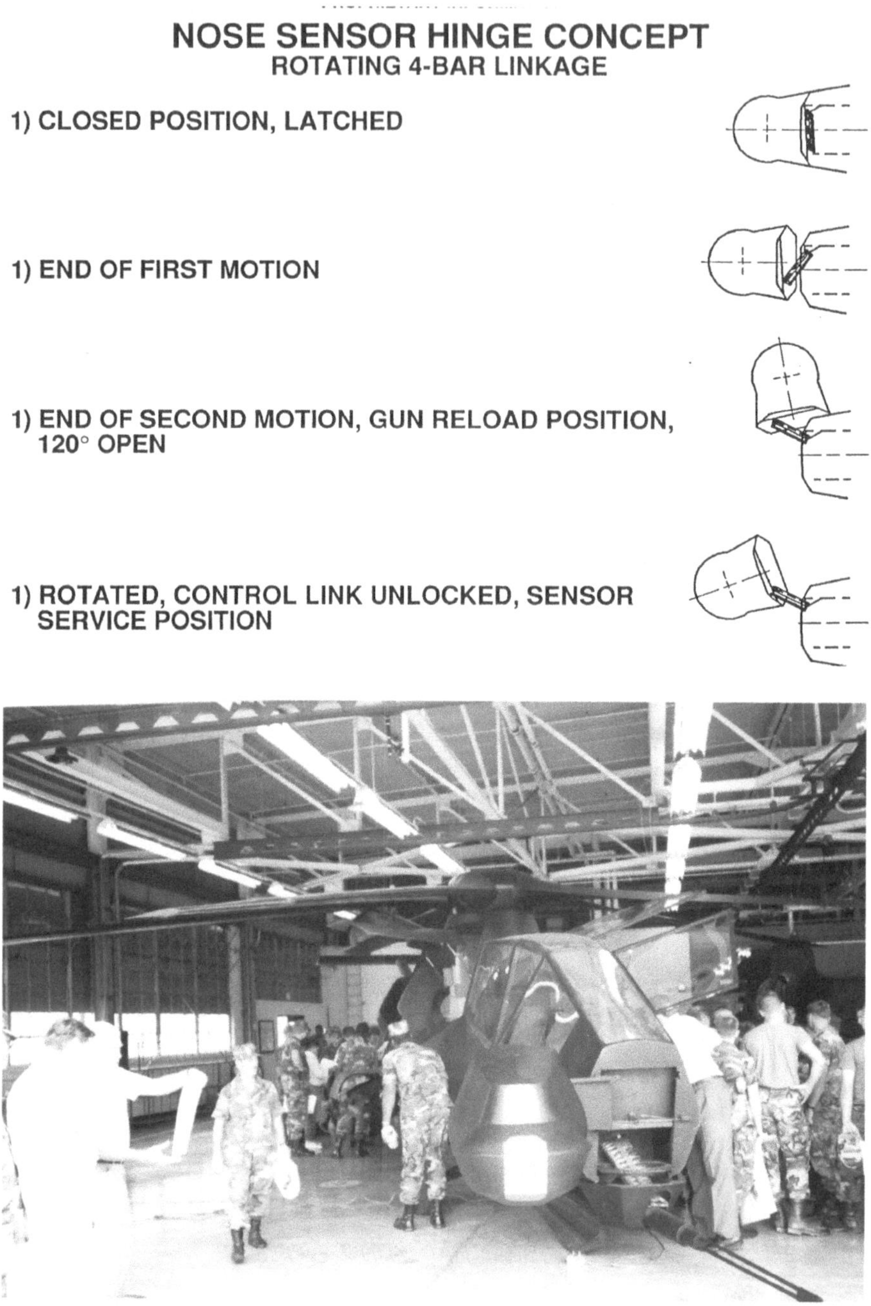

Fig. 7.13 The mockup with the swing-away nose.

and close. We went back to a fixed nose and provided other methods of access to the sensors and electronics.

SUMMARY

The Army placed a strong emphasis on supportability in its original RFP to industry. Boeing Sikorsky reacted by giving the supportability disciplines unprecedented importance as the aircraft design was developed. We were rewarded for this in the competitive evaluation between the two competing proposals. Although we thought we were well ahead of our competition on aircraft design discriminators, the evaluators gave each of us similar scores. On supportability we were head and shoulders above them. Our supportability approach was the major factor in our winning the competition.

REFERENCES

[1] David F. Brown, "Cost, Supportability Key to Boeing Sikorsky LH Award," *Aviation Week & Space Technology*, 15 April 1991, pp. 18–19.

[2] Boeing Sikorsky, "Press release from the Boeing Sikorsky Joint Program Office," April 1996, fact sheet.

[3] Colucci, F., "The Army's RAH-66 Comanche," *Army magazine*, May 1996, p. 39.

[4] Boeing News Special Edition, Vol. 45, No. 4, Philadelphia Edition, 5 April 1991, p. 2.

[5] Pratt & Whitney and Sikorsky say thanks to Research Center for support in ATF and LH projects, Vol. 12 No. 5, United Technologies Research Center, Special Feature, May/June 1991, p. 3.

Chapter 8

Flight Controls

John Occhiato
Handling Qualities Lead
Rus Stiles
Comanche Chief Pilot
Phil Gokd
Control Law Design Support

The flight control system was certainly one of the more critical features that distinguished Comanche from other helicopters at that time because it represented a huge step forward with its unique design and, more importantly, how it allowed Comanche to be controlled and ultimately perform. One of the most dramatic changes in the helicopter industry in the last 20 to 30 years was the introduction of fly-by-wire (FBW) flight control systems. With the newly established vertical and/or short takeoff and landing (VSTOL) handling requirements dictated by the military, it was imperative to go in that direction. The V-22 Osprey and RAH-66 Comanche were the pioneering programs for FBW. The experience from these two programs together put the industry at a launching point, and most new military programs going forward will inevitably employ FBW systems.

The Boeing flight controls experience was almost solely responsible for the FBW evolution in the rotary wing business. Boeing started its FBW efforts back in the 1960s with research and development (R&D) efforts on the tactical avionics guidance system (TAGS) program and continued gaining experience through the 1970s and 1980s with other programs like the advanced digital optical control system (ADOCS) on the 347 program, which was a risk reduction prototype program for the Army heavy lift helicopter (HLH) program, and with the HLH program itself. No other American helicopter company was even close to introducing FBW systems in any of their product lines. So when

Bell teamed with Boeing for the JVX program, and when Sikorsky teamed with Boeing for LHX, there was little doubt as to who would be in charge of the electronic flight control design effort—Boeing had essentially all of the industry's FBW experience to that point in time.

The new requirements for LHX were demanding much more precision and more capability across the entire operating envelope, and FBW became a necessity. There were several other factors that made a compelling argument for its inclusion. First, the Army's desire for LHX to be a single-pilot aircraft mandated a flight control interface with high-fidelity automated functions, to periodically offload control of the aircraft from the pilot and enable him to complete mission tasks. Conventional mechanical control technology, with limited capability strap-on actuators and sensors to provide autopilot functions, simply did not provide the precision or reliability needed for the demanding LHX mission environment. Second, in order to incorporate low observable features, it was apparent that some aerodynamic compromises would have to be made. Fly-by-wire offered the only solution with sufficient bandwidth and precision to compensate for the potentially reduced stability resulting from incorporating these design features. Fly-by-wire had been successfully incorporated into tactical fixed wing aircraft such as the F-16, which enabled that design to incorporate relaxed stability for improved maneuverability. In addition, the relative simplicity of the unique trim side-arm controller installation opened many options for the airframe design. For these reasons, FBW technology found its way into the early LHX design configuration, and due to the highly integrated nature of the aircraft there was no turning back.

In the concept formulation phase of the program, a fly-by-light fiber-optic flight control system was considered as the most advanced technology available. After considerable study, however, it was eliminated due to reliability, size, and the cost of the transducers, according to then–program director Lou Cotton [1]. The flight controls for LHX therefore would be fly-by-wire.

The Comanche flight control system (FCS) provided two highly augmented modes: rate command/attitude hold, which provided very responsive maneuver capability for day visual conditions, and attitude command/velocity hold, with more benign response for night and degraded visual conditions. The system incorporated a number of automated features including a flight director, which simply either aids manual or uses automatic direction of the aircraft flight path, and a fully coupled autopilot with advanced functions, including auto hover, bob-up, and return to mask. Functions that were to be added in later stages of the program included envelope cueing and integrated fire and flight control. Use of a triplex (triple redundant) electronic architecture with electrical wires in lieu of mechanical linkages provided greater flexibility in routing the wiring and reduced the probability of ballistic damage from threat weapons. The navigation function was integral to the flight control system and relied on three LN 251 (compact/lightweight fiber optic gyros) inertial/Global Positioning System (GPS) navigators to develop the navigation solution.

Also within the flight control system was the rotating air data system. This system obtained air data from blade-mounted scoops, and the data processing was contained in a card in the flight control computers. It provided the control laws with the speed of air impacting the aircraft from all angles.

The basic architecture was proven in over 500 hours of flight test on the two prototype aircraft, but the hardware was updated for the engineering, manufacturing, and development (EMD) (production configuration) aircraft. To address the potential for a common failure mode, the EMD design incorporated dissimilar processors in one of the flight control computers. Newly available Power PC chips were utilized in conjunction with the existing Pentium-based architecture. This ensured that no chip-level failure would render the system inoperative. Improvements in the flight control actuators provided better overall performance and reduced susceptibility to ballistic weapons.

Making the leap from a mechanical to an FBW system obviously expanded the opportunities for more maneuvering and precise flight path control, but the system reliability remained the first and highest architectural design priority. In that regard, the flight control systems architecture was triply redundant to provide full operational capability after a single failure anywhere in the system, and safe operation after a second failure. Ballistic tolerance was provided through the physical separation of components, which ensured that no one ballistic impact could disable the system. To accomplish this, the three flight control computers were designed with internal redundancy to eliminate the need for constant cross comparison with a companion computer. Any one of the computers could control the aircraft. Hydraulically powered generators were connected directly and exclusively to two of the flight control computers, with redundant airframe electrical power provided as a backup. The third flight control computer was provided with two sources of airframe electrical power, both with built-in redundancy. The flight control computers were connected directly to the primary rotor servo actuators via a triply redundant electronic servo control unit. Triple hydraulic power sources were reconfigurable so that any one of them could control the aircraft and supply electrical power to the flight control computer.

The flight control computers were the heart of the FCS. Figure 8.1 shows a high-level overview of the component connections to the FCCs. Figure 8.2 shows the location of these components in the aircraft.

There are three FCCs per shipset. One FCC each is installed in the electronics bay below and to the left and right of the rear cockpit. The third FCC is installed in a compartment aft of the rear cockpit. This separation improves ballistic survivability, yet centrally locates the FCCs to help minimize wire

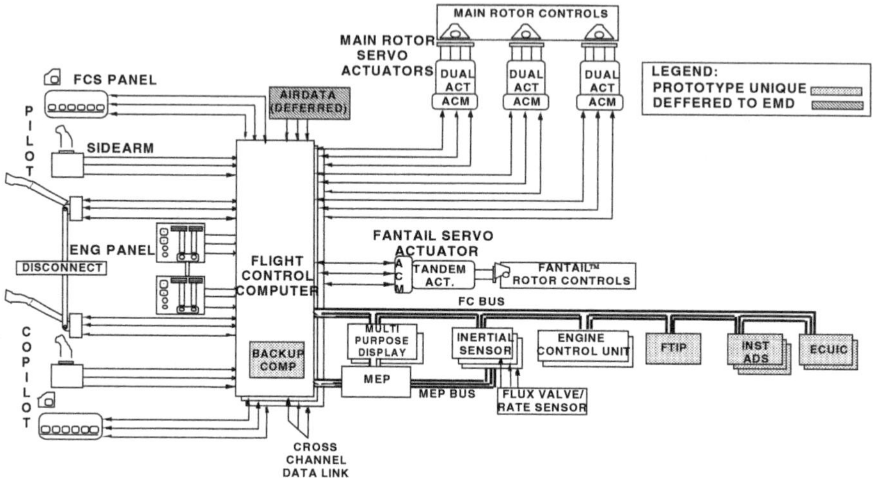

Fig. 8.1 FCS components.

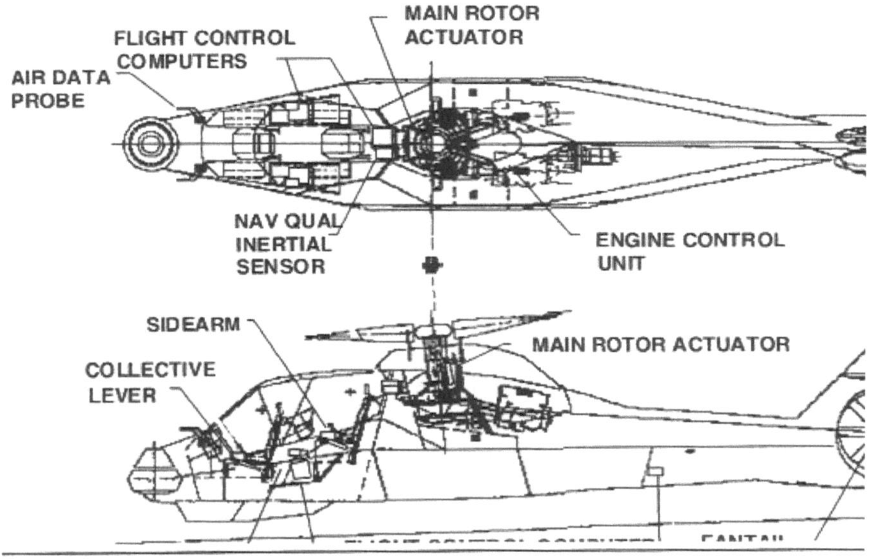

Fig. 8.2 FCS component locations.

harness lengths. The FCC architecture encompasses four 1750AE processors. Three of these processors perform flight-critical operations, and the fourth is dedicated to mission-critical operations. The three flight-critical processors are the input/output processor (IOP) and the primary flight control processors (PFCPs). The mission-critical processor is the automatic flight control processor (AFCP).

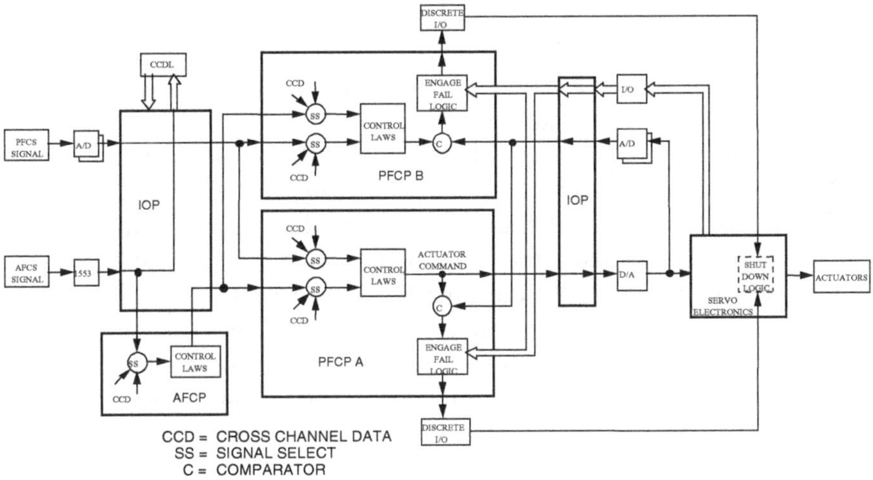

Fig. 8.3 Internal FCC architecture.

The PFCPs execute identical operational programs. This in-line pair configuration allows CPU hardware failures to be detected in a single computer. As shown in Fig. 8.3, this is accomplished by the following method:

1. Both PFCP A and B perform identical computations.
2. PFCP A's output signals pass through the IOP to the hardware I/O (D/A).
3. The output signals are wrapped back into the IOP and passed to both PFCP A and B.
4. Both PFCP A and B compare their generated commands to the wraparound.
5. If either miscompares, the processor shuts down the appropriate functions.

The FCCs communicate with each other via serial data buses known as cross channel data links (CCDLs). The data transmitted cross channel support interchannel synchronization, enhance system performance, and aid in the redundancy management of sensor input.

The Comanche was one of the first designs to incorporate full authority digital engine controls (FADEC). Instead of a segregated interface to the engine, the Comanche design embedded engine management within the flight control computers. This approach improved reliability by taking advantage of the triplex architecture and simplified the pilot interface. The LHTEC T-800 engine included two types of backup control schemes. One was embedded in the FADEC itself. The second was designed to be a manual control operated by the pilot through a beeper. The Comanche improved upon this design by embedding the manual control in the flight control computers through an independent governing algorithm. This reduced pilot workload during primary governor failures, as well as eliminating the training normally associated with manual governor operations. Because of the integrated nature of engine management, the Comanche design included multiple rotor speed modes. It also enabled automatic rotor RPM control during engine out operations to prevent excessive rotor droop due to autopilot command.

The cockpits were equipped with identical controls. A three-axis sidearm controller provided the interface for full authority control of pitch, roll, and yaw (via a twist operation). A fourth (vertical) axis was incorporated into the sidearm controller to enable the pilot to command the vertical autopilot to climb and descend. Ultimately, the aircraft could be easily controlled with one hand. The controllers all incorporated triply redundant connections to the flight control computers. Again, any one of the connections enabled control of the aircraft through the full flight envelope. The sidearm controllers were "passive" in that there was no active mechanism to move the controller for control law feedback to the pilot. Instead, the required feedback mechanisms were provided by manipulating the control laws. Springs and temperature-compensated mechanical damping were provided to maintain control centering in the absence of pilot input.

The sidearm controller configuration offered several ancillary benefits over and above improved handling qualities. First was a vast improvement in crash survivability. Modern helicopters had incorporated a crash design that focused on controlled energy dissipation. In addition to energy-absorbing structure and landing gear, the cockpit seats were designed to stroke downward in a crash, thus reducing the sudden, high vertical loads transmitted to the pilot's torso. This design was very effective, but had an unfortunate unintended consequence. When configured with a center-mounted cyclic stick, as the seat stroked downward, the stick would often impact the pilot's jaw, sometimes resulting in serious injury. By mounting the sidearm controller outside the seat stroke envelope, this problem was resolved. The Comanche design went one step further and mounted the controller to the armrest of the seat so that as it stroked, the entire controller went with it. This eliminated the possibility of arm injury due to seat stroke/stick interference.

Another improvement provided by the sidearm controller was the ability to adjust the stick location to a comfortable position independent of seat position, reducing fatigue and improving posture. It also allowed the pilot to adjust the seat to a position that optimized external visibility, without compromising relative stick position, as had been the case with conventional fixed center cyclic sticks.

Both flight control stations were normally active. The unique trim controllers were implemented such that pilot inputs were summed. Because this could result in nullification or otherwise undesirable flight control inputs during a breakdown in crew coordination or failure, a control disconnect function was incorporated. A pushbutton was installed on each stick to invoke this function. The crew assigned a "priority" controller—normally this would be the primary flying pilot. The disconnect function was active on the priority controller to allow the primary pilot to nullify inputs from the secondary pilot at any time. In the event of a failure in the primary controller, the disconnect button would become active in the secondary controller to allow that pilot to take control of the aircraft.

A conventional displacement collective was incorporated to provide proportional control of rotor thrust. The front and rear seat collectives included triply redundant transducers to convey collective position to the flight control computers. In the event of a jam, a shear pin allowed the pilots to regain control via the operational station. An electromechanical actuator was incorporated into the collective control to synch the collective position to the autopilot command, when it was in use. The actuator also incorporated mechanical dampers and a slip clutch to allow the pilot to override the autopilot.

The pilot mode select panel was simple and reflected the straightforward but versatile design of the control laws. There were only two primary modes to choose from, selected via a pushbutton directly in front of the pilot. There was one mode for agile aggressive flight and another mode for stabilized flight operations such as might be required at night or in adverse weather. The stabilized mode included automatic hover hold engagement anytime the aircraft decelerated into a low speed engagement gate. Altitude hold was further selectable via a pushbutton, but also automatically engaged when inside the hover hold gate.

A challenge to any helicopter with a full-authority FBW system is transition to and from the ground. In flight, the control laws expect the aircraft to respond to pilot inputs by movement of the aircraft (detected by inertial sensors). On the ground, however, the aircraft is constrained from movement. This confuses the control laws. Ground detection is required to tell the control laws to stop trying to make the aircraft move. In keeping with the Comanche architecture, triplex sensors were mounted on each gear and set to sense very light contact with the ground. This enabled a smooth transition from flight to ground and vice versa.

The FCCS incorporated triplex inertial sensors and interfaced to the triplex air data system. These data were used primarily in the flight control laws. The flight control computer could take over operation of one of the wing-mounted multipurpose displays to provide backup air data, heading, and rudimentary navigation capability, as well as flight-critical fault annunciations in event of a dual mission computer failure. This eliminated the need for a suite of backup instruments and wet compass. Finally, the FCS also managed the hydraulic systems to reconfigure during pump failures and leaks.

The FCC architecture ensured that the aircraft could be safely flown even after two failures of the same type component anywhere in the system, an event whose probability was categorized as extremely rare.

THE S-76 SHADOW

The Shadow was an S-76 helicopter modified with a single-pilot cockpit grafted onto the front of the aircraft (Fig. 8.4). The objectives of the program were focused on LHX to evaluate integrated sidearm flight control systems, advanced control laws primarily for low speed and for transitioning to and from ground contact, advanced cockpit configurations, and visually coupled helmet-mounted displays, and determine their acceptance by aircrews.

The test pilot controlled the aircraft from the front cockpit. The left seat in the S-76 standard cockpit is the copilot's seat. He also performed the systems operator's tasks. The safety pilot was in the right seat. He could monitor status and assume control by the push of a button or by simply resisting the movement of his conventional mechanical controls.

Behind the cockpit was a flight engineer's station for real-time control of mission simulation and data collection (Fig. 8.5).

The test cockpit was equipped with a full-authority four-axis sidearm controller that could control pitch, roll, yaw, and collective. The flight controls included a "unique-trim" function, referring to the fact that the controllers all returned to a null "trim" position upon release. They could best be compared to spring-centered joysticks, except that yaw control was accomplished by twisting the controller against a spring, and collective control was accomplished by pushing up and down against the spring.

The aircraft was also outfitted with a conventional collective, albeit with much less travel than the collective control in conventional helicopters (approximately 5 in. vs 9–11 in.). The intent of the flight control research activities was to select the best configuration for operational suitability in the low altitude terrain environment.

The Shadow also incorporated many of the advanced cockpit concepts envisioned for Comanche, including glass displays (well before they had been

Fig. 8.4 The Shadow.

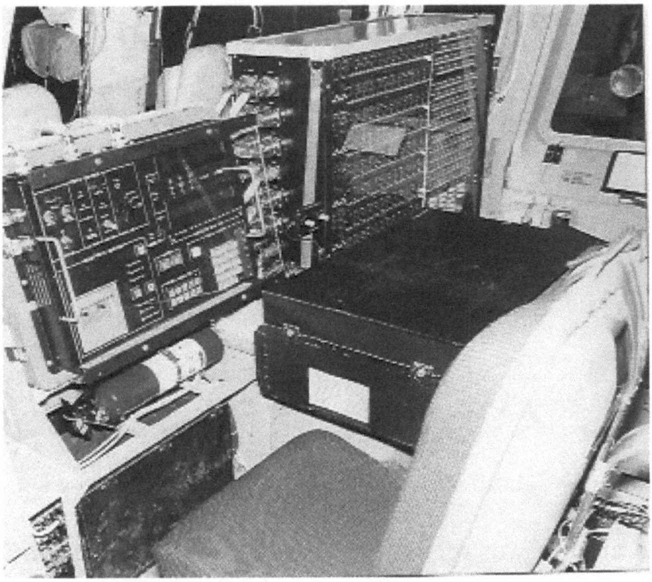

Fig. 8.5 Shadow flight engineer's station.

incorporated into production aircraft) and a head tracker helmet display with daylight symbology.

The derivation of the name *Shadow* has its own story. George Sipprell, who had a major role in the Shadow program, explained that the Shadow name was originally an acronym, the Sikorsky Helicopter Advanced Demonstrator of Operator Workload. According to George, as we were writing this chapter:

> It was cute, but most of us felt the flow of words was awkward…kind of sounded like we tried to find the words to fit the word SHADOW. … However that was its name and how it was addressed in briefings as it got initially approved. The name was used officially during the ARTI proposal period and appears in the documents and subsequent briefings fully "acronym'd." … Some of us liked the name for its connotation … I remember the discussion … something like we are going to overshadow the competition, the aircraft is ominous/scary to the competition, etc. … And people also started goofing with "the Shadow 'nose' (knows)"! Good chuckle.

(For those people younger than the authors, *The Shadow* was a 1940s/1950s radio show with a detective called the Shadow, and the claim that the "the Shadow knows.")

As time passed, the SHADOW acronym was recognized as somewhat strained, and the aircraft was simply know as the Shadow, no identifiable acronym mentioned.

Fig. 8.6 The Shadow painted black.

When the aircraft was released to conduct flight demonstrations, it was painted black and taken on the road (Fig. 8.6). Again quoting George:

> … people took notice and they were scared of what we were doing, the commitment we were making. … The aircraft was later repainted to try and paint an aircraft (LHX) within an aircraft (Shadow) using the tricky paint lines and coloring, but I ran into a buzz saw with test with their mandatory orange painted parts for safety and visibility. We reached a compromise, but I never did get my original graphic I drew up.

One not-ready-for-prime-time technology that was looked at early in the program was voice recognition. This was in its early development stages, and the thought was that by enabling control of subsystems without physical or visual interaction, it might be useful in reducing pilot workload.

Applications for voice recognition were explored in the simulator. A state-of-the-art system was purchased and integrated to control functionality of communications, navigation, weapons, and target acquisition systems. Operational pilots could choose to use voice or manual control during a variety of mission segments. The simulator offered controlled conditions, without stress, vibration, or background noise, and it revealed whether pilots would prefer the system over traditional manual interfaces.

To evaluate how reliably the system would work in a flight environment, a bread-board unit was tested on a company S-76A. Test subjects would speak words, typically those being tested in the simulator, and data were collected on recognition accuracy. The results were mixed, probably 50/50 accuracy. Most importantly, failures often resulted in misrecognition of words, rather than rejection. This could result in erroneous actions, which could not be tolerated. Mistakenly executing a "target" instead of a "tune" command could have serious consequences.

To address this, the system only listened for a few words at a time, using branching hierarchy to select the system, then function, then control. For example, to tune a radio, the pilot had to first utter the word "COMM," followed by "TUNE," after which the system would listen for numbers associated with a frequency. This made using the system training intensive.

Simulator trials revealed that even one mistake by the system, which required repeating a command, led to enough frustration to immediately abandon using it. In the two-person cockpit, it also caused confusion. The other pilot would often ask, "Were you talking to me?" This added to frustration and a higher workload.

At the same time, it was becoming normal design practice to include control of critical functions on the flight control grips [known as "hands on grip" control (HOG)]. HOG selection of radio, weapon, or target acquisition modes was being accommodated. In the Comanche design, the pilot did not look into the cockpit to control many mission functions, because symbology was displayed on the helmet.

For these reasons, voice recognition was not integrated into the Comanche design. Several years later, it was installed in Shadow, in conjunction with the Defense Advanced Research Projects Agency (DARPA) day-night adverse weather pilotage system (DNAPS) program. As suspected, when it didn't work, even once, it created a frustrating increase in workload. Then no one wanted to use it.

It might have had better acceptance in a single seat aircraft.

In fact, voice recognition has never been included in a Sikorsky helicopter since the Shadow experiments, nor have we seen its inclusion anywhere in the industry as of the publication date of this book.

Collective control on the sidearm was evaluated. During flight test, the unique trim control of the collective function on the sidearm controller was rejected for two primary reasons: first, the pilot had a difficult time staying aware of the collective position and proximity to limits, and second, fine control of collective position was nearly impossible and made takeoff and landing difficult. The fourth axis of the sidearm controller was then changed to command vertical speed (up or down) only when the altitude-hold autopilot function was engaged.

A conventional collective pitch stick was installed. At first we thought this could also be a small-displacement, spring-centered control, like the cyclic; however, testing showed that then the pilot's only cue as to power level was to focus on the torque gauge, which was inconvenient. Therefore, Shadow used a conventional displacement collective. The two collective sticks were mechanically tied together, and a trim motor moved them up and down in concert with the power demands of either pilot or with the automatic stability modes.

The sidearm controller configuration and some low-speed control laws were tested to demonstrate their effectiveness, particularly in close to the ground nap-of-the-earth operation. An interesting experiment found that experienced pilots had issues with transitioning to this new control concept, so a number of nonpilots were tested with the new system. They took to the system much more easily. Art Linden, our then–program manager and a nonpilot, was able to successfully fly the Shadow for 20 min around the Sikorsky flight field, with no help from the safety pilots. He moved sidewards and rearwards, and precisely approached hover spots for bob-ups and turns, all in winds of 15 to 25 kt, which demonstrated the ease of handling of the system. The conclusion was that young Army pilots, who had experience with video games and electronic controllers, would readily adapt to this new control scheme.

Along with using Shadow to develop control laws, it was used for evaluation and testing of the night vision pilotage system (NVPS). The NVPS incorporated a second-generation night pilotage system, including forward-looking infrared (FLIR) and image-intensified TV (IITV). A total of 60 hours of development flight tests were conducted.

The last research conducted on the Shadow platform was investigation into the takeoff and landing flight control requirements (Fig. 8.7). Unlike an airplane that develops lift and control only at forward-flying airspeed, helicopter rotors can generate substantial airframe motion on the ground as soon as the rotors are turning. Thus, the automated control laws used in flight to maintain trimmed flight state had to be altered on the ground. Boeing provided flight control engineering support, in line with its responsibilities on the program, and the requirement for sensitive weight on wheel detection was developed. This was one of the first truly cooperative flight test efforts between Boeing and Sikorsky and was approached somewhat tentatively. The Shadow was,

Fig. 8.7 Shadow takeoff and landing control law research on the front lawn of Sikorsky headquarters.

after all, an in-house-funded Sikorsky development. Ultimately, it was a very successful collaboration.

From April to June 1992, the aircraft flew 32 test sorties and completed more than 400 takeoffs and landings on surfaces ranging from hard runway to soft grass, and on slopes from level to 8-deg side slope. The program investigated several technologies for detecting ground contact, as well as developing control laws to conduct smooth takeoff and landing transitions. Two significant findings resulted from this research: first, the importance of reliably detecting ground contact, and second, the methodology required to make mode changes to the control laws during ground transition. Shadow research revealed the importance of detecting ground contact at very light weight. During landing, the desire was to retain as much of the stability as possible throughout the landing transition and retain the same control strategy during landing as was used during in-flight maneuvers. Several technologies were evaluated, including a ranging device to detect close proximity to the ground (vs actual ground contact). It was found that pilots were very sensitive to early or late mode changes, and the requirement evolved to a "weight on wheels" detector that would transition above at least 200 lb of weight on wheels, but less than 500 lb. Tolerance to side loads was essential. Although there was some speculation these values might represent percent gross weight vs absolute thresholds, because the test vehicle and Comanche were approximately the same weight, it was decided to forgo any further investigation of that question.

An additional factor noted in the ground detection system was the need for de-bounce. Once the aircraft made contact with the ground and the control moding began to transition, if the pilot bounced the aircraft in any way, it was important to hold the transition. To do otherwise created some interesting oscillatory inputs.

Control law mode changes, primarily in the areas of feedback and trim, were a somewhat more complex problem than ground contact detection. It was desirable during landing to maintain limited authority rate feedback augmentation to suppress short-term upsets during landing, while rejecting the higher authority attitude feedback. Rate feedback was additionally rejected based on an "axis constrained" criteria; that is, once the axis (longitudinal, lateral, yaw) was constrained by ground contact, the feedback would be turned off. Trim follow-up was the biggest challenge, because moding of this function would result in a change in pilot control strategy. Eventually it was found that "freezing" the trim upon ground contact actually improved pilot awareness of the control strategy requirements during landing. During slope landings, for example, the controller reverted to proportional response in the axis that was still airborne, allowing the pilot to "feel" how much input was being made against the rotation of the aircraft, which made for smoother landings and a higher degree of awareness as to the severity of the slope. The takeoff sequence also experienced important changes in control law moding. Once again, if the mode change took place too early or too late, dramatic changes in pilot workload resulted. Because trim follow-up was "frozen" during the landing, the potential existed for each takeoff to start with a different control (trim) position. This resulted in some rather large transients after liftoff, as the pilot commanded trim to the proper position. To counter this, during all landings the trim was forced to a neutral position. The controls were placed at the position for nominal hover, and trim follow-up was not restored until the aircraft was airborne. This resulted in a feature similar to the slope landing, where the pilot could bias the proportional control during liftoff, to anticipate wind for example, washing out the command to zero once achieving airborne state.

From the data collected during this testing, the characteristics of the weight-on-wheels switches, and associated control laws, were defined for the Comanche aircraft, which thankfully encountered few surprises in the takeoff and landing environment.

Control Law Design

The BMR system along with the Fantail and the empennage configuration were certainly a great starting point for eventually getting excellent handling qualities for Comanche, but to meet the detailed requirements of the specs, it was also necessary to provide a very extensive set of control laws (CLAWs) that afforded the pilot very precise flight path control "hands on" and tight stability "hands off." Once there was a commitment to an FBW system, it was almost axiomatic that detailed control laws be part of the system. That would be true for any class of helicopter, but it is even more so the case for an LHX-type vehicle where very demanding flight path control is required both from the maneuvering requirements and the specifics in ADS-33. Key features these CLAWSs made possible were:

- High gain response in automatic flight control system (AFCS) mode for aggressive maneuvering.
- Low gain mode (VELSTAB) for precise near-ground operations.
- Excellent engine integration CLAWS for maintaining rotor speed control.
- Acceleration limiting to reduce loads.
- Outstanding yaw axis capability—the ability to turn and fly sidewards at 100 kt. This provided excellent turn to target capability.
- Fully populated mixing matrix to completely decouple controls.
- Unmatched ability to turn and point quickly and accurately.
- Because it was a model-following system, there was a tendency for more similar response across all pilots.

- Complete full envelope static and dynamic hands-free stability.
- Precise pointing accuracy.
- Allowed for more precise slope landing capability.

In the traditional mechanical flight control systems, the limited authority stability augmentation and trim system, known as the automatic flight control system (AFCS) uses certain sensed parameters to basically stabilize the helicopter once the pilot makes a control input and reject disturbances once trimmed. In these conventional mechanical control systems, the electronics are limited to only about 10% authority of the control range. The pilot's inputs get summed with feedback signals, usually through a series actuator, to provide the mechanical command position of the main or tail rotors. This represents a basic set of control laws that support the pilot by providing some platform-stabilizing feedback that helps to reduce their workload to get a desired response. The AFCS is tuned to a singular set of gains that typically work the same across all corners of the envelope. They support the pilot with some stabilization with limited authority, but do not offer significant help for precisely maneuvering the helicopter. They are, however, quite adequate for most general flight maneuvers. That is what is used by essentially all the current fleet of helicopters using mechanical flight control systems. These aircraft can perform very demanding precise control, but typically the pilot workload to do so is quite high and demands considerable piloting experience. When performing to the precise flight path control, considerable, if not 100%, pilot attention is demanded.

In the full authority FBW world, the control laws do much more. They are used to get a very desirable precise and predictable response requiring minimal pilot effort. In what are called model-following CLAWs, pilot inputs do not go directly to an actuator but rather to algorithms that define a desired aircraft response, and then additional algorithms deliver that desired response to the pilot's input—the aircraft response follows the defined model. There is almost no limit to how precisely that response model can be defined. The command response can be modified as required depending on what is needed for both flying aggressively up and away and flying a precise ground track close to the terrain. The CLAWs can be defined separately for each axis of control and vary with speed. Because of this adaptability in an FBW system, it is possible to provide very desirable handling qualities across the entire envelope and for all mission task elements (MTEs).

The CLAW architecture starts with basic core system values for the primary control system and then adds capability to get to more mission-specific core AFCS and eventually a velocity-stabilized mode. There is almost no practical limit to how smart the system can be. By changing parameters inside the algorithms, the CLAWS are tuned during flight test to balance control response with ride quality and acceptable structural loads.

The Comanche system was defined as a model-following system, meaning that idealized responses were defined and the system worked to deliver those responses on pilot demand in the full up augmented modes of operation. The system had several modes of operation, starting with the primary flight control system (PFCS), which had both a core and mission capability. The core PFCS was the most degraded mode and resulted when all triply redundant sensors had failed. Its performance was similar to a typical mechanical FCS when the AFCS has failed. The core PFCS is a fixed gain system that does not rely on any feedback sensors. The mission PFCS features airspeed scheduling of parameters and yaw rate damping. Both sets of control laws feature command shaping that has been designed to be commensurate with the types of tasks envisioned for the respective degraded modes. There was still quite a bit of stick shaping and filtering, so the core PFCS mode was much easier to handle than a modern helicopter like the UH-60 with the AFCS turned off.

The mission PFCS mode had yaw rate feedback as well as some airspeed scheduled gains. The system redundancy was increased in the yaw axis with a fourth rate sensor. This was to ensure the aircraft would meet the Level 2 workload requirement in nap-of-the-earth (NOE) flight. Simulation studies and analyses showed that although pitch, roll, and heave (collective) axes were acceptable even without feedback present, the yaw axis needed rate stabilization to meet a safe level of workload in NOE. Through a series of degraded mode mission task evaluations, this premise was validated during flight test.

The core AFCS mode was the fundamental all-up mode for up and away flight and particularly for speeds above 60–80 kt. The pilot commanded an angular rate response in pitch and roll proportional to his stick inputs, and the attitudes were held when the stick was returned to detent. The stick commands were demanding specific and idealized responses, and the system was delivering them. Special altitude hold and stick force/*g* were demanded and achieved.

Attitude hold in the longitudinal, lateral, and directional axes was accomplished using a proportional plus integral control system.

Automatic turn coordination was enabled above 50 kt. In general, the turn coordination function provided longitudinal, lateral, and directional body axis rate commands that minimized lateral acceleration in a turn. The turn coordination algorithm used roll angle and airspeed to predict the desired turn rate and then modified it with lateral acceleration feedback and a roll rate signal to provide ball-centered turns. The commanded turn rate was transformed into body axis pitch, roll, and yaw rate commands and summed with the other rate commands in the AFCS.

A velocity stabilization/hover hold (VELSTAB) mode was engaged manually by pressing the VEL/HVR HOLD switch on the AFCS control panel. The pitch axis response type was nominally attitude command/velocity hold (AC/VH). Groundspeed was used as the velocity reference at low speeds, whereas airspeed was used at high speed. At very low speeds, the hover hold mode was provided automatically, which means the aircraft response to cyclic stick inputs was modified to produce a velocity command/position hold (VC/PH) response.

Hover hold was enabled whenever the VELSTAB mode was engaged. The hover hold mode engaged when groundspeed, pitch and roll rates, longitudinal and lateral linear accelerations, and pitch and roll stick commands were all small. Altitude hold automatically engaged when hover hold engaged, if it wasn't previously engaged. The velocity command response was provided at groundspeeds of less than ±5 kt, making it easier for the pilot to precisely position the aircraft in degraded visual conditions.

When VELSTAB was engaged, the yaw axis control laws provided automatic low speed turn coordination (LSTC), which was enabled above 15 kt groundspeed. Longitudinal groundspeed and commanded bank angle (instead of airspeed and actual bank angle) were used to calculate a commanded turn rate. A feedback signal proportional to lateral groundspeed was also calculated and summed with the commanded rate. This implementation provided ground-referenced coordination by driving the aircraft lateral speed to zero so that the aircraft heading aligned with its ground track in a turn. This function had been optimized with pilot-in-the-loop testing and significantly reduced pilot workload when maneuvering in NOE flight.

The altitude hold mode was engaged manually by pressing the Altitude Hold (ALTHLD) switch on the AFCS control panel. This mode also engaged automatically when in VELSTAB and the hover hold mode was entered. While in altitude hold mode, the pilot could maneuver the aircraft vertically using either the left-hand displacement collective stick or the limited authority vertical axis of the right-hand sidearm controller.

The pilot was able to perform all mission tasks at a Level 1 rating, which meant he was able to meet performance demands in the mission tasks with very low workload. This was critical to allow the Comanche pilot to perform precise maneuvering while attending to additional piloting tasks as required, such as weapons delivery.

For low-speed, very close to terrain operation, a VELSTAB mode allowed for precise and vernier type control of the flight path. This was especially needed in degraded visual environments and where mission operations may demand divided attention. Here, stick pressure controlled groundspeed. In this mode of operation, hover hold capability would automatically engage once stick pressure was released when near hover.

SUMMARY

The Comanche flight control system clearly pioneered the way forward in the industry from both an architectural design and performance perspective. It changed the manner in which the pilot interfaced with and controlled the helicopter and hugely impacted favorably the pilot's ability to get a more desired aircraft response. Sadly, Comanche never made it through to a production stage, but the experience gained in this unique design and the successes borne out through the development stage are clearly finding their way to flight control designs in follow-on programs.

Reference

[1] Stanley W. Kandebo, "Boeing Sikorsky LH Technologies Will Increase Army Combat Capability," *Aviation Week & Space Technology*, 15 April, 1991, p. 20.

Chapter 9

THE COCKPIT

RUS STILES

Although the Army initially wanted the LHX to be a single-seat aircraft, it was eventually convinced that, even with automation, workload during the scout/attack mission could exceed the capability of a single pilot. To allow two flight crews to share mission tasks without the restrictions found in legacy aircraft such as the AH-64 Apache, the Comanche incorporated identical crew stations. All flight control and mission functions could be accomplished from either cockpit without exception.

The Comanche cockpit layout was a balance of external visibility, multifunction display area, crashworthiness, and comfort. Using mission and task analysis design techniques, it was determined that two multifunction displays (MFDs) would be required to provide efficient presentation of mission functions such as target acquisition imagery and digital map data. The two large color displays were centrally located side by side in front of the pilot. To declutter the primary displays of control menus, a smaller multipurpose display (MPD), which would reconfigure according to the menu of functions selected by the pilot, was mounted outside and slightly below the multifunction displays. A keyboard was mounted below the displays on a retractable hinge. This was used by the nonflying pilot to enter message data for mission reporting. In the event of complete mission equipment package (MEP) failure, MPDs provided "limp home" capability directed by the flight control computers (FCCs).

This layout resulted in easy viewing of the display suite and comfortable access to the interfaces for data entry, while maintaining the 25-deg over-the-nose visibility required by the military specification for scout helicopters. A wide-field-of-view helmet-mounted display (HMD) provided the wearer heads-up performance data necessary for aircraft control.

Side stick controllers were used in both tactical fighters and commercial airliners, and experience with those installations was exploited to design the sidearm controllers for the Comanche. A four-axis sidearm controller was mounted on the right-hand side of the crew seat, with fore/aft adjustments for pilot comfort. The controller resembled a joystick. An internal spring/damper resisted movement; when released, it returned the controller to the central detent position. Yaw control was accomplished by twisting the grip.

Initially there was considerable skepticism regarding the incorporation of yaw control into the sidestick, because this had never been done before in a production aircraft. This design decision was supported by piloted simulation studies, in both ground-based simulators and airborne surrogates such as the Sikorsky S-76 Shadow aircraft. There were two apparent reasons why this configuration worked: first was the incorporation of automatic turn coordination that essentially eliminated the need for yaw inputs in forward flight, and second was the sidearm mounting arrangement. Because the controller was aligned to the normal relaxed pilot's arm position, twisting around the wrist joint presented no difficulty. Many pilots reported that at low speed there was a natural connection between the position of the hand and direction of flight, as if there was an arrow projecting from the front of the hand that pointed in the direction of the desired flight path. The pitch, roll, and yaw control laws incorporated automatic trim, which was invoked any time the controller was in the detent position, giving it the moniker "unique trim controller."

To allow the pilot to control the aircraft in all axes with only one hand, the sidearm controller also included vertical command via a force transducer in the base of the controller. This worked in conjunction with the altitude hold autopilot. When altitude hold was engaged, a vertical speed of up to 600 ft/min could be attained by pulling up (or pushing down) on the controller grip. The rate of climb (or descent) was proportional to the force applied, with a maximum of 6 lb. When the force was relaxed, the altitude hold reference would capture the new altitude.

Selecting a flight control option was done through a simple three-button panel between the multifunction displays and keyboard. This allowed the pilot to select between rate command and attitude command modes, and to enable the altitude hold function. These functions were alternately controlled through pushbuttons on the stick. The sidearm controller also incorporated an innovative "two-dimentional slide switch" trim switch. The switch not only provided control of pitch and roll axis trim, but when pressed inward it activated a wing leveler and engaged the attitude command/velocity hold function (VELSTAB).

A conventional displacement collective was included, albeit with slightly less total travel. The collective incorporated a trim actuator to keep it synchronized with the autopilot when altitude hold functions were active. The front and back seat collective sticks were mechanically interconnected.

The photos in Fig. 9.1 depict the flight test configuration of the cockpit. The cockpit included additional displays above the multifunction display for data needed for specific flight test profiles.

The Comanche crew seat was mounted in a slight aft recline position for comfort, and the right-hand controller was mounted at a complex angle optimized for ergonomics. An adjustable armrest allowed the pilot to position his forearm in the most comfortable location for grasping the stick.

The Comanche seat was provided with a stroking mechanism that allowed the seat to slide downward in a controlled manner during a crash to reduce some of the loads on the occupant's back. The stroking seat had been utilized in other crashworthy helicopter designs, but had one flaw: the center cyclic

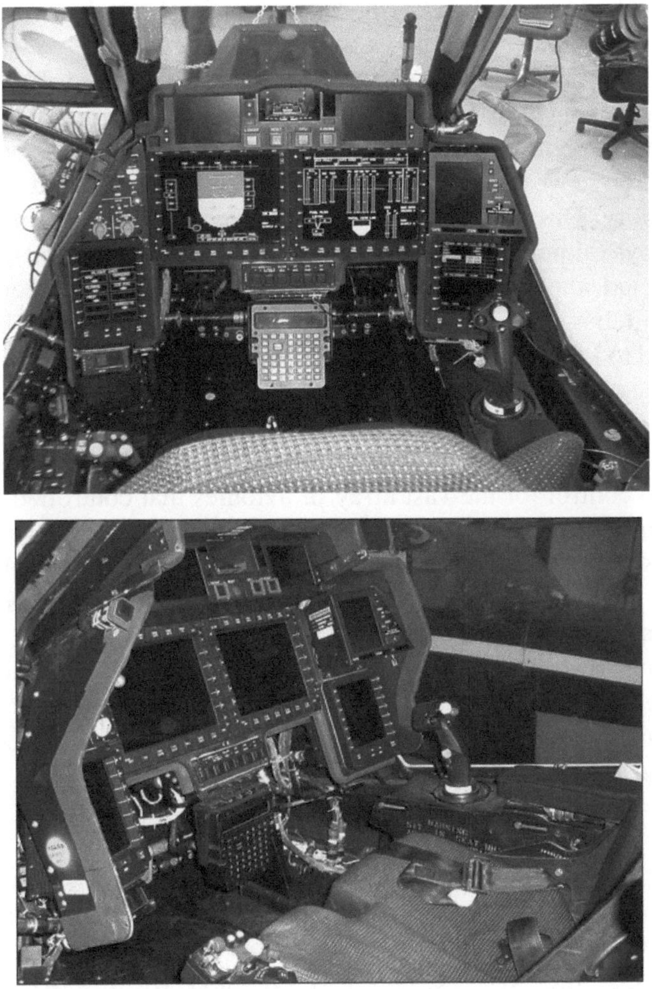

Fig. 9.1 Forward and aft cockpits enabled either crewmember to control flight via big MFDs, smaller MPDs, and sidestick controllers. Not shown is the HMD.

stick stayed in place and often impacted the pilot's jaw as the seat slid downward on the stroke track. In the Comanche, mounting the controller off to the side eliminated this interference. Additional safety was achieved by mounting the controller on the seat itself. This way, the controller stayed in position relative to the pilot's arm during seat stroking, minimizing the chance of injury to the arm or wrist.

Foot pedals were provided for pilot comfort and as a mount for the toe brakes. In addition, foot-operated pushbuttons were positioned to the side of the pedals for intercom and radio transmission. The location of the foot pedals was adjustable fore and aft. One deficiency that had to be addressed was the difficulty of using the brakes and radio transmit pushbutton simultaneously. The location of the pushbutton was altered slightly to improve this, but in the end, the pilot had to wait until finished using the foot brake before transmitting on the radio. This was considered a minor deficiency.

A wide (35 × 52 deg) field of view HMD was designed for use in both daylight and night environments. It incorporated a sunlight-readable display of flight and targeting symbology. When sensor imagery was activated for night and adverse weather flying, the same symbology appeared as an overlay on the imagery. This alleviated the need to provide flight information on the panel displays (although there was an option to do so if required).

As expected, the Comanche cockpit was environmentally controlled for crew comfort, but, more importantly, the air was filtered and a slight overpressure applied to keep the crew capsule clean and clear of biological, chemical, or radiological contaminants. The three flight control computers "lived" in the same environment as the pilots, located in bays to the left and right of the cockpit and behind the copilot seat. The same clean, conditioned air provided temperature control for the vast array of avionics that controlled the aircraft and mission systems. The environmental control system actively purged contaminants, which led to a distinctive periodic "puff" that could be heard when the rotors weren't turning. Some observers jokingly compared this to a gastric bodily function.

Comanche crew seats were armored in a conformal configuration that protected the entire crew torso without interfering with external visibility. This allowed for large areas of transparencies that exceeded all MIL STDs and provided the external view needed for nap-of-the-earth flight.

INFORMATION MANAGEMENT

The Comanche mission was a combination of reconnaissance and cooperative and direct attack—tasks that had been previously assigned to two separate platforms: the AH-1 Cobra and OH-58D Kiowa. Because of the requirement for dynamic crew workload allocation, both cockpits had to be identical so that either crewmember could conduct any mission task at any time. Through the use of the large MFDs, one crewmember could work on a task such as formulating a reconnaissance report while the other was, for example, observing targets.

This was the first time a cockpit would be devoid of instruments or displays dedicated to a single function. There simply wasn't enough room for conventional instruments. This "blank sheet" design posed both opportunities and challenges. Virtually everything was software-driven, so where in the past a crew station designer could simply pick an instrument such as an airspeed indicator from a catalog, now every aspect of the function had to be defined in excruciating detail. This focus on software also provided the opportunity for new and innovative ways to present information.

The Comanche crew station design group incorporated a very structured method to arrive at defining the information to be presented in display pages, as well as the menu structure for accessing that information. Known as a "mission/task analysis" technique, every part of the design depended on an understanding of how the mission would be conducted. The first step in the process was to develop a timeline of a typical mission. Teams of operational Army aviators were interviewed, and Boeing Sikorsky pilots visited Army units and flew on missions with them. The product of this effort was a "worst case" definition of a mission timeline, incorporating the most complex combinations of mission tasks envisioned by operational aviators, including response to threats.

The mission was then dissected into segments and analyzed to determine what information and control of which systems would be required at any point in the mission. The primary objective was to find which parts of the mission posed the highest workload to the pilot. At which points in the mission were the crewmembers bombarded with the most information and at the same time required the most response?

A secondary analysis examined which senses were utilized in these critical mission segments. Pilots have essentially three channels of information input: visual (things they see), auditory (things they hear), and psychomotor (things they feel and touch). Similarly, a pilot can use these same channels to transmit information or control tasks: visual (head-tracked displays can detect pointing angles), verbal (the pilot can speak to the system or transmit over the radio), and psychomotor (the pilot can manipulate controls). But there is a limit to how much the pilot can accomplish simultaneously, which was determined by the next step in the task analysis process.

A study of the Comanche mission, filtered by the information requirements, translated into a series of tasks the crew had to accomplish to achieve the mission timeline. The intent of the task analysis was to ensure the cockpit was organized into control and display protocols in such a way as to never overload the crew.

After completing the mission/task analysis, it was concluded there was still the potential for crew overload. Again, because of the software-intensive nature of the design, the next step in the design process was to determine how much of the mission tasks should or could be allocated to the computer. This so-called “man–machine allocation” defined how much automation would be incorporated into the design.

There were numerous opportunities for automating the functionality provided by the avionics computers. Some candidates were obvious, having been done to a certain degree in legacy aircraft. For example, rather than requiring the pilot to continuously monitor gauges indicating the health of basic aircraft systems such as engines and transmissions, the Comanche avionics system easily duplicated the monitoring job in the background and informed the pilot via a caution/advisory annunciator when a parameter went out of normal range.

To help organize the analysis required to complete the man–machine allocation, the cockpit functionality was broken down into major subfunctions, including aircraft systems, communication, tactical situation, targeting, and threat detection and response.

For each subfunction, the level of embedded automation was initially assessed. For the aircraft system subfunction, for example, there was a standing requirement under the reliability and maintainability umbrella to detect nearly 100% of potential faults in the systems that supported basic aircraft operation. This would include electrical, hydraulic, flight control, navigation sensors, lubrication, propulsion, environmental control, and landing gear monitoring and operation. Exploiting that capability led to a “display-by-exception” design for systems control and health monitoring.

Unlike legacy monitors, the Comanche display-by-exception system included an automated scenario monitor that determined when subsystems should be in operation and either activated them automatically or displayed a message to the pilot when something was wrong. For redundant systems, such as hydraulics, the monitor would automatically deploy a backup system when necessary. This eliminated the need for dedicated switches to control generators, hydraulic pumps, backup lubrication systems, and so forth. By first checking to see if a system should be in operation prior to announcing a problem, automation eliminated “invalid” fault annunciations such as low oil pressure prior to engine start. It looked for transient faults and remembered them but did not interrupt mission operations unless the fault persisted to a point where action was required. Not only did this automation clean up the cockpit and create room for dedicated mission controls, it greatly simplified the checklists required to operate the aircraft and reduced the emergency procedures to simple one- or two-step responses—usually the only decision after a failure was to continue or abort the mission.

Piloted simulation was relied upon heavily to assess the effectiveness of automated mission functions. Because everything from the environment to threat systems to target types and locations was simulated, it was possible to automate virtually any operation associated with the mission. For example, a completely automated target search, identification, reconnaissance report, weapon selection, and weapon deployment was possible in simulation. During studies as to the feasibility of using a single pilot, high levels of automation were attempted. At one point in the design investigation, all the pilot had to do was fly the aircraft through the mission profile; everything else was done automatically. Not only was this perceived as improbable in the real world, the automated systems often surprised the pilot by conducting a weapon engagement before he even knew there was a target.

The question most often asked by operational pilots during a simulated mission of this type was, "What is it doing now?" This led to a reduction from full automation to a system that would conduct mission operations in the background, but at a certain point in the mission task, ask for the pilot to consent to actual weapon deployment. Even this level of automation was unrealistic, often bombarding the pilot with decisions without context because he was still not participating in much of the mission task processes. This led one crew station designer to declare that the design created the "burden of automation." The pilot spent most of his time trying to understand what the system was asking him to do and where he was in the mission.

Ultimately, the man–machine allocation studies led to a balanced approach. First was a realistic reflection on just how much the avionics would be able to do without pilot assistance. For example, although an automatic target recognition (ATR) system was originally envisioned, the Comanche ATR function became known as an *aided* target detection and classification system (ATD/C). The technology, understandably, just couldn't support a full identification of every target possible in the world. Referring back to the task analysis, the avionics and crew station designer then took this knowledge of what automation was possible, and found those areas where assistance from the automatic systems could lower the crew workload to an acceptable level.

Once the information required to accomplish each mission task was understood and the level of automation determined, the task of developing an interface to access easily the information and control the systems began. Looking at legacy systems, the Comanche designers wanted to improve on the typical tiered menu structure. Careful analysis revealed the shortcomings of legacy systems—control and display were structured by system, not task. For example, everything associated with a targeting sensor was embedded under a tier of menus. But to launch a Hellfire missile, the pilot needed information from several systems, including the targeting sensor, the missile seeker, communications with team members, and the navigation system. This resulted in the requirement to provide access menus that were several layers deep, with few dedicated controls and many "soft" controls, which change function depending on what system has been accessed.

In the traditional system orientation menu, everything associated with the communication system was under one menu. But there are several important functions associated with the communication system, including voice and text, both clear and encrypted. One menu had to serve all those functions. This tended to be confusing and difficult to learn and then remember where to access a particular function. Constant, repetitive training was required, or the skill in cockpit operation was lost.

The integrated nature of the Comanche avionics system and the high level of automation possibilities provided the opportunity to do a better job of providing the pilot with a logical menu structure. The answer was envisioned by Dr. Bruce Hamilton, the PhD human factors engineer with Army experience who led the crew station design team. He suggested that rather than structure the cockpit mission system control architecture by system, we design it by mission task. After all, Army aviators are generally not engineers; they are warriors that understand their mission and the tasks required to complete it. Why not provide a cockpit that communicated with the aviator on a task basis? It would be easier to remember where to find a particular function, and it would allow continuity in task completion for better mission timelines and fewer errors.

A document called the Pilot–Vehicle Interface Mechanization Specification (PVIMS) was created to describe all of the Comanche cockpit displays and controls. As discussed earlier, because the cockpit was largely software driven and lacked hard-wired instruments, every aspect had to be described in excruciating detail so the software engineers could write code to convert the mission capabilities into reality.

COCKPIT LAYOUT

The result was a cockpit with a "breadth vs depth" philosophy. Breadth meant that there were numerous controls and menu buttons to permit high-level access to mission functions rather than systems. For example, in the original Comanche cockpit design, there were five menus for the communication system (serving the core function of reconnaissance) to allow for radio selection, frequency tuning and planning, textual report formulation, and encryption configuration. The final design included 20 hard-coded buttons along the top and bottom of the multifunction displays, each providing direct access to a functional menu. Although this initially sounded like a lot to learn, because they were grouped by function, pilots found it relatively easy to remember how to perform a task by functional association. The cockpit displays are shown in fig. 9.2.

- Group 1, on top of the left multifunction display, provided access to all communication functions.
- Group 2, on top of the right multifunction display, provided access to imagery and other helmet display options, as well as management of the navigation functions.

- Group 3, along the bottom of the right multifunction display, provided access to all of the functions associated with the tactical situation display (based on a digital map).
- Group 4, along the bottom of the left multifunction display, provided access to the targeting functions, system status, and backup flight instruments.

Instead of taking up valuable display area for second-layer menus, labels for these functions were located on a dedicated display to the left of the multifunction displays. Originally, this display was to be touch-sensitive, but the technology had not yet matured, so pushbuttons were added. It was anticipated that touch technology would one day catch up and allow expansion of this display's utility. The basic layout is depicted in Fig. 9.2.

In comparison with the then-current scout aircraft, the OH-58D Army Helicopter Improvement Program (AHIP), the number of button presses to perform a mission task in the Comanche was greatly reduced, as shown in Fig. 9.3.

Due to their criticality, panels with specialized controls, such as pushbuttons and rotary knobs, provided direct access to the functions associated with the target acquisition (to the left of the multifunction displays), weapons, and laser (to the right of multifunction displays). This was partly due to the importance of these functions, but was also a required safety initiative to ensure there would be no software bug that could result in inadvertent weapon launch or laser emission. In addition, direct access to weapons was provided through buttons located on the flight control grips.

The astute observer will view these figures and wonder—where is the information needed to simply fly the aircraft? Early in the design process it was recognized that the extremely close proximity to the ground and obstacles during nap-of-the-earth flight required intense and continuous scrutiny of the outside world. One of the most major technology initiatives of the Comanche

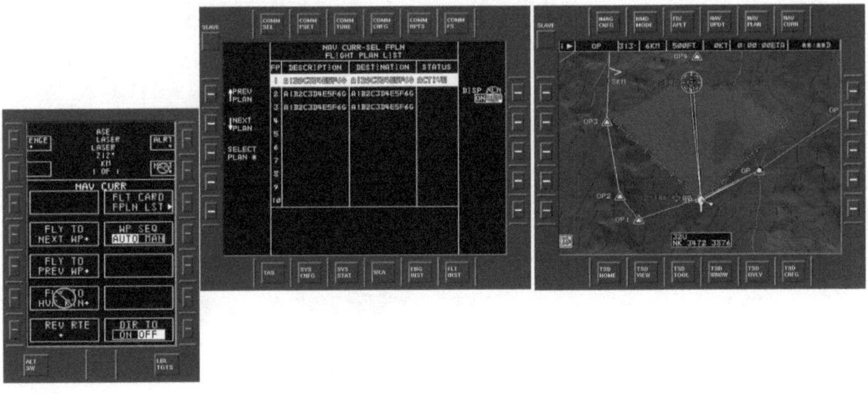

Fig. 9.2 The Comanche cockpit display layout arrayed switches by functional groups rather than the systems they operated.

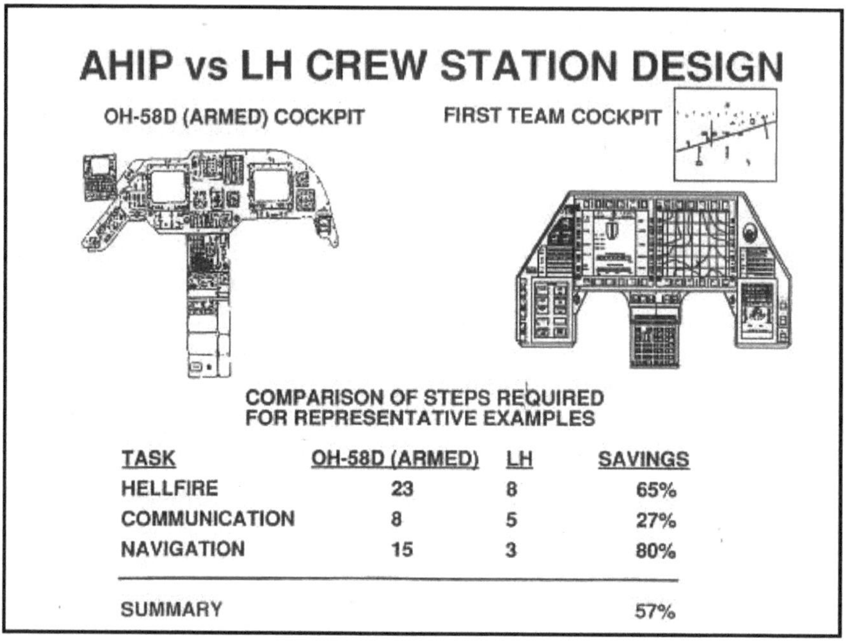

Fig. 9.3 AHIP vs LH step reduction.

program was development of the helmet-mounted display that had the capability to provide flight performance information to the pilot while keeping eyes outside the cockpit.

It was widely acknowledged that maximizing the time the pilot spent looking outside the cockpit greatly increased both the safety and effectiveness of

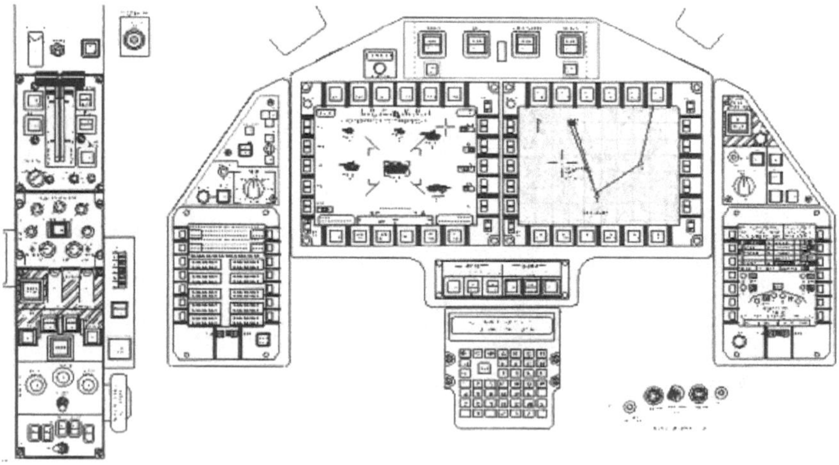

Fig. 9.4 The Comanche cockpit layout used bezel switches instead of touch displays to access mission functions.

nap-of-the-earth (NOE) flight. Flying close to the terrain and obstacles reduced the visual detection range of air defenders and improved the probability of flying through a threat area undetected. Requiring the Comanche pilot to look inside the cockpit for information regarding aircraft performance, navigation, and weapons sighting detracted from visual flight path control and increased the probability of an obstacle strike.

The objective, then, was to provide the information needed to fly, navigate, and deploy weapons while the pilot looked outside the aircraft. As the demand for 24/7 operations increased, the need for effective sensors to provide a visual depiction of the outside world resulted in development of image intensifier and infrared imaging systems. Aside from night vision goggles, the only fielded helmet display system was the integrated helmet and display sight system (IHADSS) installed in the AH-64 Apache. This was a narrow field of view (30 deg circular) monocular device that required extensive and perishable training to operate effectively. There was a strong desire to push the technology towards a wide field of view (FOV), binocular helmet display. The development of this system is described in Chapter 3, The Mission Equipment Package.

Embedded Training

The Comanche cockpit also had an embedded training capability. Essentially, this allowed the pilots to operate all of the systems, even the laser and weapons, without actually launching anything from the aircraft. The pilots utilized procedures to search and prosecute targets that were identical to those that would be used in combat. Once a simulated weapon was launched, the aircraft could assess a probability of kill using on-board sensors, or communicate with other military vehicles such as might be found during a joint training operation. The aircraft training mode was a global software command, which reconfigured the systems for safe operation. After shutdown, the aircraft automatically reverted to a full operational configuration. The Comanche program was canceled before this training function could be implemented on the aircraft.

Summary

As with many other technologies associated with Comanche, the cockpit design was truly a transformational program, moving from round dial "boiler gauges" to "glass cockpit" multifunction displays. The fact that it was able to combine the mission capabilities of both scout and attack aircraft is a testimony to its success.

Although not many aviators were able to fly the actual Comanche, hundreds experienced the cockpit design during simulation research studies, formal evaluations, and marketing displays at trade shows. The overwhelming response was positive. Many pilots said, "We need to get this into the field as soon as possible." The large majority agreed that, as measured by workload reduction, mission effectiveness, and safety, the Comanche would have been a huge benefit to the Army aviator. Further, we now see that many of the Comanche technologies have migrated to other modern cockpits. This would not have happened if the design had not offered innovative but successful concepts for others to exploit. For that, everyone associated with the Comanche cockpit design is deservedly proud.

Chapter 10

Software

Frank Schrader

Software development was in its infancy when the LHX program started. Microprocessors had just been developed, and engineers were learning to program them. The first digital navigation computers used heavy core memories with fixed programs. Core memories needed to go back to the factory whenever they needed reprogramming.

The use of microprocessors in desktop computers and the miniaturization of programmable memories led to many highly complex functions being available for everyday use. Airborne computers generally used a single microprocessor with an associated cache memory for speedy operation and a remote memory to store data and to store the programming elements loaded to the cache. As a note of scale, Commodore 64 computers used 64 KB of memory; so many complete airborne computers would be programmed by a handful of programmers. The operational software of the 1980s' AH-1 Cobra, for example, contained 10,000 lines of code. Comanche software estimates at the time were over 1 million lines of code. This grew to over 4 million lines at the time of the engineering, manufacturing, and development (EMD) proposal, and was still growing.

Initially, Sikorsky leased supercomputer time to run our radar cross-section (RCS) codes. Over the Comanche development time period, technology was advancing so rapidly that when we got cell phones in 2000 they had almost as much processing power as the Comanche mission equipment package (MEP).

Processor speeds and memory density growth led to an explosion of capabilities; however, programs written by one company were not easily transported to other machines. The military desired transportability and transparency in future software. Among other reasons was the desire not to be forever bound to companies with proprietary software.

So the LHX software requirements included that any major software be created with a standardized microprocessor and with a standardized, military-approved software programming language—Ada. Additionally, to foster transparency and to avoid any ability for an enemy to corrupt or destroy the mission computer with hidden programs, the LHX mission computer operating system was to be built from scratch as a "trusted" operating system.

Teaming with Boeing

Boeing Wichita was lead for the central computer and central data processing, including software development; Sikorsky was lead for cockpit controls and displays and also for air vehicle interfaces with the MEP and thus also was responsible for any software related to these two areas. Any software within dedicated electronic units was the responsibility of that unit's product development team. Flight control software was Boeing's lead, although Sikorsky was responsible for the control laws, which defined the requirements to a certain extent.

Sikorsky had developed the Pilot–Vehicle Interface Mechanization Specification (PVIMS) to explain how the pilot would operate his controls during various missions and what functions would be automated (by software) to lower pilot workload. A Sikorsky team of human factors engineers worked with simulation engineers to test and refine the PVIMS, and the result was incorporated into Sikorsky's motion base simulator's Comanche cockpit for team and customer evaluation. The software to accomplish the PVIMS within the central computer was called Cockpit Interface Management (CIM).

The automation via intelligent software extended to the pilot's interface with most of the air vehicle subsystems. The subsystems were monitored automatically by software; the pilots were not informed of their status unless something was out of tolerance, and even then, software usually provided corrective action without pilot intervention. The software within the central computer to automate the air vehicle subsystems was called Aircraft Systems Management (ASM).

These two packages of software were controlled by the CIM and ASM computer system configuration specifications. They turned out to be the two largest software packages on Comanche.

The Early Days

Before teaming, Sikorsky had planned to employ a multiprocessor central computer and partitioned software environment to allow graceful degradation and recovery from multiple processing failures. This architecture was developed in part by TRW's experience with satellite designs, where in-outer-space repairs were impossible. This approach was not normally done with the segregated systems of the 1970s. The only hint at fault tolerance in most digital system architectures was simply to install dual, identical, segregated computers running the same program. If one failed, the other simply did the same work.

Boeing concurred with the centralized, parallel processing approach, especially because the centralized (Boeing-led) architecture collected all the data to be shared and processed. Also, Boeing Wichita was probably aware of the similar architectures from PAVE PILLAR programs.

Boeing selected the Intel I380 chipset (the predecessor to the first Pentium chip) as the baseline computer. This chipset had a full complement of instructions, which went in line with the military's desire to ensure some transportability of code across different platforms. Sikorsky's technical experts would have preferred a Reduced Instruction Set Computer (RISC), which was faster and could be tailored for more efficient coding. But standardization, and Boeing as lead, won out. The planned availability of an Ada compiler, separately funded, was also a factor.

SOFTWARE DESIGN USING ADA

One of the mandates of Comanche was that the software for the central computer be done in the Ada software language. Ada is a modern programming language designed for large, long-lived applications—and embedded systems in particular—where reliability and efficiency are essential. It was originally developed in the early 1980s. [This version is generally known as Ada 83 (for 1983), so it was "new" when Comanche started.]

Ada has a simple syntax, structured control statements, flexible data composition facilities, strong type checking, traditional features for code modularization (subprograms), and a mechanism for detecting and responding to exceptional run-time conditions (exception handling).

Unlike languages based on C syntax (such as C++, Java, and C#), Ada allows the programmer to simply and explicitly specify the range of values that is permitted for variables of scalar types (integer, floating-point, fixed-point, or enumeration types). The attempted assignment of an out-of-range value causes a run-time error. The ability to specify range constraints makes programmer intent explicit and makes it easier to detect a major source of coding and user input errors.

As mentioned in the MEP chapter (Chapter 3), MEP resets were a continuing nuisance. These were a function of several questionable decisions in the early program to stop all processing and display of MEP data (including aircraft instrumentation) if a single error occurred (so nothing erroneous was displayed to the pilot). This Ada feature would cause run-time errors whenever an input was out of range, so it is clear that many of these resets were caused by operator input values incompatible with the scalar ranges (as in a bored copilot on first flight). In part, the aircraft problems were due to incomplete software packages—we were always waiting for additional functionality to appear.

A key to reusable components (which would speed coding) is a mechanism for parameterizing modules with respect to data types and other program entities. Ada met this requirement through a facility known as *generics*. Sikorsky software engineers were adept at using this feature, and it facilitated rapid code generation for the displays.

SOFTWARE PROCESS IMPROVEMENT

In the 1980s Sikorsky was instituting process controls (similar to Six Sigma) to reduce errors throughout the design process. The software organization was a leader in mandating process control and understanding the full set of requirements before initiating coding.

Software process improvement efforts were based on the Capability Maturity Model Integration (CMMI—Systems Engineering) and the Software Acquisition Capability Maturity Model (SA-CMM) process improvement frameworks developed for the DoD by the Software Engineering Institute (SEI). Sikorsky enthusiastically endorsed the process and achieved early certification under it. The lead engineer for this massive effort to retrain all software engineers to a standardized process was Leigh Grey, and she did an excellent job.

CMMI allowed us to track design progress, error rates, and many program variables. The bane of software developers was requirements changes (volatility). But with concurrent designs of the airframe and mission systems, the ability of system engineers to exactly specify interfaces and software requirements was limited, if not impossible. The requirements stabilized fairly quickly on the airframe side, but the cockpit control and displays were constantly "improving" due to better ideas evolving through simulation of the mission.

Boeing also supported early standardization of software development. It transferred Taylor Warren to lead the software effort. He had been recently awarded a government medal for developing software program controls and accurate reporting of software metrics.

SOFTWARE ORGANIZATION

Both Sikorsky and Boeing recognized that software engineers had a specialized skill, and each company had functionally grouped software developers as a separate organization. But software engineers need to receive several "boundaries." At Sikorsky, there was an avionics systems group that established the initial overall desired operation of a system and its software, partitioning down to individual black boxes. Then there was a software systems group that worked out the operational partitioning between a processor and software within a "box." Within Comanche, the software systems group and the software engineers were one group, led by Rocky Albano. The mission equipment group, led at Sikorsky by Frank Schrader, was the equivalent of the avionic systems group.

At Boeing, there was one overall leader for MEP development, but at Sikorsky there were two. This led to a lot of boundary disputes, especially when CMMI metrics were tracked.

The key boundary metrics were defined by the quality and volatility of requirements definition. Changes in requirements caused software rework (programming, peer review, testing, and integration retest). Mistakes in interpreting requirements also led to "that's not what I wanted" rework. There were two schools of thought (Boeing's and Sikorsky's) on how to most efficiently develop MEP software.

Boeing preferred to have "top-level" specifications from its systems engineers handed to the software engineers. These computer software configuration items (CSCIs) were also the data deliverables for Army review and approval. By keeping them somewhat general, Boeing reduced volatility in Army-approved requirements; however, this required continuous hand-holding and clarification at the software requirements level between the systems and software engineers to get the desired operational software. This was normal at Boeing and didn't ripple through its metrics. It did mean that there was typically a long time between approval of a CSCI and generation of software code.

The Sikorsky software group took a different path. We didn't want "top-level" requirements that were subject to interpretation. We didn't accept vague requirements. So the group bounced back anything lacking detail. This significantly improved its metrics, to the detriment of the systems engineers' work. Sikorsky CSCIs were extremely detailed. Unfortunately, because there was no single group leader for MEP and software, there were constant unresolved arguments about when requirement details were "too detailed" or "not detailed enough."

SOFTWARE DEVELOPMENT

Sikorsky had a bunch of talented software engineers. Software development started with our Advanced Rotorcraft Technology Integration (ARTI) LHX simulator, where the pilots and Bruce Hamilton's human factors engineers developed the single-pilot concept and identified a lot of automation requirements to allow the pilot to concentrate on the mission phase (attack and so forth) while the aircraft took care of itself. The simulator provided detailed requirements to allow the mission computer to identify what the pilot was doing and what he might do next. The mission computer was to anticipate his next need in the context of what was happening inside and outside the cockpit. The mission computer was also required to react to any changes in equipment status and do ordinary reactions to changes in equipment status (failures, incoming messages, changes in power requirements). Conveniently, these automation functions primarily were in areas that Sikorsky controlled through software in the controls and display CSCI and the aircraft management CSCI. Thus, the simulator (and pilots) helped draft the CSCIs. To define the control and display operation, Bruce Hamilton's crew wrote the PVIMS before we even won the contract. This was constantly iterated and expanded during the contract phases. The aircraft systems management CSCI was built with the help of the electrical systems group, which did all the electronics for aircraft systems. Mike Bradford was the key contributor to this CSCI. The volatility of this ASM CSCI was driven by "newer and better" ideas emanating from the diagnostic group, which in fairness didn't know which failures to detect until the supplier's hardware went through critical design reviews and environmental testing.

In general, because of the extensive simulation work, these CSCIs were very detailed and allowed accelerated development of our portion of Comanche software. Unfortunately, the Boeing-led mission computer operating system development lagged.

The operating system was a difficult task. It had to be built from the ground up as a trusted (no extraneous code) OS. It also had to operate multiple parallel processing cards and pull programmed software packets to available processing cards and prioritize which software packets would run at a given time, based on priority of tasks and availability of processing elements. This was also complicated by the need for "graceful degradation" of capabilities in backup modes due to failures.

Boeing had a staff of three to five software engineers writing the operating system, but they had to start from scratch—writing code for addition, multiplication, and trigonometric functions as well as all the transfer functions of a multiprocessor environment. Basically, the whole software development was paced by the ability of these engineers to provide OS capability.

So Sikorsky could generate code, but had to wait for the OS to integrate with it.

Boeing was surprised at how much code Sikorsky could generate. In the display area, Sikorsky took early advantage of macros to magnify "lines of code" generation.

Because Boeing was the MEP lead, all software had to be tested in its integration lab before being issued to the aircraft. This sometimes took weeks of manual testing of all possible inputs and outputs with hardware in the loop.

Finally, Sikorsky software engineers developed automated scripts to simulate all the inputs and outputs and greatly reduced the turnaround time at Boeing.

Because the program was delayed by funding in the later years, much equipment went unfunded, and the corresponding mission software was subbed out. This added to the software iterations and schedule delays. At the time of termination, Operational Flight Program (OFP) drop 10.4, which supported electro-optical target acquisition and designation system (EOTADS) and night pilotage functions on prototype two was nearing completion to support the late April flight. Also in process was OFP drop 11, which supported the first flight of the EMD, aircraft planned for June 2005. The Army was amazed that the first time the OFP 11.00 CIM and ASM CSCIs were brought together in the lab, within three days, the software was integrated and controlling tactical subsystems in a hardware-in-the-loop configuration. Considering the complexity of the system, this was a remarkable achievement, attributable to effective execution of the common process from the system engineering and software levels all the way through system integration and test.

A Year 2000 Tri-Services Assessment found fault with the lack of a weapon system–level systems engineering organization within Sikorsky, the fact that there was no weapon system–level end-to-end integration facility, the requirements volatility in the remaining fault detection and Portable Intelligent Maintenance Aid (PIMA) systems, and a dependency on the Joint Strike Fighter (JSF) program. In December 2000, the Joint Program Office (JPO) and Sikorsky went to brief Walt Hollis on the Tri-Services assessment. Hollis felt that although the Comanche software represented a significant challenge, everything that could be done was understood by Sikorsky, and efforts were in place to address the issues identified by the Tri-Service. The real issues were recruiting qualified software people; software tools within Sikorsky to support and manage the integration of such a large software program; multiple software languages, formats, and compiles to be used by team members; and lack of a clearly defined fault detection isolation architecture for everyone to design to.

The final planned OFP drop was drop 15, which was scheduled for completion in 2001. With drop 15, all the features and capabilities would have been in place. One highly advanced feature that was completed in this future release was the Tactics Expert Function. This function would assist the crew in making time-critical decisions on how best to employ the aircraft while maximizing its survivability against a host of potential threat systems.

Chapter 11

THE MAIN ROTOR

BRUCE KAY AND JOHN OCCHIATO

Advances in rotor technology were available for the LHX that were not feasible for earlier generation helicopters. One of these was the bearingless main rotor (BMR). This was a simple composite flexbeam that replaced the conventional flapping, lead-lag, and pitch change bearings of then-current helicopters. Consideration for using a BMR over a conventional articulated rotor was strongly influenced early on by the LHX handling qualities requirements for aggressive, precise, and quick response characteristics. The tradeoff between an articulated rotor and a bearingless rotor is described in Chapter 6, Handling Qualities.

THE BEARINGLESS ROTOR

Early helicopters had mechanical bearings for all three motions, which were grease lubricated. The need to purge the grease on a daily basis made for a high maintenance burden; this led to oil-lubricated bearings to reduce maintenance requirements. In the 1960s, a concept was developed to replace these conventional bearings with elastomeric bearings; these were first used on the Sikorsky CH-53D and then on the UTTAS/Black Hawk main rotor.

The Black Hawk tail rotor was the first production use by Sikorsky of the bearingless flexbeam concept. The Black Hawk tail rotor uses two graphite flexbeams connecting two rotor blades, clamped together for the four-bladed rotor (see Fig. 11.1). All motions—flapping, lead-lag, and pitch change—were accomplished by twisting or bending of the flexbeam. A torque tube around the flexbeam was used to transmit blade pitch control inputs to the blade.

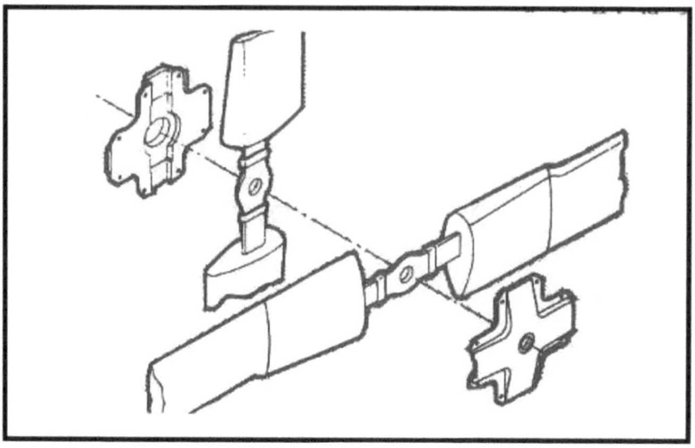

Fig. 11.1 The UH-60 Black Hawk tail rotor using the bearingless concept.

The LHX rotor was the first BMR main rotor design for Sikorsky. It had not been even prototyped previously. (The LHX design tradeoff study between the bearingless main rotor and a conventional articulated rotor is discussed in Chapter 6, Handling Qualities.)

After the decision was made to use the BMR, there were all the concerns that typically accompany nonarticulated rotors regarding potential vibration and structural concerns as well as the potential for ground resonance and a regressive lag mode. There was no time before proposal submittal to build a prototype and conduct a risk reduction program to resolve these potential risks.

To address risk, Messerschmitt Bolkow Blohm (MBB) in Germany was brought in to help perform an independent assessment of the flexbeam design. Its BO-105 had used a metal flexbeam main rotor system and had been in production for a number of years. MBB reviewed the analysis and endorsed the Boeing Sikorsky BMR design.

Next, the rotor geometry needed to be determined. This was done using standard performance methodology based on meeting hover and high speed performance requirements. The performance folks established a rotor diameter of 39 ft and selected an SSC-A09 blade section airfoil with a high Cl_{max} and drag divergence Mach number.

Rotor Solidity

Next, the rotor solidity (blade area) had to be determined, and it was based on LHX meeting the 300-ft obstacle avoidance maneuver at 4000/95 conditions and V_h. In order to meet those requirements, the rotor must be able to sustain 1.95 g for 2.5 seconds. The chord that produces that level of sustained Nz is performed at the rotor incipient stall point, which is typically at approximately $Ct/\delta=0.15$. The task of determining the total blade area was the responsibility of the handling qualities group, who used a GenHel (Generalized Helicopter simulation program) version of the LHX. The process is a not a closed-form solution but rather a trial-and-error process. The pull-up maneuver, using nonpiloted simulation analysis, is performed over and over until the aircraft achieves the necessary profile while operating at the sustained required Nz with a blade area that places the rotor at a defined incipient stall condition.

The technique used by the handling qualities section defines stall using a proven technique that was originally developed back in the early 1960s. It defines stall using a parametric formulation related to the rotor blade drag coefficient when the blade is at the 270-deg azimuth position. It turns out that there is a very good correlation between a rotor stalling and a specific buildup of that blade's drag data. Sikorsky had developed a flight test procedure that essentially measures a similar phenomenon by measuring the blade push rod load buildup as the rotor approaches stall. This measure is referred to as the ERITS (equivalent retreating indicated tip speed) calculation. The technique of determining rotor stall in the GenHel program correlates almost exactly with the test ERITS data, which gives good credibility to the analysis prediction. The rotor solidity question was resolved.

NUMBER OF BLADES

The next tradeoff study was on the number of rotor blades. Engineering Vice President Dr. Ken Rosen was pushing for four blades, similar to the Black Hawk, and keeping the total rotor parts count down to a minimum. After much debate, acoustician Phil LaMessurier convinced Ken that at least five blades were necessary to meet the acoustic requirement. Helicopters are notoriously loud, and in many cases they can be aurally detected before they become visible. Because great measures were being taken to make LHX/Comanche stealthy, something obviously had to be done to reduce its aural signature. The primary source of the main rotor slapping noise is from the blade tips, and in addition to blade thickness, this noise is a direct function of the tip shape and tip speed. This noise source was reduced by increasing the number of rotor blades (thus reducing blade thickness) and doing two additional things. The blade included a tapered tip that was optimized both for acoustics and to maximize vertical rate of climb performance. In addition, a "quiet mode" operation was enabled by reducing rotor rpm by 10%. The result was an approximate 15-dB reduction in audible noise compared to existing H-60 and AH-64 helicopters. Both of those helicopters were already quieter than the infamous "wop-wop" H-1 Huey noise signature.

The trade then boiled down to comparing five- and six-bladed configurations. The magnitude of vibratory forces that shake the helicopter is directly dependent on the number of rotor blades. By reducing these vibratory

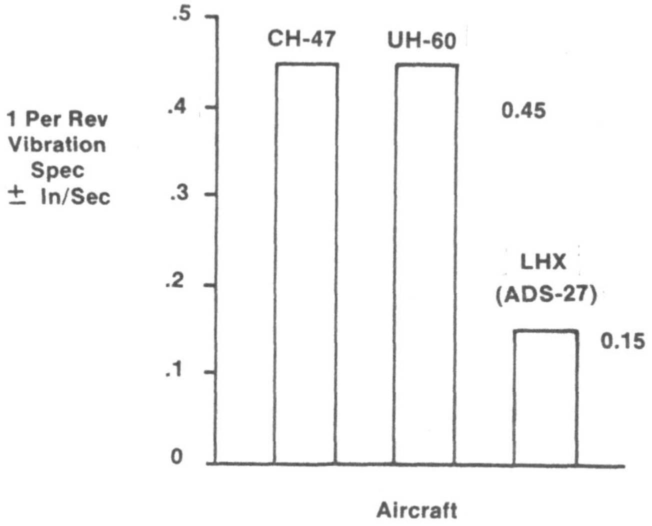

Fig. 11.2 Vibration spec comparisons.

excitation forces, the need for supplementary vibration absorbers could be minimized or eliminated. Low vibration levels would enhance crew comfort and increase the reliability of avionics equipment. A very low vibration level requirement was reflected in the Army's LHX specifications (Fig. 11.2).

Clearly, six blades would be better than five; however, there were mitigating circumstances that made a six-bladed configuration impractical. The biggest obstacle was ballistic survivability. With six blades, the structural spars became so small that damage from most ballistic threat rounds would destroy the spar. Loss of a major portion of a rotor blade would create large imbalance forces that would destroy the helicopter. To preclude these imbalance forces, jettisoning an opposite blade was considered for a while. As far-fetched as this seemed, the high speed NASA/Army/Sikorsky S-72 Rotor System Research Aircraft (RSRA) had rotor blades that could be jettisoned to enable the aircrew to bail out in an emergency. However, inadvertent jettison rendered this option to be unsatisfactory.

The decision was made to use five rotor blades. Compared to four blades, this reduced individual blade forces and moments and improved the rotor's radar and acoustic signatures, without making the blade chord so small as to introduce new problems. As hoped for, fuselage vibration levels were very low, and no auxiliary vibration absorbers were needed.

With five blades and the rotor solidity previously determined, the rotor chord was 15 in. We called it the "Pentaflex" rotor.

So the chosen Comanche design used a 39-ft-diameter, five-bladed rotor using the bearingless rotor head (see Figs. 11.3 through 11.5). Blade twist was –11.1 deg, or –13.5 deg on a thrust-weighted basis. A Boeing VR-12 10%

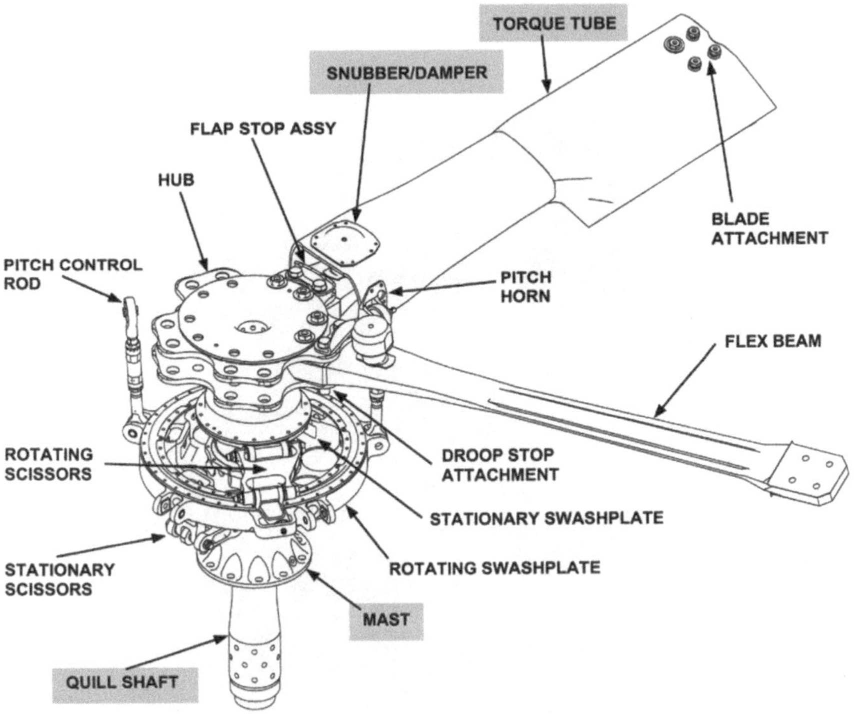

Fig. 11.3 Main rotor head components.

Fig. 11.4 Rotor on the Propulsion System Test Bed (PSTB).

thick airfoil was used out to 85% radius. This transitioned to a Sikorsky SSC-A09 9% thick airfoil at 90% radius. The outer 8% of the blade was swept aft and tapered. Equivalent hinge offset was 10%.

Fig. 11.5 Rotor head and swashplate.

To address risk, a full-scale whirl test project was initiated under Sikorsky's supporting Independent Research and Development (IR&D) program. This was sized for the Sikorsky S-76, with a 44-ft-diameter rotor, and used existing S-76 composite rotor blades and a new five-bladed hub, employing design features similar to the rotor proposed for the LHX. The whirl tests were completed in August 1991. In total there were more than 10,000 hours of wind tunnel tests on models ranging from one-tenth to full scale. The full-scale testing was conducted in the NASA Ames 40-ft×80-ft wind tunnel with the objectives of investigating dynamics and stability, rotor structures and loads, handling qualities, aero performance, and acoustics. The rotor was flown to 200 kt with rotor thrust up to 19,000 lb [1].

The tests were conducted from June to October 1992. According to a 1993 American Helicopter Society technical paper [2], the results included the following:

> In the dynamics area, rotor stability trends and rotor natural frequencies were generally close to predictions. The SBMR [Sikorsky bearingless main rotor] demonstrated positive stability in the regressive lag mode at

all conditions tested in the wind tunnel. In addition the rotor produces low flap moments at 4, 5, and 6 per rev frequencies at high velocities, which should lead to low vertical 5P vibration in an aircraft application.

During rotor structural testing, the SBMR was found to be quite robust resulting in low accumulated fatigue damage.... Spanwise bending moments were found to correlate well with predictions in both flapwise and chordwise direction.

For handling qualities, generally good correlation of control derivative data with predictions was found. Rotor frequencies response methods were used to provide a large database for comparison with predictions. Generally good correlation was found in the area of control system modeling and rotor system natural frequency identification.

In the aero performance area, rotor performance analysis and test results were in good agreement after accounting for actual hub and torque tube tares. Fairing of the torque tube/blade joint had a measurable effect on rotor performance.

For acoustics, high speed impulsive trends...were generally consistent with predictions....

ONE-PIECE PENTAFLEX ROTOR

It is interesting to note that during the LHX proposal phase, Sikorsky was considering a one-piece Pentaflex rotor head instead of five individual flex-beams. The concept was based on using a fiber placement machine to lay up multiple Pentaflex components on a large, 12-sided dodecahedron mandrel (Fig. 11.6). The objective was to reduce parts count and exploit automation. Although a mockup was built, fortunately the approach was discarded

Fig. 11.6 One-Piece Pentaflex Rotor.

because of the high technical risk. With perfect hindsight based on the individual flexbeam manufacturing experiences, the one-piece Pentaflex would have been unbuildable and become a show stopper.

1991 DEM/VAL PROPOSAL

It was assumed that the competing SuperTeam would use a more conventional rotor design with a lower offset and four blades, similar to the Bell 680 rotor. Thus, the Boeing Sikorsky proposal stressed the advantages of higher offset. The charts in Figs. 11.7 and 11.8 were included in the proposal.

HINGE OFFSET FOR VARIOUS AIRCRAFT

	% Offset	
UH-1, AH-1	0	Teetering rotor
-680 rotor	2½	Super Team LHX
SH-3, HH-3	3	Transport, ASW helo
AH-64	4	Apache
UH-60A	5	BLACK HAWK
First Team BMR	10	High agility LHX rotor

Fig. 11.7 Hinge Offset for Various Aircraft.

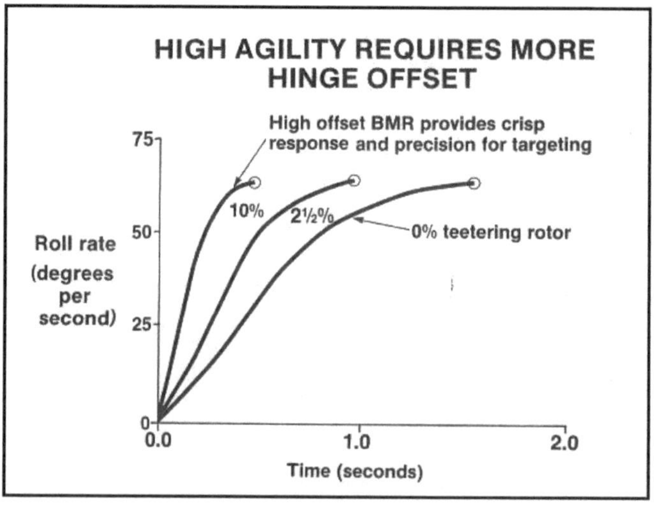

Fig. 11.8 High Agility Requires More Hinge Offset.

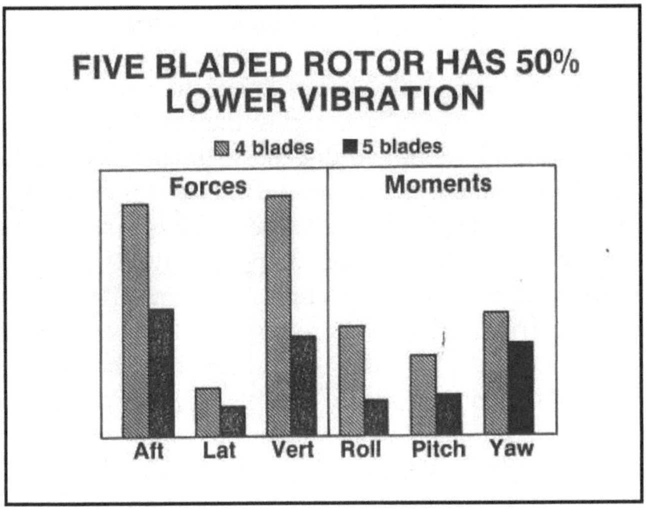

Fig. 11.9 Five Bladed Rotor has 50% Lower Vibration.

The impact on effectiveness with offset is apparent. The higher offset rotor allows tighter terrain following and more roll agility as well. These same factors carried over into an air combat scenario and could have a profound effect on the outcome.

The Boeing Sikorsky proposal also stressed the advantage of the five-bladed rotor on aircraft vibration (Fig. 11.9).

In a verification of this conclusion, Dr. Kenneth B. Amer of the RAND Corporation presented a paper on the subject at the 1989 American Helicopter Society Annual Forum [3]. Titled "A Comparison of Four Versus Five Blades for the Main Rotor of a Light Helicopter," he concluded that the five-bladed rotor is superior, primarily due to lower vibratory forces, which don't require vibration absorption or cancellation devices and thus leads to a lower rotor system weight and cost.

The Flexbeam

Although the basic flexbeam geometry appeared pretty straightforward, the structural design and analysis was extremely complex due to the number of constraints the design had to satisfy. The cross-sectional area was minimized to keep it flexible. Torsional and bending stresses, induced to accommodate blade twisting and flapping motions, had to be accommodated. The torque tube surrounding the flexbeam transmitted pitch control inputs to the rotor blades. Keeping the virtual hinge offset within a predetermined range limited the ability to minimize bending and torsional stresses by lengthening the flexbeam. Another constraint was keeping natural frequencies away from rotor harmonics. Finally, strengths and stiffness of available materials further constrained the analysis. Due to cyclic and ground-air-ground loading conditions, various fatigue failure modes also had to be contended with. All of these constraints left very little design space to work in. Painstakingly detailed design included carefully tailoring the graphite and fiberglass ply layup angles and overall width and thickness dimensions to allow bending and twisting motions without excessive strain. This was necessary to achieve a point design that had the necessary fatigue life. Extensive fatigue testing was then performed to substantiate the component lives.

The flexbeams were press molded using matched metal tooling (Fig. 11.10). In theory, manufacturing rectangular cross-section flexbeams would seem to be straightforward. In actuality, they were anything but. Difficulties arose because of the need to maintain precise thickness tolerances in order to control stiffness. Stiffness variations between flexbeams would cause each blade to "fly" differently, resulting in large vibratory forces. Although the graphite and fiberglass used to lay up the flexbeams had stringent specifications controlling thickness and resin characteristics, there were still tolerances that had to be contended with. Because the flexbeams were very thick and contained hundreds of plies, very small variations in ply thickness could accumulate to create unacceptable overall thickness tolerances. Ply layups also had to be controlled to preclude unacceptable step and gap conditions between plies. Arduous techniques for compensating for these conditions were developed. In addition, resin viscosity could vary somewhat between material batches and be affected by temperature variations within the laminate. These variations could create internal laminate pressure differentials during the cure process that caused fiber wrinkling, or voids that adversely affected laminate fatigue strength. Stringent quality control procedures and ultrasonic inspection processes were implemented. In spite of these measures, initial production yields were relatively low, and many expensive flexbeams had to be scrapped.

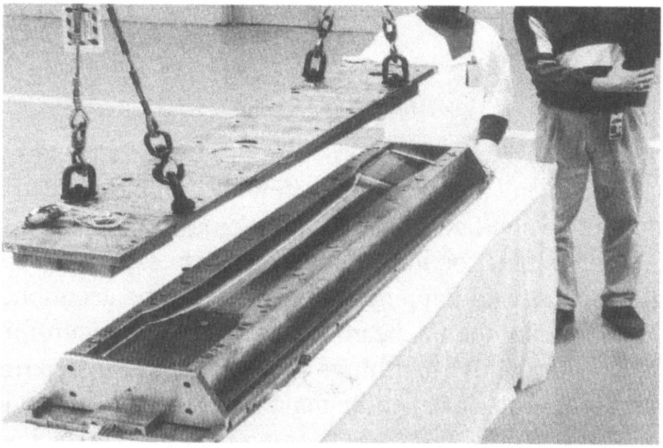

Fig. 11.10 Flexbeam matched metal press molds.

One concern with soft inplane bearingless rotors is inplane damping. Neither Boeing nor Sikorsky had any experience flying this particular rotor, and there was some apprehension about potential issues like the regressive lag mode. At the time, our analytical tools were indicating that this mode could have very light damping, which could create disturbing roll coupling issues at some point in the flight envelope. Therefore, a snubber/damper was included at the inboard end of the blade, between the flexbeam and the torque tube.

THE CONTROL SYSTEM

One of the few systems on the LHX that did not offer any innovation was the rotor control system. It contained the same type of swashplate system used on virtually all Boeing and Sikorsky helicopters. The swashplate transfers hydraulic servo inputs from the fixed system to the rotating blade pitch control links.

The technology for electric motor servos and/or airfoil shape modulation was briefly considered but was felt to be too great a leap for the LHX time period.

THE MAIN ROTOR BLADES

The main rotor blades were all composite, based on proven low-risk Boeing production designs (see Fig. 11.11). They had also been proven to be ballistically tolerant to 23-mm rounds.

During early preliminary design studies, the main rotor blade configuration was debated. Boeing proposed a configuration derived from its CH-46 and CH-47 helicopters. The primary load path in the Boeing blade spars was constructed from fiberglass. The Sikorsky proposal was a more advanced concept that incorporated fiber-placed graphite epoxy. Sikorsky argued vigorously that graphite was a stronger material and was a more environmentally resistant material. Boeing countered that it had tens of thousands of hours of flight time, and the fiberglass spars were perfectly suitable for the application. In regard to manufacturing cost, which was the other prime consideration, Sikorsky claimed that automated fiber placement was inherently cheaper than the multipiece spar that Boeing was proposing. Boeing

Fig. 11.11 Dem/Val rotor blade.

countered that it could methodize its detail part manufacturing and be lower in cost than Sikorsky's approach. This debate went back and forth to the point that outrageous claims were being made on both sides. In the end, the debate escalated through various levels of management, and the program manager decided to step in and make a decision. He selected the Boeing design based on lower technical risk. By then, the work split between the companies had been decided, and Boeing would be responsible for building the main blades, so it made sense for it to stick with something that it was comfortable with.

Aside from the usual structural dynamic conditions, the LHX blades had some new requirements. One unique requirement was the ability to cut through small tree limbs without compromising structural integrity. This capability was validated by test. No additional spar reinforcement was needed.

By far, the most challenging requirements for the rotor blades (Fig. 11.12) were associated with the leading edge. Here, there was a set of conflicting requirements for anti-ice, erosion, low radar cross section (RCS), and lightning protection. Erosion requirements by themselves are conflicting. Soft elastomeric coatings used for rain protection are susceptible to sand erosion, whereas hard coatings used for sand protection are susceptible to rain erosion. For anti-ice protection, blade leading edges needed to be heated, and this meant embedding electrical heating elements. Rotor blade tips were also the primary attachment point for lightning strikes, and the blades had to carry very high electrical currents to the rotor head and fuselage, where they exited the aircraft. Finally, blunt leading edges caused high radar reflections, which necessitated special treatments. Concepts were proposed, and risk reduction tests were conducted, to deal with all of these conflicting requirements. Full development was deferred to the post-Dem/Val phase of the program. The prototype YRAH-66A blades did not incorporate these features.

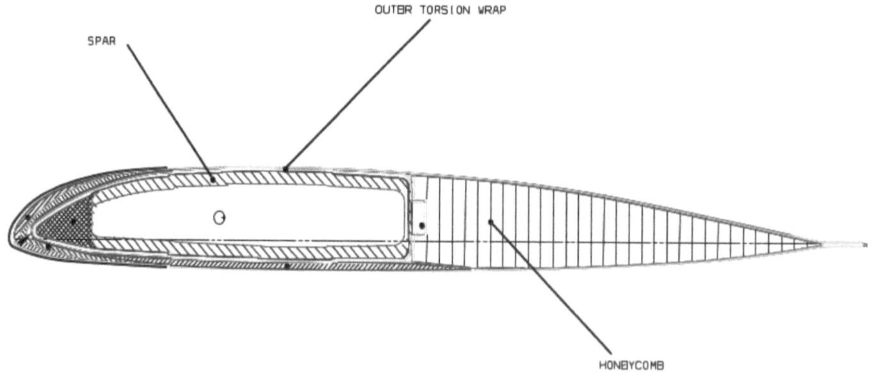

Fig. 11.12 Main rotor blade cross section.

FLIGHT TEST

The prototype aircraft rotor was designed and built based on the initial wind tunnel testing. It accomplished over 500 hours of flight on the two prototype aircraft. There really were no major technical issues with the rotor blade in terms of performance or vibration. On aircraft 1, once track and balance evolution was completed, no adjustments were made, and the rotor flew through the entire flight envelope with no objectionable vibrations. Interestingly, Mach 1 at the blade tip was inadvertently exceeded on one high speed flight with no ill effects.

Ultimately though, one of the design concerns did materialize: the regressive lag mode appeared at times during elevated *g* level turning flight. A more detailed discussion of this issue is contained in Chapter 6, Handling Qualities.

THE GROWTH ROTOR

As the aircraft weight grew over the years due to the various added features and systems and the resolution of technical issues, the rotor needed to be redesigned. Mission gross weight had increased from 10,088 to 12,350 lb. Concerns with the rotor at this higher weight included rotor weight, the regressive lag mode, blade removal time, flight control system limits, snubber damper reliability, and a few other more minor issues.

These issues were big enough for the Army to ask Boeing Sikorsky to conduct another tradeoff study between the BMR and an articulated rotor (Fig. 11.13). This was done in 2000. As with the 1989 study, the difficulty with the articulated rotor was incorporating the high hinge offset required on Comanche for maneuverability and agility. An 8% offset was selected as the best compromise between head moment generation and physical size.

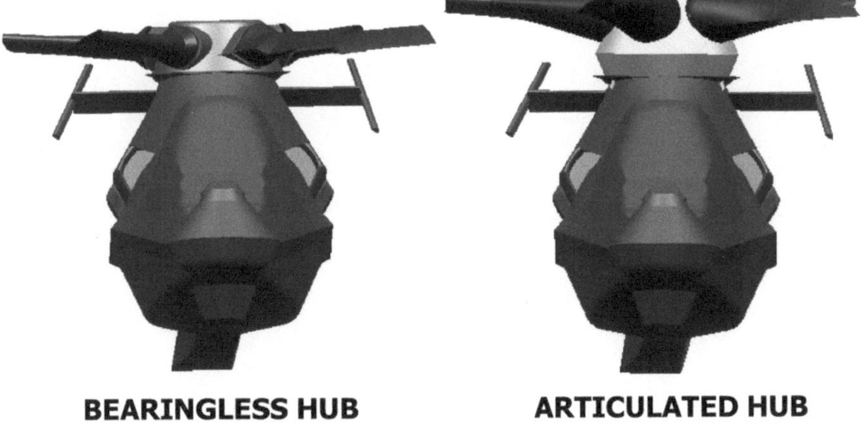

Fig. 11.13 The 2000 tradeoff study for BMR vs. articulated rotor.

The result was an articulated rotor that was bigger and significantly heavier than the BMR and had adverse effects on aircraft performance (Vertical Rate of Climb (VROC) and dash speed). It was also concluded that the 8% offset rotor would have to be raised 13 in. over the BMR position for acceptable rotor-to-fuselage clearance.

As with the earlier study, the conclusion was that the BMR was the better rotor type for Comanche, and a growth BMR rotor was introduced in the engineering, manufacturing, and development (EMD) phase. This increased the rotor diameter from 39 ft to 40 ft and the chord from 15 in. to 16 in. It introduced the anhedral swept tip for performance improvement and improved acoustic signature. The effective hinge offset, which came out to be 9% in the Dem/Val rotor, was increased to 11.3%. To provide more flapping range, the flexbeam length grew from 71.4 in. to 82.5 in. The rotor blade was also redesigned for better ballistic tolerance.

At the time of program termination in 2004, the assembly of the first growth rotor was nearing completion. The combination of the higher hinge offset and highly responsive flight control actuators continued to provide high agility, but also higher load generation. All the rotor system components, including hub, flexbeam, and blades, had undergone redesign to accommodate load increases and tuning to match stiffness.

REFERENCES

[1] Stanley W. Kandebo, "RAH-66 Comanche Rolls Out," *Aviation Week and Space Technology*, 29 May 1995, p. 75.

[2] Norman, T. R., Cooper, C. R., Fredrickson, C. A., and Herter, J. R., Sikorsky Aircraft. "Full-Scale Wind Tunnel Evaluation of the Sikorsky Five-Bladed Bearingless Main Rotor," presented at the American Helicopter Society 49th Annual Forum, St. Louis, MO, May 19–21, 1993.

[3] Amer, K. B., "A Comparison of Four Versus Five Blades for the Main Rotor of a Light Helicopter," presented at the American Helicopter Society 45th Annual Forum, Hollywood Florida, January 8–11, 1989.

Chapter 12

THE FANTAIL

BRUCE KAY, ART LINDEN AND JOHN OCCHIATO

The Army defined a very high level of yaw control power for the LH to overcome the limitation of its then-current light helicopter fleet when flying at high altitudes. The Bell UH-1 and AH-1 and the Hughes OH-6 were all designed for a maximum crosswind of 17 kt. In many conditions requiring high performance, those helicopters simply ran out of tail rotor control. The UH-60 Black Hawk and the AH-64 Apache had a maximum allowable sideslip of 35 kt. The LH requirement was to perform a 180-deg yaw turn while flying in a 45-kt sidewind within 5 seconds (Fig. 12.1).

This led to a more powerful antitorque/yaw control system than the industry was accustomed to. In addition, the Army required the antitorque system to be protected in order to prevent personnel accidents where soldiers ran into the tail rotor under stressful combat conditions. Another important factor was to have an antitorque system that could operate more safely and freely in congested close-to-the-ground operations. A conventional tail rotor was not acceptable. And of course, it had to have a low radar signature.

Although competitive secrecy was strong during the initial design and proposal phases of the program, it was assumed that the McDonnell Bell SuperTeam would propose the use of its NOTAR (no tail rotor) antitorque system. This was flying on an OH-6 and receiving much publicity, and it certainly solved the problem of soldiers running into a spinning tail rotor. Thus, this concept was evaluated as one option for Boeing Sikorsky.

The NOTAR required power levels above those of conventional tail rotors. Boeing Sikorsky was favoring a ducted fan design, which we called the "Fantail." This design also was thought to required more power than a tail rotor, although later tests with the H-76B Fantail demonstrator showed that

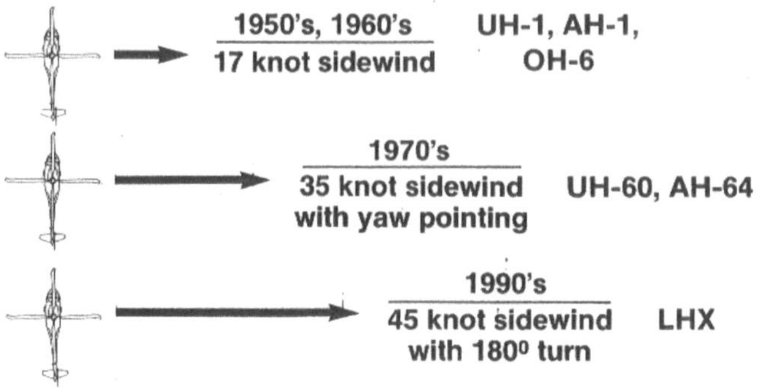

Fig. 12.1 Tougher antitorque requirements.

the Fantail power required was quite similar to that of a conventional tail rotor. Because of the extreme maneuverability requirements, however, the maximum continuous power for an LH-sized Fantail was over 900 hp, and it provided 7400 lb of thrust. For perspective, this is more than the total installed engine power for typical light helicopters.

The NOTAR required even more power due to its internal fan and ducting losses. At Sikorsky, the NOTAR power calculations were based on very optimistic assumptions, because the team wanted to consider its "best-case" attributes. Our conclusions were that not only was the NOTAR not as efficient as the Fantail when sized for small helicopters, but also its weight and power penalties grew as it was scaled up to LH-sized aircraft.

The conclusion of tradeoff studies on the antitorque systems led Boeing Sikorsky to select the Fantail design (Fig. 12.2). Our analysis showed that the

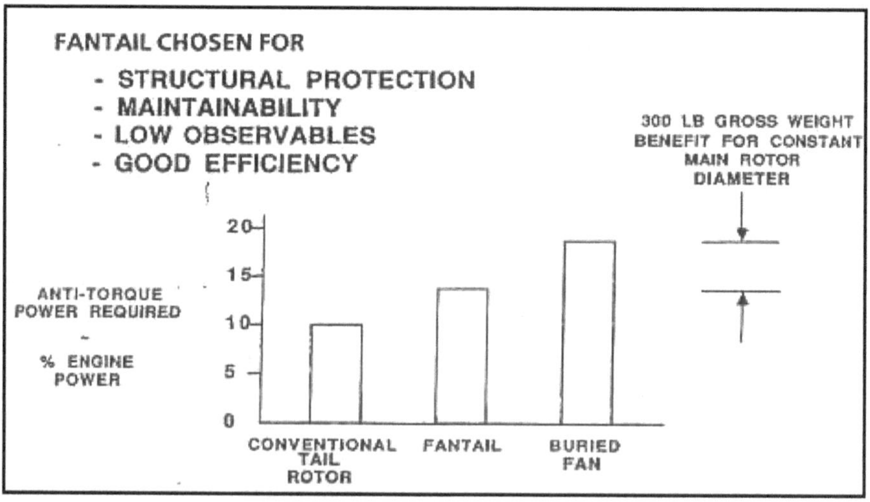

Fig. 12.2 Dem/Val proposal tradeoff results.

NOTAR approach resulted in a 300-lb gross weight penalty at constant rotor diameter.

This same point was made in a First Team advertisement saying (Fig. 12.3):

> The heat of battle is no place to pay the high price of an inefficient design for LHX. Efficiencies in the First Team's FANTAIL system increase payload by 200–400 pounds—equal to 2–4 Hellfire missiles. The First Team's LHX fights better...with more missiles.

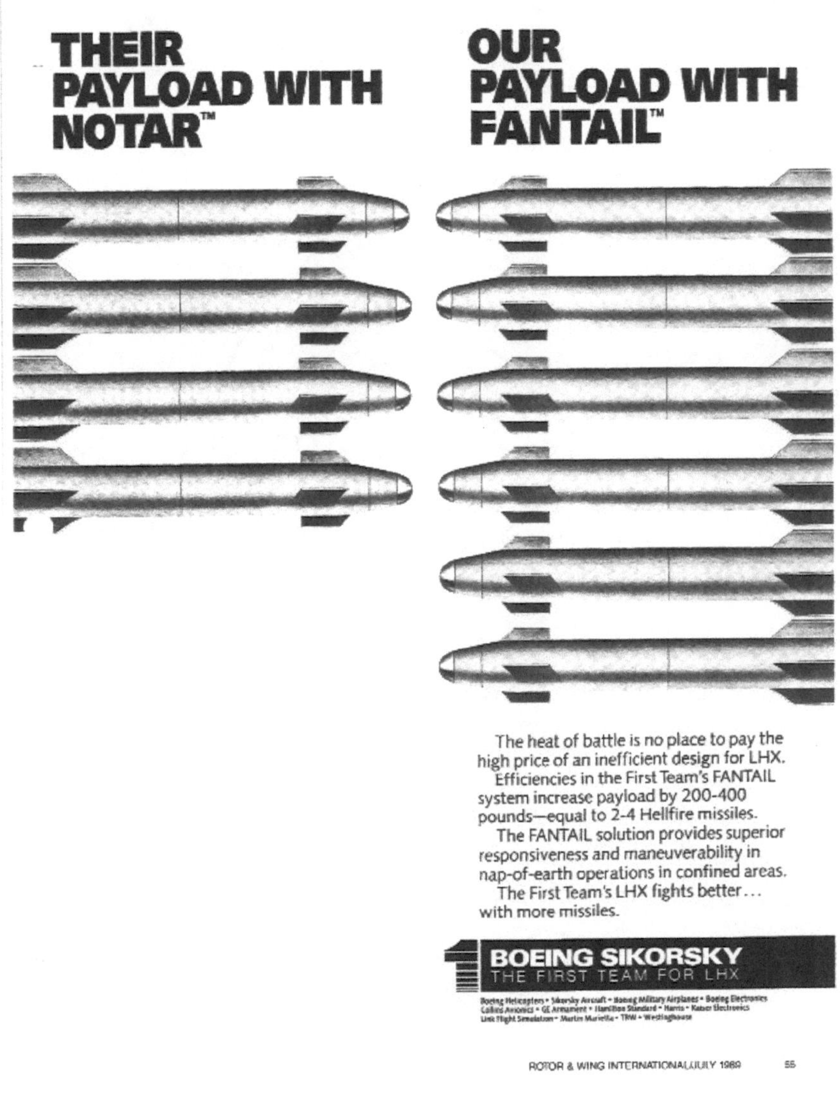

Fig. 12.3 Boeing Sikorsky advertisement.

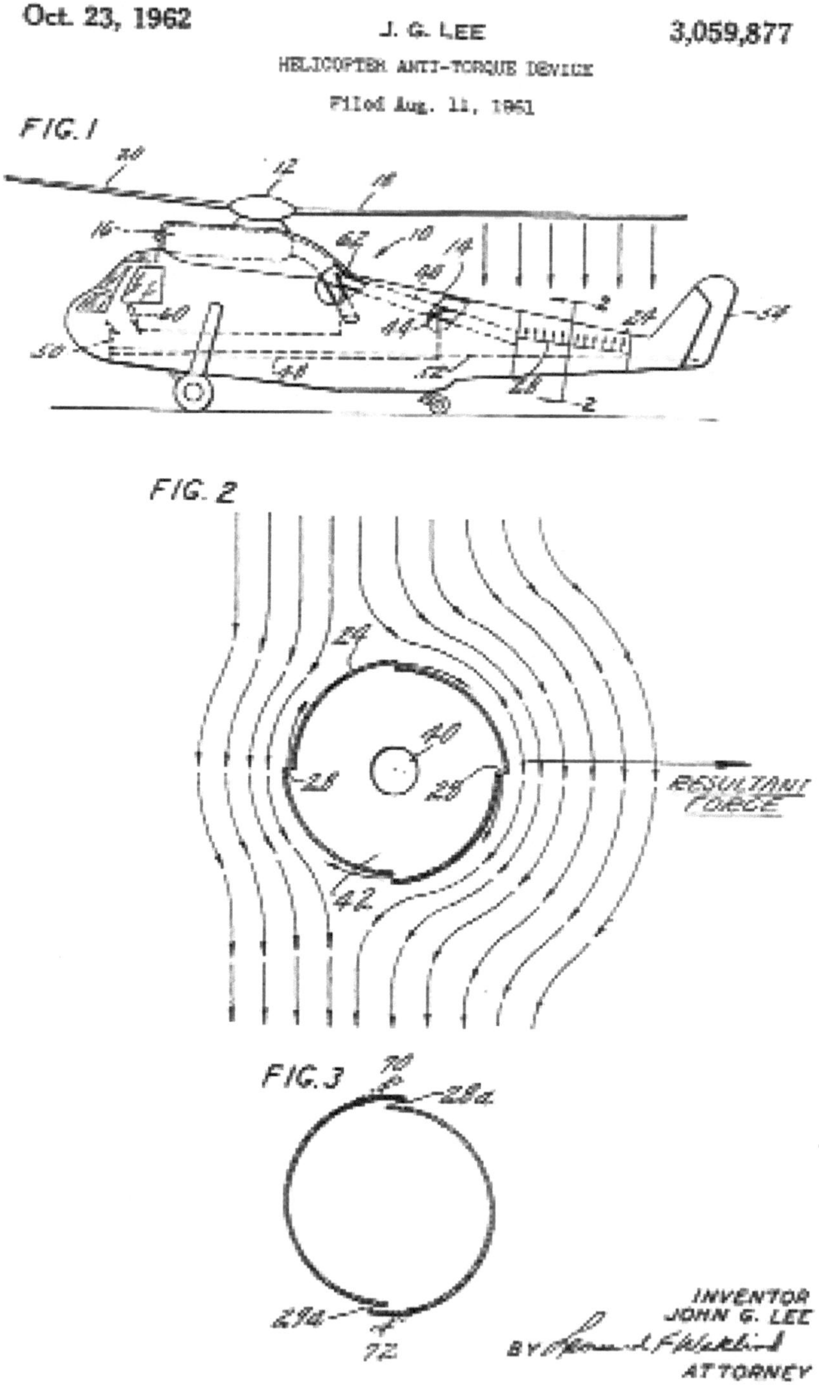

Fig. 12.4 Original United Aircraft Corporation NOTAR patent.

During one of the numerous reviews with United Technologies senior management, this conclusion was challenged by Stanley Hiller, a consultant to UTC. Although 35 years younger than Igor Sikorsky, Hiller was a contemporary of Sikorsky's in the early days of helicopter development. Igor Sikorsky first flew his VS-300 in 1939 at age 50. Stanley Hiller first flew his coaxial helicopter in 1944, at age 19, and he went on to form the Hiller Helicopter Company, which was moderately successful in the early days of helicopter production.

Hiller made a comment that the design team had an NIH (not invented here) attitude when evaluating the NOTAR. This was put to rest when the original patent for the concept was shown. The idea of generating sideward lift on a helicopter tailcone using the airflow from the main rotor was invented by engineers at the United Aircraft Research Laboratories, United Aircraft being the parent corporation to Sikorsky.

The 1991 patent drawings showed the concept applied to a Sikorsky SH-3 helicopter (Fig. 12.4).

In 1975, Sikorsky evaluated a Fantail on the S-67 gunship demonstrator, a modified S-61, to compare it to a conventional tail rotor (Fig. 12.5). This gave us valuable experience and baseline data on the Fantail configuration.

Fig. 12.5 Sikorsky S-67 Blackhawk Fantail demonstrator.

S-76 FANTAIL

We decided that it was necessary to acquire quantitative data on the performance of the Fantail, to ensure the design would achieve the required capabilities. Both whirl stand tests and a risk-reduction flying prototype were funded. A military version of the S-76B, known as the Eagle, was chosen as the host vehicle. This aircraft was quite close to the proposed LH in size (Table 12.1), negating any issues with scaling the concept to vastly different gross weights, a problem the SuperTeam had with its OH-6 Notor demonstrator. First flight was on 6 June 1990 (Fig. 12.6).

The Fantail Eagle demonstrator provided an abundance of technical data about performance, handling qualities, and acoustics that were used to refine the final design configuration and supported the formal LH proposal. Just as important, the full-scale flying demonstrator strongly established the technological superiority of the Fantail over the competitive concepts.

The time from S-76B Fantail demonstrator decision to the first flight was a remarkably short 18 months. Flight testing took only 6 weeks, was concluded in 29 hours of flight, and included a full maneuver envelope. These included:

- Right sideward flight at 70 kt
- Left sideward flight at 68 kt
- Rearward flight to 70 kt
- Maximum level flight speed of 163 kt
- Maximum dive speed of 189 kt
- 180-deg hover turns in 4 seconds

Also included were +2.5-*g* pullouts, –0.5-*g* pushovers, roll maneuvers in both directions, hammerhead stalls, split S maneuvers, extreme side-slipping maneuvers at speeds up to 155 kt, and high-speed turns and autorotations.

The testing demonstrated an acoustic signature 15 dB below that of the baseline S-76.

The fan was designed for full pedal application up to 160 kt, although it was limited on the test aircraft to 120 kt due to the structural limitations of the S-76 tailcone.

TABLE 12.1 THE S-76 EAGLE VS THE PROPOSED LHX

	LHX	S-76 Eagle Fantail Demonstrator
Gross weight	10,024 lb	10,500 lb
Yaw inertia	17,500 slug-ft^2	24,776 slug-ft^2
Main rotor diameter	39 ft	44 ft
Tail rotor diameter	4.5 ft	3.937 ft
Aircraft power available	2053 hp	1605 hp
Maximum fan power	944 hp	615 hp
Maximum thrust	2347 lb	1624 lb

Fig. 12.6 S-76B Eagle Fantail demonstrator, first flight.

The Fantail provided exceptionally high control power in hover and low speed. In addition, unlike a tail rotor, the ducted design masked the rotor from high airloads normally encountered by conventional tail rotors in high speed sideslips.

An interesting fallout of the unique Fantail attributes resulted in the creation of a new flight maneuver. One requirement by the Army was to do a quick turn-to-target maneuver at moderate speeds. This was expected to be a roll into a 90-deg turn, turn, then roll out and line up with the target. Flight testing showed that the Boeing Sikorsky LHX could simply yaw 90 deg at speeds up to 70 kt to meet this requirement. No extreme rolling required. This maneuver was named the "snap turn" and became a favorite of pilots flying the aircraft.

The First Team Program Director Lou Cotton said the Fantail demonstrator's performance was validation of the years of engineering development, wind tunnel, and whirl stand testing that preceded the current flight test program.

Aviation Week & Space Technology flight tested the aircraft and reported the results in its 27 August 1990, issue [1]. It reported that at startup the low noise was noticeable, as was the absence of medium-frequency vibrations associated with conventional tail rotors. In flight, a maximum yaw rate of 100 deg/s was achieved. The magazine also did numerous timed yaw maneuvers against the Army's requirements, and the aircraft accomplished them all within the required times. Sideward flight was demonstrated to 70 kt right and 65 kt left. The article commented on the lack of roll-yaw coupling, due to the fan being located lower than a conventional tail rotor.

Quoting the pilot, Nicholas C. Kernstock [1]:

> The most impressive capability was the aircraft's unlimited yaw pointing ability at speeds up to 80 kt. Flying down the runway at about 20 ft. altitude, I applied full left pedal and adjusted the cyclic control to maintain ground track. I was able to point the nose of the aircraft to any direction at will.

> Following our maneuvers on the runway, we climbed to 1,500 ft. to evaluate the yaw pointing characteristics in out-of-ground-effect flight. One of the LH maneuvers we performed was the coordinated 90 deg. turn to target at a constant airspeed and altitude. LH requirements specify the turn be completed in less than 6 sec. This maneuver would be used to quickly bring the aircraft into firing position after identifying a threat at its 90 deg. position. The conventional maneuver is a steeply banked turn to target, and we were able to execute this turn in 3.32 sec. However, using the Fantail's unrestricted yaw capability, we made a turn using only the anti-torque pedals in 2.28 sec. The 1-sec. advantage would be of significant value in an air-to-air combat situation.

In conclusion, the article said:

> The Fantail far exceeds the abilities of conventional tail rotors in several respects, and many of the maneuvers the system is capable of doing do not yet exist in combat flying doctrine. The selection of a Fantail-equipped LH would provide Army and Boeing/Sikorsky test pilots with some unique and interesting challenges.

In its December 1990 issue, David S. Harvey of *Rotor & Wing International* also flew the aircraft. He provided these details [2]:

> We then flew backwards, which at 70 knots give anyone other than a test pilot an apprehensive feeling. What's behind? But the Fantail was doing its work well. [Sikorsky test pilot Nick] Lappos placed his feet in the floor, essentially steering with cyclic. Deceleration in this flight mode is an interesting experience; it involves pointing the nose directly at the ground, while *climbing* backwards, the reverse of a conventional rapid flare maneuver.

Harvey also commented on a quick-stop capability from 90 kt forward speed:

> Lappos performed a conventional, nose-up deceleration, and the aircraft covered some 1,500 feet (as indicated by runway markers) before stopping. To halt the aircraft a second time, he applied a maximum-rate yaw input to the left as a primary means of inducing drag. The result was impressive. N3124G was stopped and hovering neatly within a distance of only 800 feet. And Lappos reported no control difficulties at all.

He then commented on the efficiency of the Fantail design, where the fan duct shape is integrated into the total thrusting system. Again quoting Sikorsky test pilot Nick Lappos: "Our flight-test results show that in an OGE hover, Fantail is using about the same percentage of power for thrust produced as does a conventional tail rotor...".

After the flights at West Palm Beach, the aircraft went on a 30-hour demonstration tour to Ft. Rucker and the Army Aviation Systems Command in St. Louis. Reporting on those demonstrations in the October/November 1990 issue of *Defence Helicopter*, Harold Coghlan reported [3]:

> Fantail controllability is vastly superior to that of conventional attack helicopters. One would not attempt this manoeuvre [360° pirouettes at 30 knots] in current generation attack helicopters.
>
> The Fantail's ability to turn 90 degrees so quickly is frankly amazing.
>
> During nap-of-the-earth flight, the Fantail H-76 behaved in a very conventional way, that is, it required no additional skills. Not only was the Fantail easy to fly in a tactical mode, I had a very secure feeling knowing the tail rotor was fully enclosed... This secure feeling allowed me to fly more aggressively than I would normally when flying a new helicopter for the first time.
>
> I also noted the power required to hover at 5 ft. was 80 to 81%... It is interesting to note that the Fantail H-76B required the same amount of hover power as a stock H-76B with conventional tail rotor... I had expected the Fantail H-76B to use slightly more hover power than the standard Eagle at both hover and cruise....
>
> I noted during the preflight that the Fantail fan was readily accessible from the ground, enabling maintenance to be performed more easily than on conventional high-set tail rotors....

In summing up, the author said:

> First, the radar cross section (RCS) is much lower, with fewer sharp angles and makes use of radar absorbent materials. The LH Fantail will be canted to deflect radar return away from ground-based air defence radars, and the broad-chord fan blades help reduce the flicker signature that makes conventional rotors so noticeable to radar. Second, LH has a classified acoustic requirement already met by the development Fantail on the Eagle. LH will also have a Whisper Mode to reduce tail rotor speed and noise in stealthy tactical situation. The Fantailed Eagle is about 15 dB quieter than the unmodified H-76B.

The aircraft was next demonstrated at the 1991 Paris Air Show (Fig. 12.7), participating in the daily air show presentation. The snap turn maneuver created quite a media buzz. The headline in the 8 July 1991 issue of *Aviation Week & Space Technology* was "Boeing Sikorsky Fantail Demonstrates Unprecedented Maneuverability" [4].

This show was particularly interesting considering its presence in the home of Aerospatiale, the manufacturer of the only other ducted fan antitorque system, the Fenestron. The relatively low acoustic signature and high yaw control power frustrated the French manufacturer, which was displaying a

Fig. 12.7 Paris Airshow Patch.

Fenestron-equipped helicopter during the air show. The Fantail prompted it to attempt to add a similar maneuver to its routine, but it couldn't manage to conduct the maneuver at anything but low speeds, only showing it once before abandoning any more attempts to match the Fantail capability.

Design Details

The Boeing Sikorsky antitorque trade study conclusion was to use a 4.5-ft diameter Fantail for the LH.

A basic tenet of the Fantail configuration was to cant the tail (Fig. 12.8). This was done for two reasons. A canted tail provides lift, allowing the aircraft center of gravity to be behind the rotor center, resulting in a shorter aircraft, and it enables a larger center of gravity range. This approach was first successfully employed on the UH-60 helicopter. It also contributed to a lower lateral radar signature, because incident radar beams would be reflected up, or down and away, from the radar receiver.

The next design details to be defined were related to the fan and fan duct configuration. A considerable amount of time was spent on optimizing the inlet and exhaust lip radii of the fan duct. These radii had a considerable impact on efficiency and radar signature. The variable lip radii around the circumference can be seen in Fig. 12.9. About 1400 hours of wind tunnel testing were conducted to support this effort (Fig. 12.10).

Fig. 12.8 Canted Fantail configuration.

Fig. 12.9 Close-up view of Fantail.

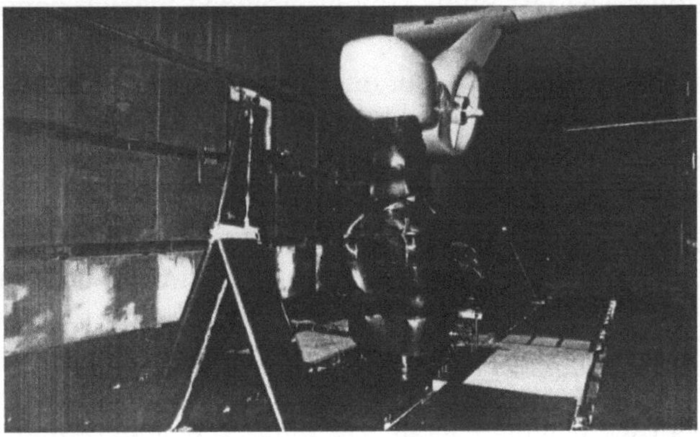

Fig. 12.10 Fantail wind tunnel test.

Boeing did 518 hours of wind tunnel testing in its wind tunnel from July to September 1989 with a 3/4 scale Fantail. This validated the fan's figure of merit and aural signature, and showed a maximum thrust 10% greater than predicted.

To reduce the parasitic aspect of using tail rotor power to counteract torque in forward flight, modern helicopters incorporate a vertical tail that develops adequate sideforce in forward flight to allow the tail rotor to null pitch (zero thrust) when at cruise speed. The Comanche design was no different. The vertical fin was designed to offload tail rotor thrust at speeds above approximately 120 kt-indicated airspeed (KIAS); however, this is not the only requirement for the vertical fin. In addition, sideforce is needed to maintain balanced flight in gusts or turns (a handling qualities attribute known as directional stability). Conventional tail rotors, which are in the free stream air in forward flight, tend to act as airfoils themselves and provide much of the directional stability required to maintain balanced flight at forward speed (essentially augmenting the effectiveness of the vertical fin). The Comanche vertical fin lacked adequate surface area, and because of the duct, the Fantail contributed nothing to directional stability unless commanded to change pitch by the flight control system. In addition, there was a deadband in the thrust vs Fantail pitch response. It was thought that the fly-by-wire system would provide fast, correct commands to maintain directional stability. It turned out, however, that when the fly-by-wire system sensed the need to provide Fantail thrust to counter an out-of-balance condition due to a turn or gust, there was a lag in the response while the pitch moved out of the deadband area. There was a tendency to overshoot the required pitch command, which resulted in a subtle but noticeable fishtailing at higher speeds, and a lag in achieving coordinated flight in a turn. In Chapter 19, The Flight Test Program, there will be a discussion on how this tendency was corrected. This phenomenon was not observed in the S-76 Fantail demonstrator aircraft because it incorporated a simple stability augmentation system without automatic turn coordination or heading hold capabilities. The test pilots simply provided manual control inputs in turns or gusts.

The ducted fan approach was not immediately seen as an advantage due to the use of this concept on other production helicopters. Eurocopter trademarked the design as the "Fenestron." Fenestron helicopters were noticeably louder than tail rotor helicopters and produced an annoying high-pitched sound. A simple analysis showed that the noise was due to supporting struts being located too close to the fan, in essence creating a siren effect. With sufficient clearance between the fan and the struts, the noise was very acceptable. The Fantail design was different from the Fenestron design in several other aspects. The Fenestron incorporated more blades of much smaller chord, and operated at much higher rpm than the Fantail. The reduced Fantail rpm further contributed to a less objectionable noise signature. Whereas the Fenestron-equipped aircraft sound like an old-fashioned siren, the Fantail sounded more like a conventional tail rotor.

Figure 12.11 shows the sources of noise from a ducted fan. The major acoustic discriminators that drove the design were the number of blades (eight) and the spacing between the blades and the stators that support the fan hub. Dr. Charles A. Yoerkie, Jr., chief of acoustics at Sikorsky, was quoted in the 27 August 1990, issue of *Aviation Week & Space Technology* [1]:

> Another important noise factor…is the placement of the support and drive-shaft struts on the tail. Struts that are oriented radially to the center of the fan tend to accentuate fan noise. The Fantail's vertical support strut is placed off axis to the center of rotation and therefore dampens sound waves coming off the blades.

An even number of blades was chosen to meet ballistic survivability requirements. In the event that a blade was lost due to a ballistic hit, the fantail hub and gearbox were designed to resist the resulting imbalance forces.

Boeing was responsible for the design and fabrication of the entire empennage, including the Fantail. The mechanical design of the fan hub was unique (Fig. 12.12). Concentric bearings provided a rigid mount to resist flapping and lead-lag motions, while allowing rotational freedom for pitch control. Centrifugal loads were resisted by tension straps inside the bearings that attached the blade to the hub. A shaft-mounted spider provided pitch control.

The fan blades were of all-composite construction with a foam-filled airfoil section and integral tubular root end pitch bearing mounting surfaces. The composite layups were optimized to minimize radar signatures. The centrifugal tension strap was secured to the blade via a retaining pin in the root end (Fig. 12.13).

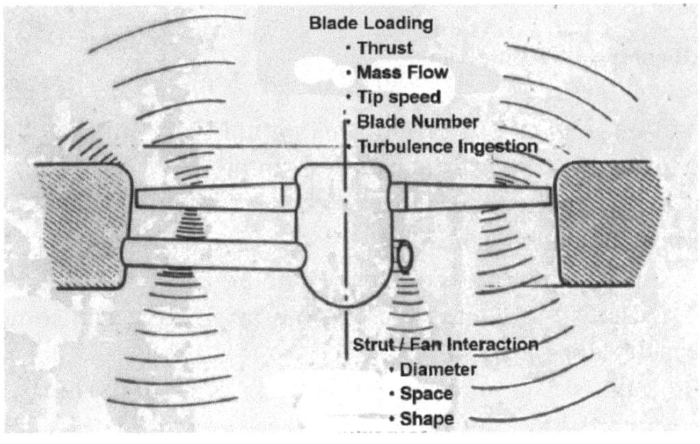

Fig. 12.11 Ducted fan noise sources.

Fig. 12.12 Fantail hub design.

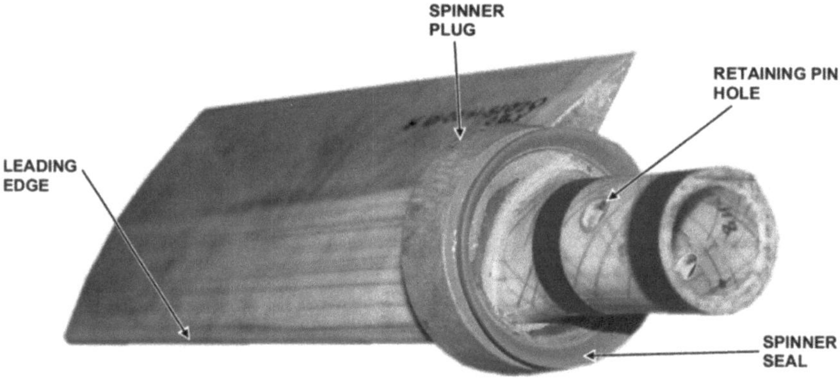

Fig. 12.13 All-composite Fantail blade.

Concerns about hazards to ground personnel were mitigated by a safety guard on the inward-leaning surface of the Fantail (Fig. 12.14). The overhanging slope of the opposite side along with the fan setback and stators were sufficient to preclude any hazards.

The end result was that the Corhanche Fantail configuration was very different from then-existing ducted fan configurations that had some perceived negative characteristics.

The Boeing Sikorsky team was confident that we had the better antitorque/yaw control system when compared with the SuperTeam's NOTAR. To people observing the competition, the tail configuration was a major difference in the designs.

Fig. 12.14 Comanche Fantail with personnel guard.

In the November 1990 issue of *Interavia Aerospace Review*, author/pilot Robert Ropelewski reported on flying both the OH-6 NOTAR demonstrator and the S-76B Fantail [5]. He was positive on both aircraft and stressed how they both had yaw control power greatly superior to conventional helicopters. In closing, he wrote:

> Considering the extensive data on both systems available to the LH Source Selection Board, it will not be easy to decide between the two teams if the tail configuration becomes the deciding factor. Mission equipment packages, main rotor systems and costs will also be influential to the final selection. From the exterior, however, it is the tail that seems to make the difference and both competitors are striving hard to show that theirs will do the job.

In actuality, after the downselect, the Army told us that the tail configuration was not a large driver in the final selection.

REFERENCES

[1] Nicholas C. Kernstock, "Boeing/Sikorsky Fantail Allows Radical Maneuvers," *Aviation Week and Space Technology*, 27 August 1990, p. 32.

[2] Harvey, D. S., "Flying The First Team's Fantail-Equipped H-76," *Rotor & Wing International*, December 1990, p. 42.

[3] Coghlan, H., "Flying the Fantail," *Defence Helicopter*, October/November 1990.

[4] David Hughes, "Boeing Sikorsky Fantail Demonstrates Unprecedented Maneuverability," *Aviation Week & Space Technology*, 8 July 1991, pp. 45–49.

[5] Ropelewski, R., *Interavia Aerospace Review*, November 1990.

Chapter 13

THE FUSELAGE AND LANDING GEAR

BRUCE KAY

The LHX/Comanche fuselage was not just a new design for a new mission. It also capitalized on the materials technology advances that had been made since the previous helicopters were designed, and it had difficult requirements for low observables, supportability, and life cycle costs.

It was recognized early on that the fuselage would be made from composite materials. The belief was that composite structures are 15–20% lighter than aluminum structures, and if designed properly could also be less expensive. Designed properly meant making parts as large as possible with integral cocured details. This approach reduced assembly labor to such an extent that the higher composite raw material costs were negated. Data supporting this premise were based on experience from the Sikorsky Advanced Composite Airframe Program (ACAP, see Chapter 1), as well several Sikorsky commercial ventures producing composite parts for general aviation applications. Furthermore, it was felt that composite construction was necessary to meet the low radar signature requirements. As a result, there was never a tradeoff between composite and metal construction.

In the work split with Boeing, Sikorsky was responsible for the main fuselage and main landing gear, and Boeing was responsible for the empennage, tailcone, and tail landing gear.

THE CENTER BOX BEAM CONCEPT

After many conceptual design studies, the Boeing Sikorsky team chose an unusual design for the fuselage. Unlike virtually all other helicopters (and fixed wing aircraft, for that matter) they decided not to use the exterior surfaces of the aircraft for the primary fuselage structure. Because of the need for a low radar signature, the aircraft weapons had to be stored within the fuselage and then extended out when they were to be deployed. A retractable landing gear was required for both low radar signature and high speed. Additionally, the Army's emphasis on supportability required many access panels to simplify maintenance tasks. To maintain structural integrity with most of the surface cut away meant that conventional semimonocoque construction would be very inefficient and heavy. What emerged as a solution was a more elegant design in the form of a central box beam primary structure. The box beam was a continuous structure that could carry the major fuselage bending and torsional loads. Hard points for attaching the main landing gear, main rotor, nose gun, weapon launcher supports, and the semimonocoque tailcone were all incorporated into the box beam. Having straight, continuous load paths also yielded a very stiff, easily tunable structure that enhanced vibration control. The outside skin was unloaded from the primary structural loads. Doors and access panel could then be placed in the optimal positions, improving both radar signature and ease of maintenance for the internal parts.

Longitudinal keel beams and a lower skin, running from nose to tail, formed the sides and bottom of the box beam. Individual bulkheads and decks then closed out and framed the structure. Seldom removed components comprising the fuel and environmental control systems were located inside the box beam. Bolted structural access panels were provided to access these components. Hardpoints were included for attaching the major components (Fig. 13.1).

The external skins attached to the box beam were still considered semistructural because they were capable of carrying some loads. Their structural role was to react to air pressure loads and provide support for doors, lights, sensors, antennae, and so forth.

Early in the program, this concept started to be described as the "fuselage within a fuselage" with some derogatory implications that it must be too

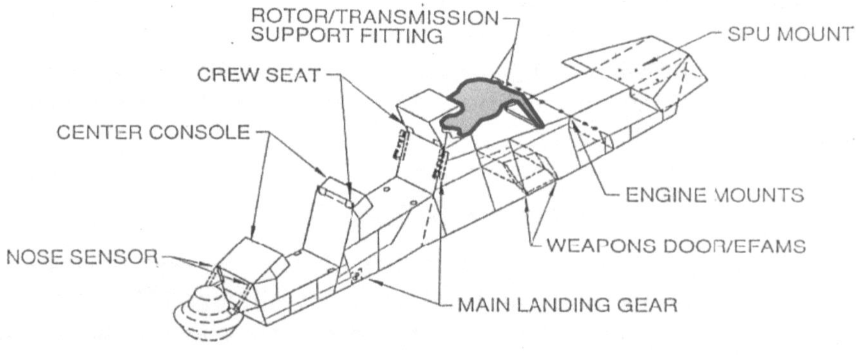

Fig. 13.1 Box beam hardpoints.

heavy. In actuality, it became one of the most positive attributes of the Boeing Sikorsky design.

Shaping the fuselage was relatively straightforward and became a tradeoff between radar signature and aerodynamic performance. Both of these characteristics were readily determined by analysis and small scale model tests, so it came down to selecting a good balance.

Design Criteria

From the outset, weight was one of the most important attributes that had to be met. Both the government and Boeing Sikorsky understood the challenge and were fully committed to doing their best to meet this objective. One of the extraordinary measures that the government took to step up to this challenge occurred during an LHX spec negotiation session, Charlie Crawford, Aviation Systems Command's (AVSCOM's) technical director, pulled Sikorsky Chief of Design Bruce Kay aside for a discussion on design criteria. Crawford wanted to take a fresh look at structural design criteria to remove any conservatism. He noted how Sikorsky was proposing to use extensive NASA structural analysis (NASTRAN) finite element analyses to design the structure, and would be conducting building block tests to validate strength. This was anticipated to be more accurate than previous structural design methods, so he suggested exploiting this benefit by tweaking design criteria. Accordingly, it was agreed to use a 1.4 ultimate design load factor instead of the usual 1.5 factor. This effectively reduced the safety factor by 9% compared to industry standards. However, a proviso was added to the spec stating that structures designed to this criterion had to be tested.

As it turned out during the airframe static tests, there were a few premature failures, but none of them was attributed to the use of the 1.4 load factor. They were due to analysis oversights or other anomalies.

Crawford also pointed out that airframe structures tended to fail at higher loads than predicted for crash conditions. This occurred because analyses did not account for large deformations and secondary structures that absorbed energy in crashes. So, it was agreed to use a 0.85 ultimate load factor for crash conditions. These were bold steps on the part of the Army, and set them apart from other government certifying agencies.

Structural Analysis

Sikorsky's commitment to meet the weight challenge was embodied by assigning the company's top two stress analysts to the program: Darryl Toni and Bruce Barr.

Darryl was given the responsibility to support the IPTs by sizing the structures and approving the detail design drawings. Bruce was given the responsibility for structures technology, which included developing analysis procedures for configurations that could not be analyzed with standard methods, and also developing internal loads. In addition to giving this duo different roles and responsibilities, their reporting structure was different. Darryl worked with the IPDTs, whereas Bruce reported at a higher level to the chief of design. This was intentional because when the IPTs would fall behind schedule, there would be a lot of pressure to draw in the team working on technology to help. Having the technology group report to a higher level protected them and avoided compromising their mission.

Darryl automated Sikorsky's Airframe Structures Manual into a collection of software routines. To do this, he first had to learn how to use the Java programming software. When he showed the programs to his functional engineering colleagues, there was push back. They questioned the lack of extensive validation. Darryl's rebuttal was that it was more important for him to be able to review analytical results, instead of getting immersed in arduous calculation checking. If the results did not look right, he would go back and recheck the software. This explanation was good enough for the chief of design, and his DTLAM programs were used exclusively going forward. The Army's certification agency, the Aviation Engineering Directorate (AED), also embraced his programs, and they turned out to be a very valuable design tool.

Bruce Barr was responsible for building the NASTRAN finite element analytical models that were used to establish internal loads for the structure. These models were very important in order to provide accurate loads for the stress analysis. Structural stiffnesses had to be modeled precisely, because they affected not only the load distribution, but also how the structure responded to vibration and how it affected weapon bore sighting. The latter requirement was necessary because precise alignment between target acquisition sensors and weapons had to be maintained during severe maneuvers. Bruce did an outstanding job in this regard.

Bruce also managed the structures technology effort. For composite structures that contain novel or unique design features, building block test programs were used to substantiate structural integrity. The building block program is essentially a pyramid of increasingly more complex tests to validate structural integrity. It starts with material characterization and property testing, followed by small element tests, then larger subcomponent tests, and culminates with full scale tests. Bruce Barr was responsible for the element and subcomponent testing part of the pyramid. There were a lot of new, unproven design features in the airframe. Therefore, dozens of element tests had to be conducted to reduce risks. The tests were used to determine failure modes and then correlate them to analysis. Larger subcomponent tests addressed more complex structural interactions. Subcomponent testing covered phenomena like canopy bird-strike resistances and fuel enclosure ballistic damage tolerance.

THE INTRODUCTION OF DIGITAL DESIGN

From the 22 June 1992 issue of *Aviation Week & Space Technology* [1]: "Sikorsky is one of the first aerospace companies to rely almost exclusively on IBM workstations, instead of IBM mainframe computers, to run CATIA [Computer Aided Three-Dimensional Interactive Applications] software. . .".

Solid modeling was used for spatial integration, electrical wire and plumbing routing, and interference checking. This was the first time that Sikorsky used 3-D solid modeling, although Boeing had used the system on the V-22. Based on Boeing's experience, Sikorsky fully committed to using the system as the primary design tool.

However, there was a learning curve price that had to be paid. First and foremost was the necessity to train the entire design team and their integrated product development team (IPDT) colleagues on how to use the system. This was no small task, because designing in 3-D is very different than conventional orthographic drawing, which was the customary process that designers had used for their entire careers. New design procedures and a different thinking process had to be developed and adapted. Initial estimates of model size and computer processing speeds were grossly underestimated by the IT department, and it was a constant battle to enable the computers to keep up with designer needs. This was not entirely the fault of the IT department, because a new conundrum became apparent. Almost as soon as the computer network was upgraded to satisfy design needs, the design team figured out how they could do even more with the system, which inevitably required even more processing and storage space. As Tom Sheehy, then Sikorsky chief engineer, put it, "Given the choice of living cheaper or living better, the design teams always elected to live better." So like an arms race, the ability to do a better job with more computer power still continues to this day.

There were two other somewhat unexpected fallouts related to using 3-D graphics for design. First, the attractiveness of 3-D graphics created a renewed interest from recent college graduates to engage in design. Up to this point, the best and brightest preferred to be analysts and were turned off on design, which focused on mechanical drafting, which they perceived to be not worthy of their education. Yet in spite of this proclivity to design in 3-D, the complexity required to be proficient in using CATIA resulted in some of these engineers becoming CATIA specialists instead of creative designers. They tended to zero in on CATIA modeling nuances and had difficulty focusing in on "big picture" design issues. Over time a new culture evolved in which the different skill sets would work harmoniously together.

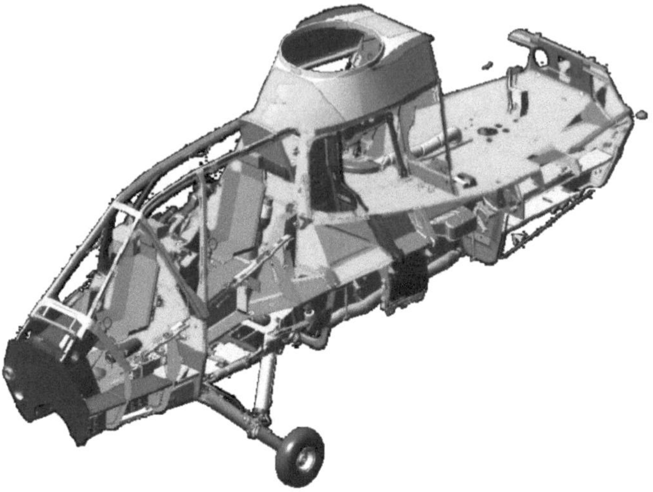

Fig. 13.2 Electronic mockup of the forward fuselage.

Initially a three-dimensional model or electronic mockup (EMU) of the aircraft was used to allocate space (Fig. 13.2). These models were used during the preliminary design phase. This was a very powerful tool for visualization, but it had some unique challenges. Ideally, the EMU would consist of the actual detail models used to define the parts; however, this was not the case because of model size limitations. Very large models would take up too much computer space and would take a very long time to manipulate on the computer screens. Therefore for the EMU, simplified versions were used. The simplification was accomplished by tessellating (tiling of a plane using one or more geometric shapes, called tiles, with no overlaps and no gaps) surface features. This reduced the model size to manageable levels. This worked in two steps. Initially the EMUs were built with simple block models, and then they were refined as more detailed part definitions were developed. Configuration control to keep up with these iterations was very difficult, and the reliability and accuracy of the EMU was always in question. Personally double-checking with model authors was the only way that this quandary was overcome.

One of the most immediate and profound benefits of using 3-D models was the visualization provided to the IPDT members responsible for building the helicopter. The Operations Department assigned a general foreman (Joe Gagliano) and assembly foreman (John Benvenuti) to the IPDTs at the outset of preliminary design. Both of these men were very experienced, capable, and outspoken. They wallpapered their office with 3-D models of the structure and spent most of their days visualizing and discussing how they would put the aircraft together. As a result, they were able to provide very valuable insights and inputs on how to optimize the design for producibility. This also gave them "skin in the game" so that even when there were problems, it was not "engineering's fault," but rather something that just had to be worked out.

Because they built the helicopter several time in their minds, this in effect provided learning curve benefits, and when the actual helicopter was built the labor hours were significantly below the manufacturing budgets.

MOCKUPS

The digital engineering mockup precluded the need for full-scale hard mockups that were traditionally used for systems integration on previous programs. Digital mockups were much less expensive and were easier to maintain than hard mockups. They also provided greater fidelity. However, there were certain aspects for which hard mockups were still required. Even with anthropomorphic model simulations, some evaluations still needed real human feel and touch. For these purposes, two hard mockups were constructed. An engineering "soft mockup" and a "marketing mockup."

The engineering soft mockup was made from wood and foam board (Fig. 13.3). These materials were easily modified to accommodate changes. It was used to validate equipment accessibility and human factor ergonomics. Although it lacked fidelity, it was more than adequate for these applications and was used in conjunction with the EMU.

A hard mockup (Fig. 13.4) was fabricated more as a marketing tool to show the aircraft to customers and also to demonstrate certain features, such as the

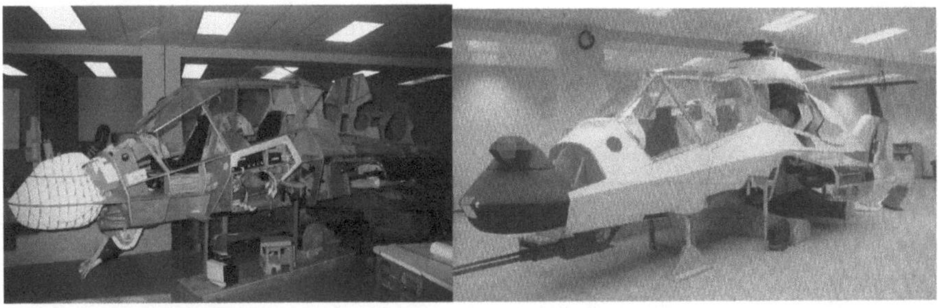

Fig. 13.3 Soft mockup.

Fig. 13.4 Hard mockups with external ferry tanks and being loaded into a C-130.

cockpit arrangement, armament, and external stores positioning. This mockup was also used to demonstrate how the Comanche could be loaded into C-130s for air transport.

OTHER DESIGN CRITERIA AND THEIR RESOLUTION

SUPPORTABILITY

Supportability was a major driver of the airframe design. Quick and easy access to all mechanical and electronic systems was an important factor that affected maintenance man-hours and life cycle cost. The Comanche was very densely packaged, so this requirement resulted in much of the outside surface area becoming removable panels.

To avoid the need for maintenance stands on the flight line, integral work platforms were incorporated into the fuselage design. The weapons bay door provided this function in addition to enclosing the weapons. The door could be raised to provide clear access for loading missiles. Another feature of this door was a kit containing personnel recovery nets that could be used to carry passengers in emergency conditions (Fig. 13.5).

Supportability is discussed in detail in Chapter 7.

CRASHWORTHINESS

The primary Comanche scout missions entailed a good deal of nap-of-the-earth flying, so crashworthiness was a major design consideration. Various crash conditions were defined in the LHX specification. The most severe condition was a vertical impact with a sink speed of 39 ft/s, or about 27 mph. To meet these requirements, a systems approach, developed and proven on the UH-60 Black Hawk and S-75 ACAP helicopters, was used. Extra strength was built into the structure to prevent high mass items, such as rotors, transmissions, and engines, from breaking loose in a crash. For crew safety, means

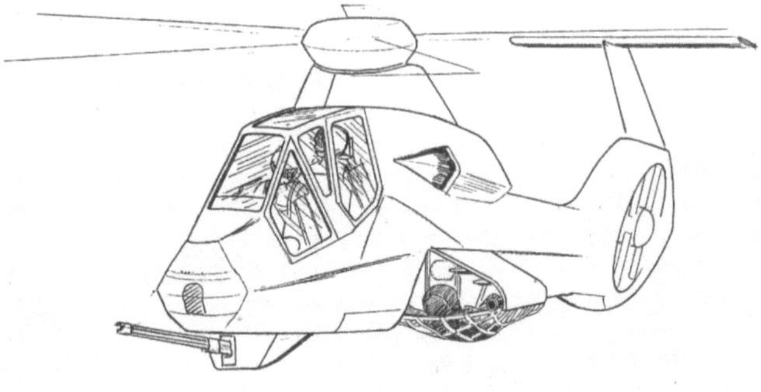

Fig. 13.5 Proposed emergency personnel recovery kit.

were provided to attenuate crash energy and reduce deceleration forces to survivable levels. This was accomplished by a crashworthy landing gear, crushable lower fuselage, and stroking crew seats. The landing gear absorbed the bulk of the crash energy. After the landing gear stroked, additional energy was absorbed by crushing the lower fuselage structure. Finally, bulkhead-mounted crew seats were designed to stroke approximately 1 ft and absorb any residual energy. The tailcone was also designed to break away in a crash to reduce mass. A leakproof, rupture-resistant fuel system prevented fires.

SURVIVABILITY

Comanche was designed to survive gunfire in the battlefield. Survivability levels ranged from invulnerability to small arms fire, up to get-home capability for larger threats. This was accomplished primarily through structural redundancy and separation of vital components. Armor was used sparingly. Crew seats were lined with lightweight composite ceramic armor. Small armor panels were also located between the engine turbines to prevent a single shot from disabling both engines. The result was a very low vulnerable area to ballistic threats. It was said that it would take a "golden BB" to take down a Comanche.

One of the more difficult areas to protect was the fuel cell. Located between the keel beams, the fuel cell was tall and narrow. Solid and explosive rounds shot into the fuel cells would create very high dynamic pressure shock waves. To prevent rupture from these threats, the keel beams contained a co-cured gridwork of unidirectional fiberglass strips. Fiberglass has a large elongation to failure and is able to absorb the shock wave energy without bursting.

The viability of this approach was successfully demonstrated by ballistic test (Fig. 13.6). Expectations during this test were mixed. With the exception

Fig. 13.6 Fuel cell ballistic test setup.

of the Sikorsky engineering representative, most of the experienced ballistic test personnel in the block house were rooting for the bullet. They were disappointed when there was no explosion, and they could hardly tell that the tank was impacted.

NUCLEAR, BIOLOGICAL, AND CHEMICAL THREATS

The Comanche was designed to operate in an environment with nuclear, biological, and chemical (NBC) hazards. The nuclear threat meant nuclear radiation, and that implied shielding the avionics equipment against electromagnetic interference. Shielding was provided at the component and wiring levels. For the airframe, the biological and chemical threat was gas warfare. First and foremost, this required protecting the flight crew to prevent lethal chemicals from entering the cockpit. A twofold approach was used to accomplish this goal. First, the air necessary for breathing was filtered to remove any contaminants. Filtering was done by a pressure swing adsorber (PSA) filter built into the environmental control system (ECS). Its operation is detailed in Chapter 14. In addition to filtering the air, the cockpit was pressurized to prevent gas leakage into the living space. The avionics mission equipment package compartments were also pressurized. Some mission equipment package (MEP) components had to be accessed after each flight, and it was important to make sure they were not contaminated. The overpressure level was fairly low, set just above the magnitude of positive aerodynamic pressures enveloping the fuselage. Thus, any leakage would exit instead of entering the aircraft.

Besides protecting the flight crew, it was important that the helicopter should be available for continued operation in NBC threat areas. This meant the aircraft materials had to be resistant to damage from, and absorption of, these noxious chemicals. All of the aircraft materials had to be evaluated for their chemical resistance. Data were available for some of the materials, but any new materials had to be tested. This was done at the Dugway Proving Grounds in the Utah desert.

One more step was needed after protecting the occupants and helicopter from deleterious gas attack effects, and that was decontamination. Any trace of residual chemicals left on or inside the helicopter had to be removed to protect personnel working on the machine. Several decontamination approaches to wash down the aircraft were available. Some of these remedies used harsh chemicals, and materials had to be resistant to them. However, the most insidious problem was decontaminating all of the nooks and crannies inside the helicopter. The Army's design guide suggested having the structure be smooth and self-draining, but this was impractical. That meant decontamination would be a very tedious operation. On the other hand, the likelihood of large-scale, long-term repeated forays into contaminated war zones was very unlikely. So, this was not considered a serious deficiency.

BIRD STRIKE

Comanche was the first production helicopter with a cockpit enclosure designed to resist bird strikes. This was necessary because a good deal of its mission would be flown at high speed and low altitudes. Bird-strike tests were conducted on windshields mounted to representative canopy framing. The laminated windshield was designed to deflect when impacted to absorb energy; however, the most critical conditions were hits adjacent to the windshield posts and in the corners. To meet these conditions, rugged canopy framing was required. The canopy framing was also designed to act as roll bars to protect the aircrew in rollover crash conditions. Besides the obvious requirement for strength, the post cross sections had to be small to minimize visual obstructions. Previous helicopter difficulties in getting good fits between the structure and its windshields and doors led to a decision to manufacture the canopy enclosure as a one-piece co-cured assembly. The need for close-fitting canopy apertures was even more critical for low radar signature.

Meeting structural and visibility requirements led to a box-beam cross section for the framing members. In order to manufacture this framework in one piece, a unique co-cure process was developed. This process entailed laying up the inner surfaces of the box-beam members as a network of interconnected U-channels. The channels were then filled with foam and allowed to harden. Finally, the outer closure surface was layed up and cured onto the foam and U-channels to complete the assembly. A technique was also developed to remove the foam after curing, to minimize weight. A patent diagram showing how the post cross section was manufactured is provided in Fig. 13.7.

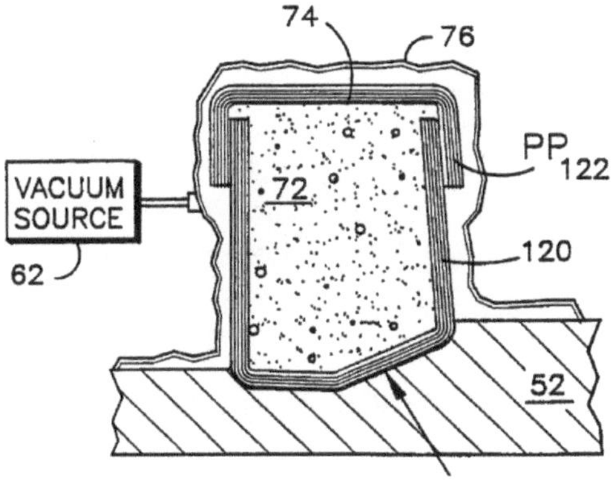

Fig. 13.7 Patent diagram of canopy post.

MATERIALS

Comanche utilized the latest state-of-the-art material technology. For the primary structure, the graphite epoxy material incorporated toughened epoxy resin systems that significantly improved damage tolerance. Other basic composite materials, including Kevlar, fiberglass, and even quartz epoxies, were used mostly in secondary structure applications. High-temperature composite materials were used in the engine compartment hot zones. In addition, improved metallic and nonmetallic honeycomb cores, adhesives, and sealant materials were used throughout the aircraft. An extensive series of building block tests was performed on all of these materials to characterize mechanical and structural properties, co-curability, compatibility with environmental fluids, and so forth.

Comanche set an all-time record for the amount of new materials used in the air vehicle. The entire external surface was composite material. All the composite materials used comprised 51% of the air vehicle weight, with another 30% being metal. The remaining miscellaneous materials included the transparencies, wiring, fuel cells, tires, and so on (Fig. 13.8).

From a March/April 1992 article in *Vertiflite* by Sam Garbo and Ken Rosen [2]: "Part count reduction was dramatic for the Comanche fuselage compared to the Black Hawk. Comanche's 350 composite parts replaced about 6000 metal. Comanche used around 6,000 fasteners compared to 60,000 on the Black Hawk."

Lightning strike protection, antenna ground planes, and electromagnetic shielding were provided by applying perforated copper foil and conductive coatings to the composite materials in selected areas. Electrical equipment was grounded through dedicated copper bus bars that ran through the helicopter. On aluminum helicopters, equipment is normally grounded to the airframe, but the graphite material is not conductive enough to meet this requirement, so there was a small weight penalty incurred for the bus bars.

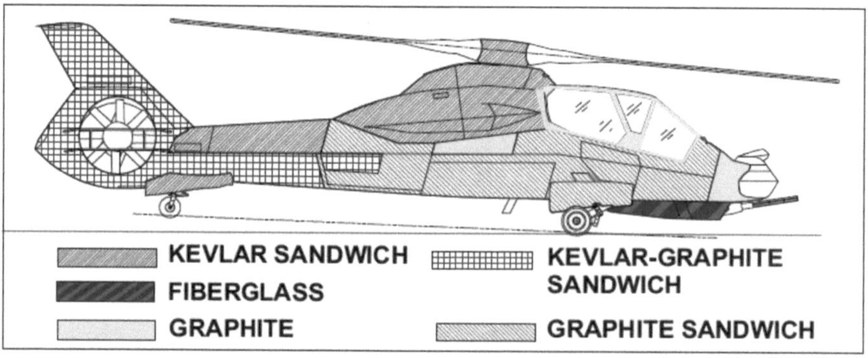

Fig. 13.8 Comanche external materials.

A number of special materials were used in a few locations to minimize radar signatures. Where shaping could not deflect radar waves, radar-absorbing materials were used. This situation occurred on the main rotor pylon and vertical fin. Honeycomb construction was used in these applications, where the honeycomb core was designed to absorb the radar energy. This was a much lighter solution than the heavy magnetic radar-absorbing coatings typically used on stealth airplanes.

Metallic materials used on Comanche also exploited the latest technologies that were available. For aluminum, improvements in heat treating to resist stress corrosion cracking were utilized. Corrosion-resistant aluminum honeycomb cores were employed where nonmetallic cores could not meet structural requirements. In applications where aluminum components mated with graphite epoxy, a thin layer of fiberglass was co-cured on the graphite parts to isolate the metal and prevent galvanic corrosion. Other metal alloys employed included high strength titanium in the airframe and rotor systems. Beryllium copper castings were even used in the electro-optical target acquisition and designation system (EOTADS) because of its very high stiffness-to-weight ratio.

A tremendous amount of work was required to qualify all these new materials, including the preparation of comprehensive procurement and process specifications. Thousands of coupon and specimen tests were conducted. This activity was primarily orchestrated by Bill Fallon, as lead materials engineer for the program.

DESIGN AND MANUFACTURING

Ken Kadezebek was responsible for the Comanche loads and criteria. When the Sikorsky team was moved from Stratford, Connecticut, to a new office building in Trumbull Connecticut, Ken was given only a desk without a reference table, because he was only going to be there a short time until the preliminary design review (PDR), when all the loads would be frozen. As it turned out, he stayed on for virtually the entire program, simulating complete maneuvers, resulting in hundreds of load conditions compared to a couple of dozen peak load conditions used to design previous helicopters. Also, whenever designs did not look right, one of the first things reviewed was the loads, which were rerun and frequently tailored. Fine tuning was required to close the gap. These never-ending iterations kept Ken fully occupied for the duration of the design activity, and then well into the test program when flight loads were validated.

A low radar signature was very important, and there was a constant struggle to achieve pragmatic design solutions. Shaping the fuselage was the easy part and fairly well understood because radar reflectivity was pretty much a function of geometry. The geometry was also validated with pole model radar cross section (RCS) tests (Fig. 13.9). However, designing radar-absorbent

Fig. 13.9 Comanche RCS pole model (Boeing Photo).

structures and mitigating signature returns from traveling waves got into very detailed nit-picking. This became a battle of theory vs practicality. RCS criteria affected steps and gaps, ply layups, and part terminations. The problem was that this was the first time the RCS engineers had to work in a "real world" with designers to integrate all of their requirements to meet the stringent Comanche spec requirements. The group lacked the experience to put forward credible design solutions with confidence. Unfortunately, even if something had been used successfully on another program, security classification restrictions prohibited revealing that information. In an environment with veteran designers, the RCS engineers were essentially uncalibrated. So there were many debates. Analyses were viewed with a jaundiced eye. Small-scale building block tests were performed when possible. Decisions were eventually hammered out, but the final proof of what worked would not be accomplished until a full-scale radar model of the helicopter with all its details would be tested.

What was not so easy was meeting step and gap criteria for fitting panels and operable doors to the skin surfaces. The step and gap requirements were almost an order of magnitude more stringent than traditional helicopter practices.

Rick Isaacs, the manufacturing engineer responsible for fuselage assembly, had been working on this problem for several years when he was offered a higher position in Sikorsky's Huntsville, Alabama, office. Upon leaving, he wrote a memo listing his open action items. In that memo he stated that the door step and gap requirements were beyond the current state of the art, and that his successor should attempt to negotiate more lenient criteria. This opinion was based in part on previous helicopter experience where doors never fit

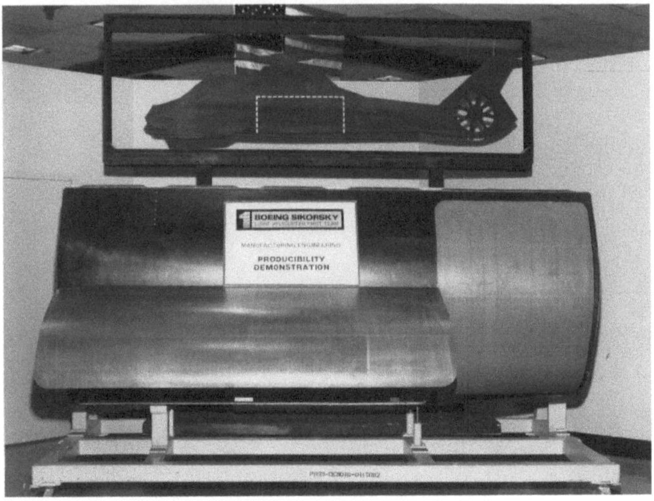

Fig. 13.10 Fuselage producibility demonstrator.

properly, as well as early-on results from building a producibility demonstrator section of fuselage (Fig. 13.10). Although the demonstration article did not completely achieve the desired results, it did identify the problems and actions needed for success. The team persevered, and the requirements were ultimately met by paying close attention to design details and tooling.

Several approaches were implemented to achieve tight-fitting doors and panels. Precise tolerances were established to control part contours and trims. When manufactured, composite parts should take the shape of the molds they are laid up on. So the first step was to have stable, dimensionally exact molds that do not change shape when they are heated during the composite cure process. To accomplish this goal, Comanche molds were precision machined to close tolerances from invar steel plate (Fig. 13.11). The invar material has a coefficient of thermal expansion very close to that of the graphite material the composite parts were made from. That meant when the composite parts were cured at approximately 350°F in the autoclave, there would be very little differential thermal expansion. The part surfaces would replicate the mold surfaces. To further ensure good fits, integral doorjambs were molded directly into the skin panels. This minimized tolerance buildups.

Much tighter tooling tolerances were possible because tool design data and tool component machining data were downloaded directly from the design database. Tolerances on the Comanche's 4 ft $\times$ 23 ft-long bottom skin mold were kept at $\pm$ 0.007 in., and door molds measuring about 3 ft $\times$ 6 ft maintained tolerances of $\pm$ 0.005 in.

The next step was to prevent the parts from warping when they were cured. Composite parts are layed up by stacking individual plies of fabric, or unidirectional tape, on top of each other. These plies contain graphite fibers oriented

Fig. 13.11 Invar steel mold for the fuselage side skin.

in certain directions based on the loads that the part has to resist. Fiber orientations in typical laminates are 0 deg, 90 deg, and ±45 deg. To avoid warping, laminates have to be stacked symmetrically. An unsymmetrical layup will warp, because the thermal expansion characteristics of the material are based on the fiber orientation. This phenomenon causes the top and bottom plies to expand at different rates, which causes the panel to warp. In symmetrical layups, the expansion is balanced, and there is no warpage. During detail design, great pains were taken to provide symmetrical, balanced composite ply layups.

Invar tools have very long lead times, and the data sets for these tools were released right after the preliminary design review (PDR). This was extremely risky because each invar mold cost approximately $1 million. Any changes to the outside contours would be extremely expensive to reconcile. Fortunately there were no changes, and the risk paid off.

Although it was possible to maintain and preserve the outside contours or shape of the fuselage, developing inside surfaces of the skins, to which the structure mated, was another story. The inside surface of the skin is developed by subtracting a set-back equivalent to the skin thickness. The skins were composite construction made up of multiple plies of composite material, layed up over honeycomb core material. Ply thicknesses were variable, being tailored based on local loading conditions. It was originally believed that Dassault CATIA's Composite Ply Development (CPD) computer program could automate this process and accurately develop these inner surfaces, which would avoid the alternative tedious, error-prone manual calculations to otherwise develop the surfaces. However, there were numerous glitches and shortcomings with this software, and it did not work as initially claimed.

Considerable efforts were required, and the manufacturing engineers never missed an opportunity to point out how they were misled.

After laying up and curing the composite parts, they had to be accurately trimmed to final dimensions. This was accomplished with holding fixtures that secured the parts to the bed of an extremely precise, numerically controlled machine. A huge Henri Liné precision gantry NC machine was part of Sikorsky's capital investment program to support the Comanche program (Fig. 13.12).

This machine also drilled the holes that were used to join the parts to adjacent assemblies. Due to the machine's extreme precision, it was possible to drill matching holes in adjacent parts. This was a major departure from the way airframes were traditionally assembled. Traditionally, undersize pilot holes were drilled in one part, and then the parts were clamped into an assembly fixture. The pilot holes were then drilled through into the adjacent part and enlarged to full size. For Comanche, pilot holes were drilled in both parts, and then the parts were pinned together through the pilot holes. They were then opened up to full size, and the fasteners were installed.

Originally, contour boards were mounted on the fuselage assembly fixtures, and it was intended that they would be used to locate and hold adjacent parts together during the assembly process; however, it was found that the detail parts were so accurate that the contour boards were not needed. They were then removed from the fixtures and used only to check fits as a quality control measure (Fig. 13.13).

Fig. 13.12 Henri Liné gantry machine tool.

Fig. 13.13 Fuselage assembly fixture.

The attention to detail in the design and fabrication of operable doors and removable panels provided very tight-fitting assemblies when the parts were built. However, the step and gap tolerances were not just static requirements, but were intended for use when the helicopter was operating in hostile environments. To make sure gaps stayed within limits under those conditions, finite element analyses were conducted to determine how the panels would deflect when subjected to flight loads. Based on these analyses, panel stiffnesses were tailored accordingly. This attention to secondary structure was pretty much unprecedented for helicopter design.

Figure 13.14 shows a large one-piece bulkhead being lowered into the fuselage box beam. Assembly of the box beam was more like building a model airplane than the traditional built-up metal assembly methods used in the past. The amalgamation of high fidelity invar molds, precision trimming, coordinated fastener holes, and large co-cured parts had a dramatic payoff in final assembly. The first prototype box beam (Fig. 13.15) took only 38 labor hours to assemble, which was well under budget. The manufacturing team had scheduled this assembly for a two-week period.

Although the nonrecurring costs for engineering and tooling were certainly more than equivalent costs for previous helicopters, the parts fit together much better. Assembly costs were dramatically lower because there was hardly any

Fig. 13.14 Assembly of the main fuselage box beam.

Fig. 13.15 Bruce Kay in front of the composite structure.

hand fitting or in-situ trimming required. The nonrecurring investment was more than offset by reduced production costs.

Getting parts to fit together precisely was still only part of the solution for achieving low radar signatures. External surfaces had to be smooth and unblemished. For the external skins, honeycomb sandwich construction was selected. Here, the tradeoff was against monolithic stiffened skins that would be allowed to buckle under load. Allowing skins to buckle under load was pretty much the standard approach for lightweight helicopter airframes. The sandwich skins were arguably heavier, but were simpler to build, offered lower cost, and appeased the signature folks, so they won the trade on that basis.

There was also a campaign to reduce all protuberances on the outside surfaces of the helicopter. Most door and panel latches were either hidden or eliminated, and replaced with smaller screw-type fasteners that could be treated to minimize signatures. Jettisonable crew emergency escape doors were replaced with fixed windows containing explosive detonation cords. In a rollover emergency, the explosive cords would blow out the window to provide a means of exit for the flight crew.

Fastener heads were considered blemishes, even when they were countersunk, so it was important to minimize the number of fasteners that penetrated the skins. This was accomplished by integrally molding T-blade joints on the skins that were used to attach internal structures such as bulkheads, frames, and keel beams (Fig. 13.16). These parts were then bolted to the upstanding flanges of the T-blades instead of the more conventional way of bolting through structure flanges into the skins. The T-blade construction also eliminated the need for shimming between the internal structure and the skins, which reduced manufacturing labor.

Unique, patented vacuum bags were built to accommodate these details when the parts were cured in the autoclave. Called "black bags," they were made from reinforced elastomeric material and molded to contain the detailed features within the laminate (Fig. 13.17).

Fig. 13.16 Molded fuselage skin T-blades.

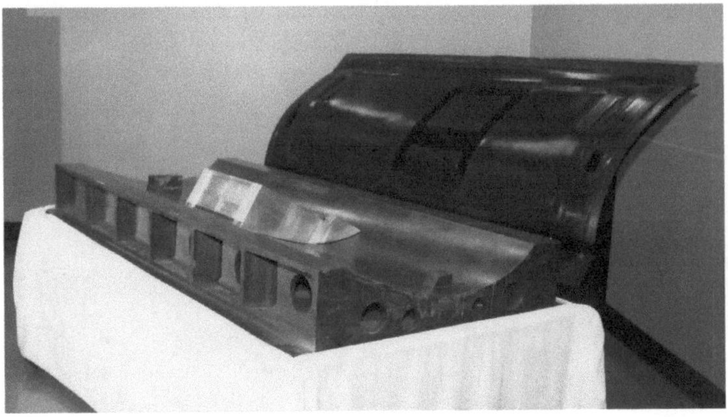

Fig. 13.17 Typical elastomeric "black bag" and mold.

It was not practical to shape the main rotor pylon that enclosed the engines and main rotor gearbox for reduced radar reflections, so this structure was constructed using radar-absorbing materials. The panels were of sandwich construction using treated honeycomb with the outside surface facings made from Kevlar epoxy material. The Kevlar was electrically nonconductive, which allowed the radar waves to enter the panels, where their energy was dissipated by the treated honeycomb. Extensive risk reduction tests were conducted to refine design details, particularly joints between panels.

Another area that required special consideration was the chines. The chines on the sides of the fuselage had a generous radius optimized for structural and producibility purposes. This resulted in a somewhat blunt surface that could adversely affect the radar signature. To mitigate this effect, radar-absorbing chine fairings were to be provided as a kit. They would be used in missions where low observables were vital. The radar attenuation requirements for these parts necessitated a very lightweight unsymmetrical construction. The outer skins were fiberglass and the inner skins were graphite, and the two skins were separated by a very lightweight core material. Processing the parts using conventional autoclave pressures and temperatures would cause warping and local crushing. To overcome this impediment, electron beam radiation was used to cure the parts (Fig. 13.18). Electron beam radiation provides a rapid cure cycle using low temperature and low pressure. Although material properties were considerably lower than conventional materials, they were more than adequate for this application. Prototype demonstration parts were fabricated by Boeing in its electron beam laboratory in Seattle, Washington. These fairings would have been the first flightworthy electron beam–cured parts had the program not been canceled.

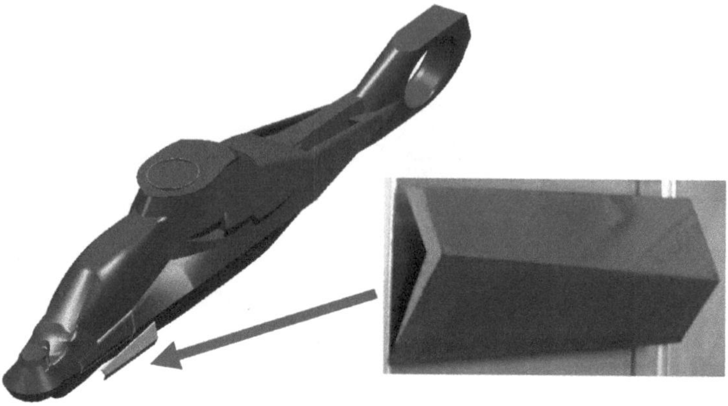

Fig. 13.18 Electron beam–cured chine fairing.

THE FUTURE

The Army's Aviation Applied Technology Directorate (AATD) sponsored another R&D program called Rotary Wing Structural Technology Demonstration (RWSTD). It used the Comanche fuselage as a benchmark against which advanced technologies improvement could be compared. As part of this program, a lower forward fuselage was built and tested. Advanced technologies incorporated in this structure included toolless assembly, paintless finish, extensive co-curing to combine parts, composite fittings, and fiber-placing the side skins with an open weave laminate to minimize weight. By skipping tows (individual fiber strands), the open weave laminate was equivalent to a half ply. For lightly loaded panels, this made it possible to go to fractional plies and effectively reduce the minimum thickness. The resulting structure was 15% lighter and used 25% less labor than the baseline. This was remarkable considering the rigorous optimizations that the Comanche went through during its gestation. The fiber-placed side skins (Fig. 13.19) were actually incorporated into the low rate initial production helicopter airframes that were built before the program was canceled.

These programs clearly demonstrated the benefits of continuous technology improvement and showed that, over time, even the most optimized parts could be further improved.

The Comanche was to have been assembled on a Sikorsky production line in Bridgeport, Connecticut. This 21st-century facility was in the same building where the Sikorsky R-4, the first U.S. military production helicopter, was built almost 60 years before. Sikorsky was to be responsible for final assembly, flight testing, and delivery. Manufacturing began with aircraft number 3, the front fuselage of which entered major assembly in 2003. The aft fuselage also started down the Boeing assembly line at the Philadelphia plant in 2003. All of this, of course, was not to be.

Fig. 13.19 Open weave fiber-placed skin.

LANDING GEAR

Comanche was the first helicopter to have a retractable crashworthy landing gear. This meant that the shock strut had two stages. The lower part of the strut was a conventional oleo shock absorber designed for normal landings, with sink speeds up to 10 ft/s. For higher crash sink speeds, additional stroke was required; this was accomplished with a second stage hydraulic absorber. The second stage actuation was initiated after bottoming the first stage. Enclosing the long shock strut was a challenge, but having an internal box-beam fuselage structure made this possible with minimal weight penalties.

The Comanche mission included a lot of nap-of-the-earth flying, so a rapid extension capability was required. The requirement was to blow down the gear in 2 seconds. A hydraulic accumulator provided the actuation force. An unexpected consequence of the rapid extension was acceleration forces that were

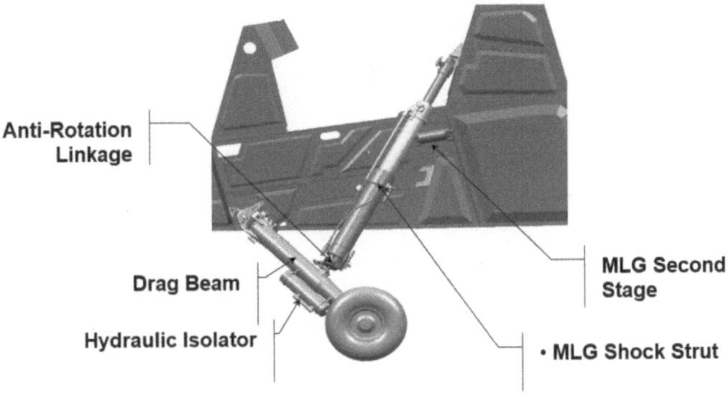

Fig. 13.20 Main landing gear.

so high they damaged the landing gear door linkage. They had to be reinforced. A kneeling feature was also built into the gear to facilitate air transport. This enabled the fuselage to be lowered so it could be easily loaded into C-130 airplanes.

Another unanticipated problem arose when the helicopter was landing on sloped terrain. Excessive roll was encountered in side wind conditions. This phenomenon occurred as a consequence of having a stiff main rotor and soft oleo struts. The solution was to add a hydraulic isolator on the gear that would stiffen the oleo for this condition (Fig. 13.20). Sensitive weight-on-wheels (WOW) switches were also built into the gear to cue the flight control system that the aircraft was on the ground. This was important because the fly-by-wire flight control laws were different for flying and for on-ground operation.

References

[1] David Hughes, "Sikorsky Exploits Workstations In RAH-66 CAD/CAM Design Work," *Aviation Week & Space Technology*, 22 June 1992, p. 52.

[2] Garbo, S., and Rosen, K., "Composite Usage on the RAH-66 Comanche," *Vertiflite*, March/April 1992, pp. 8–13.

Chapter 14

The Propulsion System

Bruce Kay

In the early 1980s, the Army had an Advanced Technology Demonstrator Engine (ATDE) program, exploring technologies for the anticipated new light helicopter. Allison and Avco Lycoming were separately contracted to study an 800- to 850-shp engine for the aircraft, which was assumed at that time to have a gross weight of 7000 lb. The engine was designated as the ATDE T-800. Pratt & Whitney was providing the gas generator and power turbines, and the inlet particle separator and blower for the Avco engines.

The aircraft size requirements then grew to a need for two 1200-shp engines. The Army issued a request for proposals (RFP) for the LHX T-800 engine in December 1984. It was assumed that the total production quantity would be 10,000 engines, and it wanted two-company teams to work together to develop the engine and then compete with each other during the large production program envisioned to follow.

In 1984, the Allison Engine Division of General Motors and Garrett independently concluded that teaming together was the best solution to meeting those requirements. They thought that codesign and codevelopment was the best means of ensuring the technology exchange necessary for two companies to effectively compete with a single design for production. The teaming was formalized in October 1984. They called themselves the Light Helicopter Turbine Engine Company (LHTEC).

Avco Lycoming and Pratt & Whitney formed a competing team, called APW (AVCO/P&W). Their engine was designated the T-800-APW. The engine was designed to produce the required 1200 shp, with a 50% growth potential, and weighed less than 300 lb. Specific fuel consumption was 0.46 lb/hp/hr. Output speed was 23,000 rpm.

In July 1985, firm fixed price full-scale development (FSD) contracts were awarded to both teams. Requirements included 1200 shp, metric format, improved reliability and maintainability, two-level maintenance, failure detection and isolation of field-replaceable items, less equipment for the field maintainer, and electronic mockup training aids.

In October 1988, the government downselected to LHTEC to continue the full-scale development program and qualify the engine (Figs. 14.1 and 14.2). At that time, the engine had a maximum contingency power rating of 1399 shp and weighed 315 lb. Its power-to-weight ratio was 4.4 hp/lb. The specific fuel consumption was a low 0.459 lb/hr/shp. It had a full authority digital engine

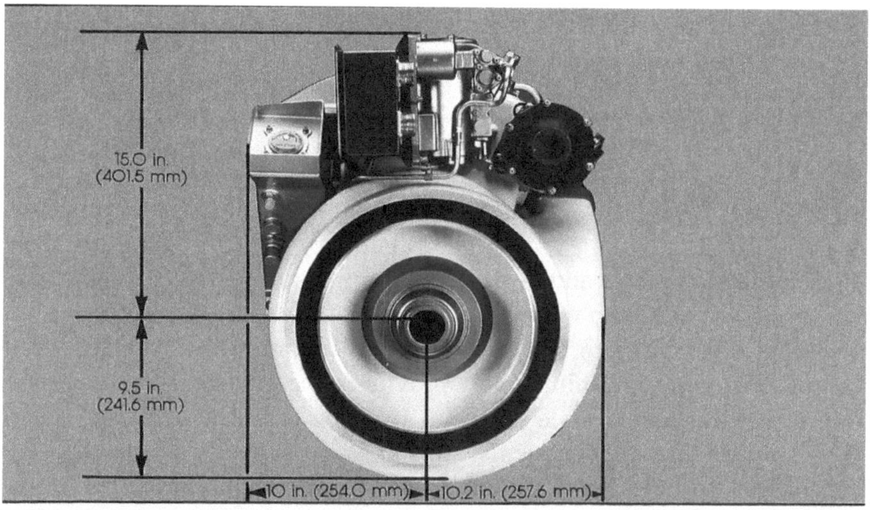

Fig. 14.1 The LHTEC T800/LHT-800 engine, and 14.2 Cutaway view of the LHTEC T-800 engine (LHTEC Photos).

Fig. 14.2 Cutaway view of the LHTEC T-800 engine.

control (FADEC) that was compatible with the Comanche fly-by-wire control system. The engine was also designed to be maintained with only a simple flight line hand tool kit.

In 1985, Garrett merged with the Allied corporation to become AlliedSignal. In 1995, General Motors sold the Allison division to Rolls Royce. In 1999, Allied Signal acquired Honeywell and adopted the Honeywell name. Thus, the Allison-Garrett LHTEC team became the Rolls Royce-Honeywell LHTEC team.

GROWTH ENGINE

Quoting *Aviation Week & Space Technology*, 19 July 1993 [1]:

> An increased power T-800 was ordered by the Army in early 1992 to accommodate a weight growth in the Comanche caused by the addition of extra systems and equipment. ... The hardware and design changes required to achieve the increased power levels will be minimal, and the engine installation envelope will remain unchanged since the engine was originally designed with planned power growth steps of 10%, 25% and 50%, [president of LHTEC, William] Schnell said.
>
> The modifications should increase the weight of the engine by no more than 20 lb., bringing it to 330 lb., Schnell said. Specific fuel consumption of the growth powerplant at cruise is projected to be slightly less than that of the baseline T-800.

Although the advanced engine was important, there is a lot more to the Comanche propulsion story.

PROPULSION SYSTEMS

Comanche was actually a three-engine helicopter. Power for the rotor drive system was provided by the two LHTEC T-800 engines. A Williams International WTS124 secondary power unit provided auxiliary power to start the main engines, provided bleed air in flight to run the environmental control system, and powered an electrical generator and a hydraulic pump.

The engine installation was driven by low observability concerns and was unlike any other helicopter before it. It featured flush National Advisory Committee for Aeronautics (NACA) inlets and low infrared (IR) signature exhausts, buried in the tailcone. Conventional ram air engine inlets have large frontal radar returns because radar waves can bounce off forward-facing turbine discs. To prevent this phenomenon, serpentine-shaped inlet ducts sucked air from flush inlets. This solved most of the problem, but the inlet cavity still allowed some radar returns, and icing protection was also required. Various schemes to deal with these requirements were evaluated, including screened inlets with bypass apertures that opened to prevent ice from clogging the screens. The prototype YRAH-66A helicopters flew with screened NACA inlets. Sand and dust filters were not required because the T-800 had an integrated particle separator as part of the engine.

Previous helicopters have used low infrared signature engine exhausts, so this technology was not entirely new. IR suppressors consisted of large appendages at the back end of the engine that diluted the hot exhaust gases with ambient air (Fig. 14.3). However, Comanche's additional requirement for low radar signature meant that the IR suppressor had to be integrated into the fuselage. This was accomplished by extending the exhaust pipes along the length of the tailcone (Fig. 14.4). The extra length created a narrow, ribbon-like

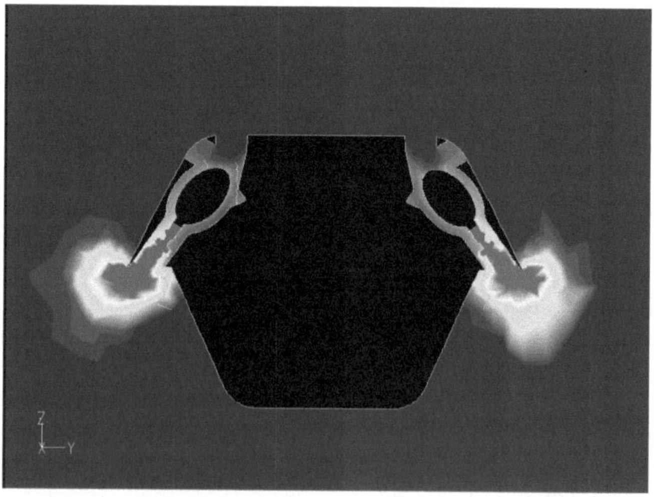

Fig. 14.3 Simulation of exhaust gas dilution.

Fig. 14.4 Low IR engine exhaust in the tailcone.

exhaust flow that could be more easily cooled to minimize the IR signature. Cooling was provided by sucking ambient air through longitudinal slots above the tailpipes. Mixing was further enhanced by fluted exhaust nozzles that increased the surface area between the hot gases and ambient air. Extensive bench testing of engine exhaust configurations were conducted at the United Technologies Research Center to optimize performance.

Engineering Vice President Dr. Ken Rosen, who had an extensive propulsion background, continually stressed to the team how important it was to keep engine inlet and exhaust pressure losses and temperature rise to a minimum while satisfying the engine's distortion limits and the Comanche's unique observable requirements. Specifically, relatively small installation inefficiencies would translate into significant payload penalties. Ken also paid detailed attention to the thermal management issues involving the engine compartment, FADEC cooling, and IR suppressor. The use of computational fluid dynamic analyses was emerging as a standard practice; however, rigorously detailed hand calculations were still the state of the art. It was always entertaining to watch Ken's emotions rise when reviewing analyst Dave Smith monotonically go over his power loss analyses or when Ken learned that the T-800 engine had no customer bleed air available, a situation that required the aircraft to incorporate and develop a unique low Specific Fuel Consumption (SFC) secondary power unit (SPU) rather than the usual auxiliary power unit (APU). This system continuously supplied the pressurized cooling air for the vehicle's environmental and filtration system as well as the cooling sink for the MEP. Aspirated wind tunnel model testing was also used to support the analyses.

LHTEC also conducted engine rig testing with simulated intakes and exhausts to evaluate installation losses. One concern was that the unsymmetrical inlets might adversely distort the airflow going into the engine. This turned out not to be the case.

The tail pipes were made from Inconel and titanium. A high temperature polyimide composite baffle box encased the fluted exhaust nozzles, where hot gas and cool air mixing was initiated. Ceramic exhaust nozzles were also developed and tested as a means to enable reduced radar reflectivity solutions.

Fatigue cracks appeared in the exhaust ducts during early testing. Rather than just increase the sheet metal thickness, which was frequently done to solve this problem, structures analyst Daryl Toni took another approach. He created a finite element model of the duct and found that it had a natural frequency that was resonant with engine harmonics. With this information in hand, he was able to tailor a local reinforcement to eliminate the cracking.

The IR suppressor eventually demonstrated an IR signature one half of the Army specification level.

Another vexing problem was air management within the main rotor pylon that housed the engines and the secondary power unit. Cooling air was required for the main transmission oil cooler, hydraulic heat exchangers, and engine control units (ECUs). The ECUs had to be located very close to the engines, exposing them to high temperatures. Another air intake was required for the secondary power unit. The T-800 inlet particle separator ducts also dumped air into the IR exhaust cooling air passageway. Maintaining adequate airflows for all these components was a complex endeavor. Inlet locations were constrained by radar cross-section (RCS) considerations and the need to maintain positive airflows under all flight conditions. Screening the inlets with open cell honeycomb core provided some protection against adverse radar reflections, but proper location of the apertures was far more important. The airflow around a helicopter's main rotor pylon is unsteady, and very unpredictable, because of its proximity to the spinning rotors. Multiple inlets were built into the main rotor pylon, but modifications had to be made during ground testing on the propulsion system test bed, and then again during flight test.

Secondary Power Unit

Medium and large helicopters often use auxiliary power units (APUs) to provide pressurized air to start the main engines. The start sequence begins by activating a hydraulic accumulator to propel the APU's hydraulic starter motor. Alternatively, a portable ground power cart can provide starting power. A buddy start capability is also available that can extract power from another helicopter. As a last resort, a hand pump can repressurize the accumulator.

The APU powers the aircraft electrical and hydraulic subsystems on the ground during maintenance. It also provides power to the avionic systems for mission planning at remote sites. Using an APU for these scenarios is more efficient, and less expensive, than running the main power plants. For these reasons, an APU was a basic LHX requirement.

During preliminary design tradeoffs, it was determined that the environmental control system would be powered by bleed air. A large amount of bleed air was required for crew comfort and avionics cooling. The T-800 engines had a very limited bleed air capacity, particularly at ground idle conditions. A much more efficient approach was to use a dedicated small engine that could provide the necessary bleed air.

It was also determined that bleeding precious air from the T-800 engine compressor stage during flight adversely affected engine performance and significantly reduced specific fuel consumption. Because the APU was already sized to provide electrical power, hydraulic power, and bleed air for ground operations, it could provide the same capability in flight without compromising the T-800 engine performance or fuel consumption. As a result, the specifications for the APU evolved into requirements for continuous flight operations. The APU was then appropriately renamed as a secondary power unit (SPU). And so, the Comanche became a three-engine helicopter.

The SPU had an equivalent power output of 176 hp. In addition to providing bleed air, a hydraulic pump and generator were mounted on the SPU. These units provided redundancy for the main gear box–mounted pumps and generators. The SPU was installed in the back of the main rotor pylon, on the right side (Fig. 14.5).

The supplier for the SPU was selected through a competitive procurement. The technical community strongly disfavored the Sundstrand proposal, but Sundstrand had the lower price. After much heated discussion and intervention by senior management, Sundstrand was given the contract based on its lower price.

Within approximately one year, Sundstrand fell hopelessly behind schedule. A high-level meeting was convened between Sikorsky Executive Vice

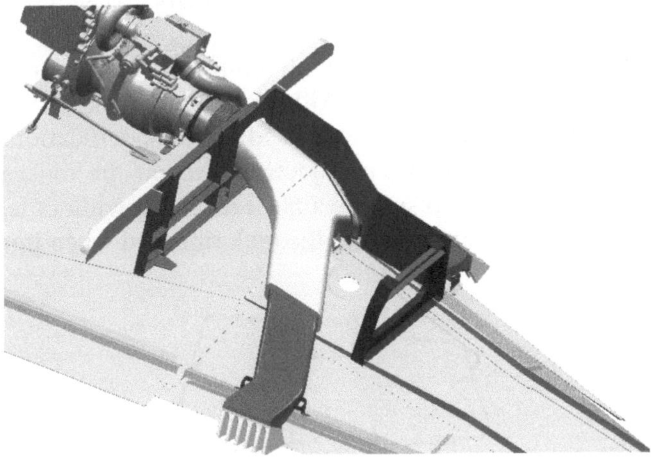

Fig. 14.5 Secondary power unit installation.

President Ray Kurlak and the president of Sundstrand. In a previous life, these executives had worked together at General Electric. In spite of this relationship, Sundstrand was unable to provide a recovery plan, and actions were initiated to change suppliers. Discussions were resumed with Williams International, a builder of small gas turbine engines, primarily for unmanned aerial vehicles (UAVs). Because of its small size, Williams offered the agility to take over the program and recoup the lost schedule time. It did, in fact, meet the schedule and delivered a spec-compliant SPU.

Fuel System

The Comanche had a very basic fuel system that incorporated state-of-the-art technology features for survivability and crashworthiness. Yet, in spite of its simplicity, it did incorporate some novel features. To save weight, a single fuel cell with two compartments was used instead of having two separate tanks (Fig. 14.6). Dual tanks provide a measure of reserve fuel capacity in the event of fuel loss from one of the tanks. On Comanche, this capability was accomplished by shaping the bottom of the fuel cell with a hump, upon which was an integral baffle. The hump was supported by a piece of rigid foam under the tank, and the raised baffle effectively provided two isolated compartments. Weight and cost savings were achieved by having only one fuel cell and eliminating the need for a dividing bulkhead and associated transfer plumbing.

The fuel cells were self-sealing to minimize leakage from ballistic penetrations. An on-board inert gas-generating system (OBIGGS) pressurized the tanks with nitrogen gas to prevent fires from incendiary ballistic threats. Fuel transfer to the engines was accomplished with suction lines, to prevent leakage if the lines were ever punctured in any way. For crashworthiness, all of the fuel lines had breakaway fittings that would automatically seal if they became detached in a crash.

The fuel cell itself was constructed with reinforced elastomeric material so it could survive the rigors of crashes. Although flexible, the material stiffness was similar to that of an automobile tire. Material specifications called for resistance to penetration by sharp objects; however, the most demanding requirement was the ability to survive a 50-ft drop test without leakage (Fig. 14.7). This was a very severe test, and the tank ruptured when initially tested

Fig. 14.6 One-piece compartmented fuel cell.

Fig. 14.7 Fifty-ft fuel cell drop test (NASA Photo).

by the manufacturer. At the time, fuel cell design was more of an art than a science, and the supplier's proposed corrective action was to just add another ply to the tank. Sikorsky considered this approach to be too crude, and structures analyst Darryl Toni created a finite element analysis simulation of the event. His analysis showed there were high local stresses in the upper crotch region of the tank. As a result of this analysis, a small local doubler ply was added, and the tank was successfully retested.

ENVIRONMENTAL CONTROL SYSTEM

Comanche requirements went well beyond the usual need for providing a comfortable environment for the aircrew. In addition to heating and cooling the cockpit, there was a major heat load from the mission equipment avionics that had to be cooled. A massive amount of electronic equipment was distributed throughout the fuselage in four avionic bay compartments, plus the targeting sensor located in the nose. In addition, there was a challenging requirement to maintain operational capability in a chemical warfare environment. That meant poisonous gases had to be prevented from entering the crew compartment. Also, to avoid endangering ground crews by exposing them to harmful chemical residues, the avionics compartments had to be contaminant free, because they were accessed pre- and postflight.

A tradeoff was conducted to select a cooling system. The options were a Freon refrigeration system or an air cycle machine (ACM). Conventional Freon systems remove heat by evaporating liquefied Freon to a gaseous state. Air cycle machines circulate cold air as the heat extraction medium. Simplistically, ACM systems create cold air by passing engine bleed air through an ambient air heat exchanger to cool it. Once the bleed air has been cooled, it is then sent to the ACM compressor and then to a secondary heat exchanger, which again uses outside air as the coolant. Finally, the air passes through the ACM expansion turbine that powers the ACM compressor. Passing the air through the turbine extracts work from the air as it expands, which cools it to the required cold temperature.

The ACM system approach was selected primarily because of its inherent reliability against leakage failures. In a Freon system, leakages will ultimately result in complete loss of coolant and system failure. On the other hand, any leakage in the ACM system is naturally replenished by new air entering the system.

The onboard computer systems for the mission equipment avionics and flight controls all contained sensitive central processing units (CPUs) whose operational performance and reliability were directly related to their temperature. They required a lot of refrigerated air to prevent overheating and degradation. Cooling was accomplished by mounting the CPUs on aluminum standard electronic modules (SEMs) that functioned as cold plates. The SEMs were installed in racks that were connected to the ECS ducts, enabling the refrigerated air to blow over the cold plates and remove the heat generated by the electronics. In the event of ECS breakdowns, backup fans were used to provide limited ambient air cooling.

Perhaps the biggest struggle in designing the ECS was the constant battle to freeze the constantly growing heat load requirements (no pun intended). Calculating heat loads for the cockpit was relatively straightforward, but avionics cooling was another story. Determining the avionics heat load was exacerbated by the fact that most of the MEP detail design was deferred to later in the program. The lines of communication with the equipment suppliers were also somewhat convoluted, with interface documents crossing various company lines to get to the actual equipment supplier. Many of the equipment suppliers were only at the preliminary design stage, whereas the ECS was in detail design. In other cases, suppliers were not even selected, and heat load predictions were based solely on preliminary proposal documents. Inevitably, this resulted in a continually growing heat load as avionic configurations matured. ECS components were constantly being tweaked to meet the growing cooling demand.

In the end, the system met its spec requirement, but there was no cooling reserve. At one point, crew cooling vests were considered to reduce cockpit heat loads. Crew cooling vests were used on the F-18 fighter for the same purpose. Cooling vests bring the cool air directly to the bodies of the flight crew, which is more efficient than having to cool the entire cockpit. Having marginal capacity also meant that on very hot days, it would take a considerable amount of time to cool heat-soaked components down to their working temperature ranges. This was a particular problem for the targeting sensor and could cause a delay in readiness if ground cooling was not provided ahead of time.

THE PRESSURE SWING ADSORBER

To filter out hazardous chemical agents, a novel pressure swing adsorber (PSA) was incorporated into the ECS. Pressure swing adsorption processes rely on the fact that under high pressure, gases tend to be attracted to solid surfaces, or adsorbed. When the pressure is reduced, the gas is released, or desorbed. PSA processes can be used to separate gases in a mixture because different gases tend to be attracted to different solid surfaces more or less strongly. This was the first time that a PSA was used in an aircraft environmental control system.

Comanche used a PSA containing two chambers that were filled with porous media. Each chamber contained a series of molecular sieve beds that were designed to filter out threat chemical agents (Fig. 14.8). Pressurized air was passed through one chamber, to adsorb the noxious gasses. The clean air was then distributed to the cockpit and protected avionic compartments. A portion of the clean air also passed through the second chamber, at a lower pressure, and then vented overboard. The lower pressure air caused the filtered gases to become desorbed, thereby cleaning that chamber and allowing it to become an adsorber again. This process was continuously cycled back and forth to provide a steady supply of clean air. The PSA also lowered the humidity of the air by

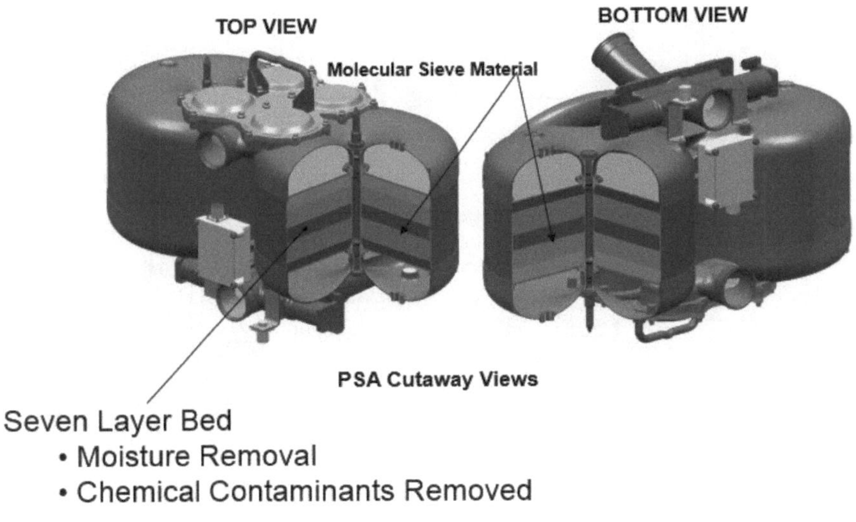

Fig. 14.8 Pressure swing adsorber.

removing water vapor. Simulants having the same molecular weights as the threat gases were used in the development and testing of the PSA.

Providing clean air to the cockpit and avionics compartments was only part of the protective measures needed to operate in contaminated environments. Means had to be provided to prevent contaminants from leaking into these compartments. This was achieved by pressurizing the compartments to a few psi, so that any leakage would go out instead of in. A dump valve was used to maintain a proper flow, accounting for some nominal leakage through the compartment door seals. However, during initial testing, there was excessive leakage through wire penetrations and other small openings in the pressurized zones that prevented the desired overpressure from being achieved. It took a lot of effort to provide adequate sealing for these sites to get the system to function properly.

ECS DEVELOPMENT ISSUES

For a helicopter, the Comanche ECS was fairly complex, containing numerous components and circuits controlled by valves and temperature/pressure sensors (Fig. 14.9). A diagnostics system was built in to facilitate maintenance and enable rapid troubleshooting. Without this capability, it would have been extremely difficult to isolate any faulty components.

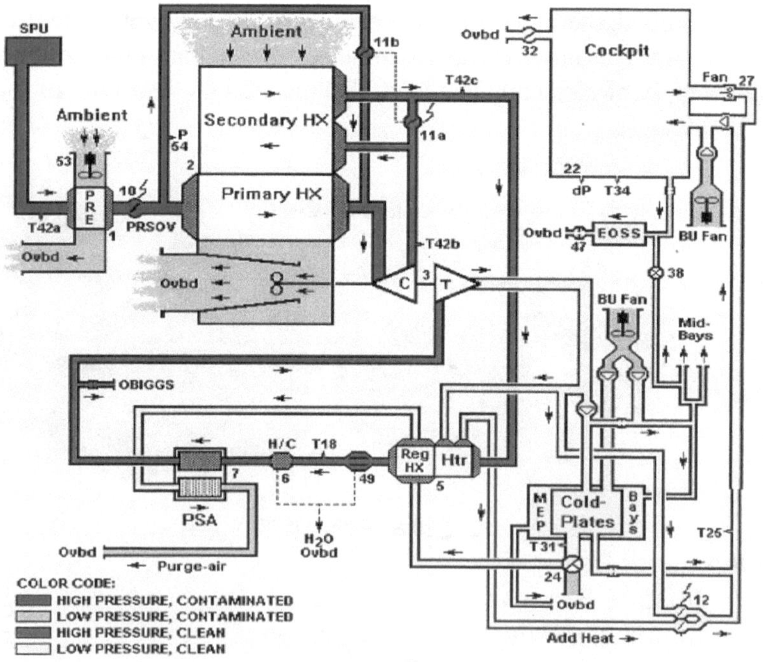

Fig. 14.9 Environmental control system schematic diagram.

Hamilton Standard was the prime subcontractor responsible for ECS design and development. It, in turn, subcontracted the PSA development to Pall Aerospace. Both companies used full-scale laboratory test rigs to develop their respective systems.

A ludicrous situation occurred when the system was being put together. A lot of flexible elastomeric couplings were used to attach the various components to interconnecting aluminum plumbing. The supply chain subcontracted some of these couplings to a fourth-tier rubber manufacturer in New Jersey. It turned out that this supplier's couplings were failing acceptance tests. A Sikorsky investigative team visited the supplier to find out why its parts could not pass pressurization tests at elevated temperatures. The supplier explained that it performed two tests—a pressure test, and then a temperature exposure test. It never conducted a simultaneous pressure test at elevated temperatures because it felt that was too severe a condition. Needless to say, its contract was terminated. Procurement was turned over to a competent supplier that was able to meet the pressure/temperature specifications.

On a more serious note, that was not the end of the coupling problems. Nuisance problems with bleed air and hydraulic line connections persisted during initial flight testing. Unrelated to the supplier coupling problem, failure of a bleed air pneumatic line coupling grounded the helicopter shortly before it made its first flight. The problem was related to fit and was rectified within a matter of days.

During this period, which was the Christmas season, there were a number of other propulsion/ECS system part issues that required replacement. To memorialize these events, the West Palm Beach test team created a "Manzolli tree" upon which they hung various propulsion/ECS components that had to be replaced. This was done in a good-natured manner, because the test engineers understood and respected the difficulties involved with putting together these complex systems. Furthermore, the tree decorations were mostly small parts with minor discrepancies that were quickly rectified.

Comanche was supposed to fly by the end of 1995. Ultimately, first flight was delayed until 4 January 1996, or in more face-saving terms, December 35th. These small hydraulic and bleed air nuisance leaks were the final straws that pushed the first flight into the next year.

Chapter 38 gets into Comanche technology transfer after the program cancellation, but we would like to say here that the failure of other organizations or programs to pick up on the pressure swing adsorber, a new and unique device to solve a difficult problem, is a real disappointment to those who brought it to a fully functioning system.

REFERENCES

[1] Stanley Kandebo, "Boeing Sikorsky Findings Underscore the RAH-66 Stealth," *Aviation Week & Space Technology*, 19 July 1993, pp. 22–23.

Chapter 15

The Transmission System

Bruce Kay

Development of the drive system, fig. 15.1, made use of major technology advances that were new to Boeing Sikorsky. Designing the final reduction stage of helicopter drive systems has always been challenging. Because the rotor turns at a relatively low speed (200 to 400 rpm), the torque in the final stage is very high. Simple spur or bevel gear reduction stages are impractical because of very high gear mesh tooth loads. This problem was typically solved by using a planetary final reduction stage with multiple planet gears, to transmit the high torque to the rotor through a number of separate gears.

A new concept of "split torque," which eliminated the complex planetary final reduction stage, was available. A tradeoff between a conventional planetary gear solution and the split torque system was conducted. The split torque approach was used on the Westland Lynx helicopter. In that approach, each of the two engines had a split path to the final reduction stage, resulting in four gears meshing with the final ring gear to transmit the high rotor torque without the use of a planetary. To help in this preliminary design trade study, Westland was given a small contract to review and critique the proposed LHX design configuration. The split torque concept won the trade and resulted in 50% fewer gears, 40% fewer bearings, and 12% less weight. Special high-strength gear steel was used to further reduce weight.

A lot of analysis was done to balance the torque loads transmitted by the second stage spur gears. Management of tolerances and stiffnesses was paramount to ensure an even splitting of the torque through the two separate load paths (Fig. 15.2).

As this design evolved, another advantage soon became apparent. Intermediate shafts in the drive system were vertical and became drives for the

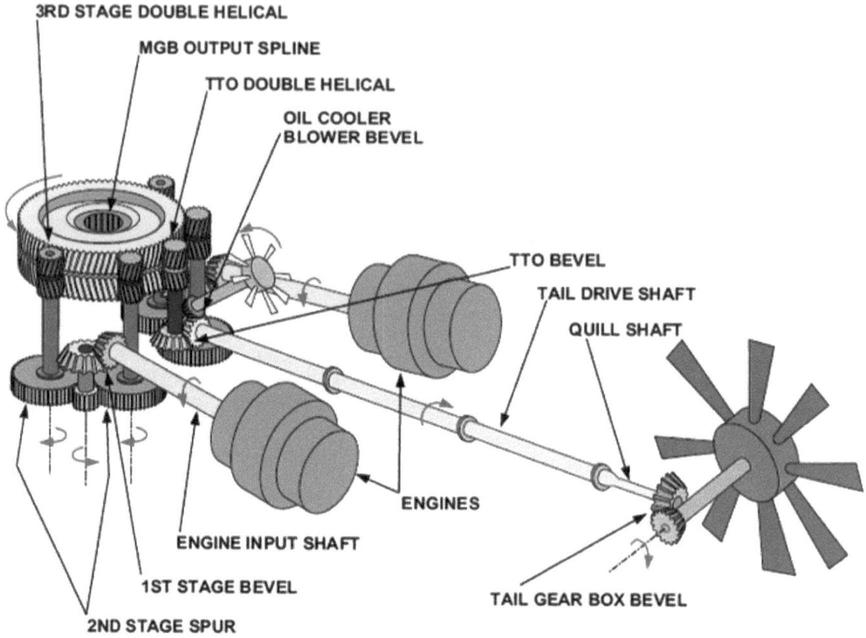

Fig. 15.1 Drive system schematic.

Fig. 15.2 Split torque main gearbox components.

hydraulic pumps and electric generators. This eliminated the need for an accessory gearbox and resulted in mounting the pumps and generators in a position where they were very accessible to maintainers. Also, because the main rotor shaft and main rotor servos were mounted on an airframe support fitting, the transmission housing did not have to react to high vibratory flight loads. The housings were designed primarily for stiffness, and to minimize weight, they were fabricated from magnesium castings instead of aluminum. A relatively new WE43 alloy was used, which was more corrosion resistant than previous magnesium alloys. The housings were relatively simple pan-shaped parts with bossed holes for retaining the bearings for the gears.

An alternative, resin transfer molded composite housings, was also studied. Prototype housings were fabricated as part of an Army MANTEC project (Fig. 15.3), but did not make it to the test aircraft. The composite housings would have saved an additional 40 lb.

Using a conventional roller ramp sprag clutch to disengage the 23,000-rpm engine output shaft from the transmission was a nonviable option. This disengagement was necessary during autorotation and single-engine flight. A new technology spring clutch evolved as a candidate solution (Figs. 15.4 and 15.5). The spring expands when the engine is the driving force and contracts when the rotor becomes the driving force. As torque is applied to the spring, it uncoils, expands, and locks against the outer housings of the input shaft and the input pinion shaft, forcing the pinion housings to turn with the input shaft. Release of torque from the input permits the spring to collapse back to the arbor, and the unit will free wheel. The spring was manufactured by laser

Fig. 15.3 Gearbox lower housing.

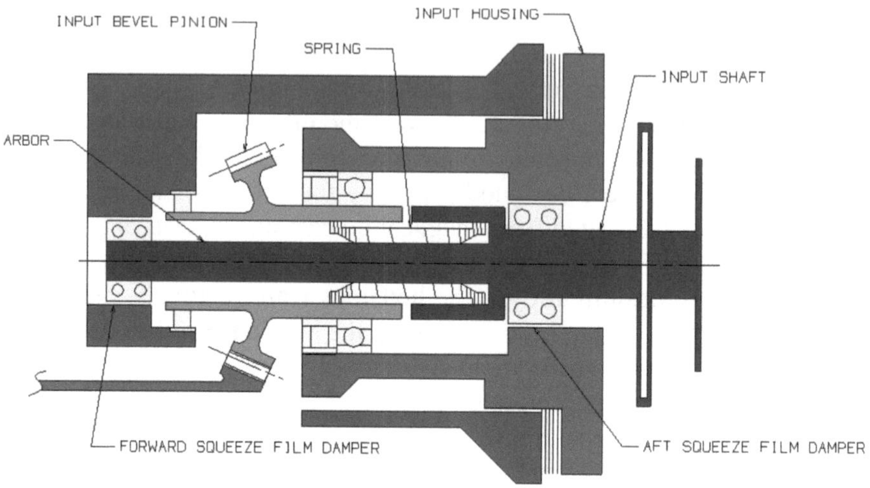

Fig. 15.4 Spring clutch freewheel unit.

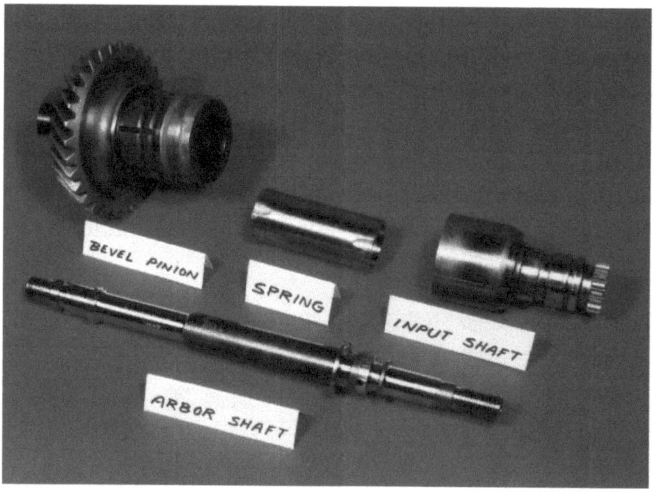

Fig. 15.5 Spring clutch components.

cutting a coil shape from tubular stock material. Although this approach was much simpler than the sprag clutch designs used on previous Sikorsky helicopters, considerable development and design iterations were required to get the system to function reliably. Weight growth over time resulted in a need for higher power, so the spring clutch then had to be redesigned.

The drive system contained a suite of sensors to monitor its health. The suite included oil temperature, oil pressure, oil level, vibration, and chip detectors located at critical sites within the transmission. Comanche also had a spec requirement that the transmissions would continue operating for 30 minutes after loss of lubrication. Theoretically, this meant that traditional preflight oil level checks were not required. The transmission sensor would alert the pilots if the oil was low, and 30 minutes of flight could be accomplished without oil. However, the maintainer community could not reconcile doing away with this historical safety check. It was never removed from their checklists, and a visual sight gauge was built into the gearbox to facilitate these inspections.

The drive shafts connecting the main transmission to the tail gearbox were made from braided graphite composite material. The major reason for selecting graphite was to match the thermal expansion between the drive shafts and the fuselage. Although aluminum could have been used for the drive shafts, differential thermal expansion would have necessitated the use of oversized couplings. The drive shafts were also designed to survive ballistic damage.

To simplify maintenance, there was a desire to allow the main transmission to be removed without disturbing the rotor system. Sikorsky traditionally integrated the main rotor shaft as part of the transmission. If the transmission had to be removed for any reason, the main rotor system went along with it, creating an unnecessary logistics burden. Furthermore, when the module was replaced, a check flight was required to make sure that the rotor was adjusted to minimize vibration. The Army's AH-64 Apache helicopter independently mounted these two components, and the Army found that approach beneficial. Comanche accommodated this request by hanging the transmission from an airframe bridge fitting, and then bolting the rotor mast to the top of the bridge fitting (see Fig. 15.6).

This approach was also a benefit for manufacturing; by preassembling the main gearbox, support fitting, rotor head, and controls as a module, and then bringing it to the aircraft on final assembly, it would simply be bolted into place. That was the intent. As the design matured, and subsystems were more fully defined, accessibility constraints for installing the dynamic module became more evident. Manufacturing started to raise concerns about installing some of the fasteners and requested some modifications. Because drawings for the long lead fittings and structure were already released, these requests for changes were met with resistance. Manufacturing continued to press, and raised even more concerns about the overall accessibility for installing the module. The module weighed over 700 lb, and it became evident that complex motions were required to slide it into place. The attachment between the bridge fitting and the fuselage included structurally efficient, tight-fitting, bayonet joints. New concerns about worker safety and damaging the adjacent structure were brought forward. Manufacturing was requesting changes that would require significant redesign as well as weight increases.

These concerns initially fell on deaf ears. Ed Fabian was the design lead for the mid fuselage, and he was naturally defensive. In this case, Ed had done everything by the book; he had met his weight targets and had the

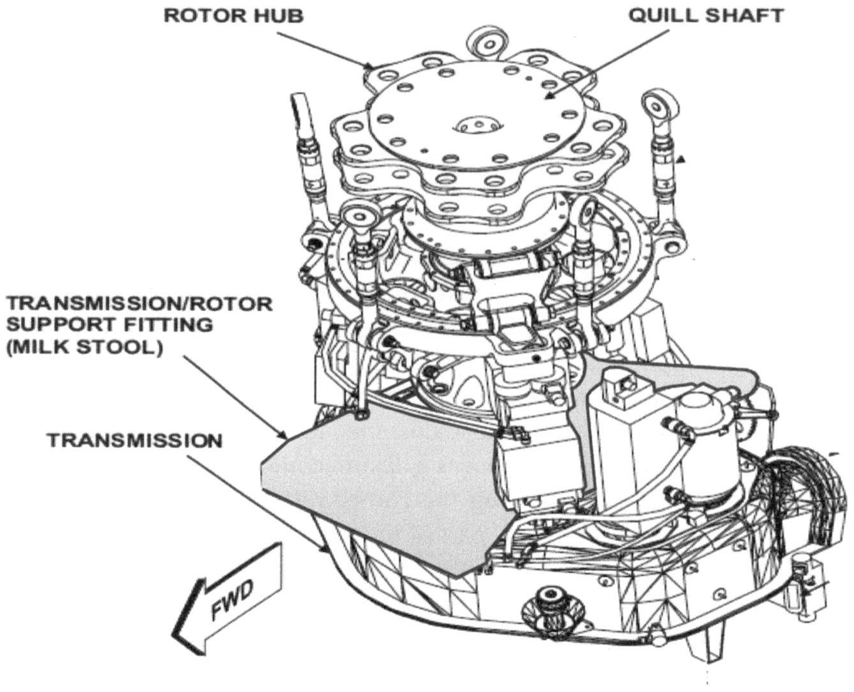

Fig. 15.6 Main transmission and rotor head module.

manufacturing folks sign off and approve his design. He had also used up any schedule slack and had no time or resources available for a redesign. Emotions ran high as the team polarized into factions. Escalating the matter to sympathetic functional managements further aggravated the situation. Comments such as "too late, too heavy, and impossible to build" proliferated. Engineering perceived the issue as a manufacturing inconvenience, whereas manufacturing viewed it as a show stopper that they would have to live with forever.

At this point, the issue had risen to the company president, who declared that the integrated product development teams (IPDT) had to solve its own problem. The IPDT then agreed to conduct a tradeoff study in earnest. It was agreed to start the tradeoff study using a configuration that manufacturing deemed optimum; however, this configuration was found to be very heavy. Eventually, though, synergism started to occur, and the design was iteratively refined toward lighter solutions. At the same time, detailed analyses of the manufacturing task, including crew loading, and labor hours were prepared. Shop demonstrations were also conducted to show the limitations of moving heavy objects with overhead cranes. These actions quantified the subjective manufacturing concerns, and they were then bought into by the entire team. Ultimately, some nominal changes to the fitting and structure were agreed to

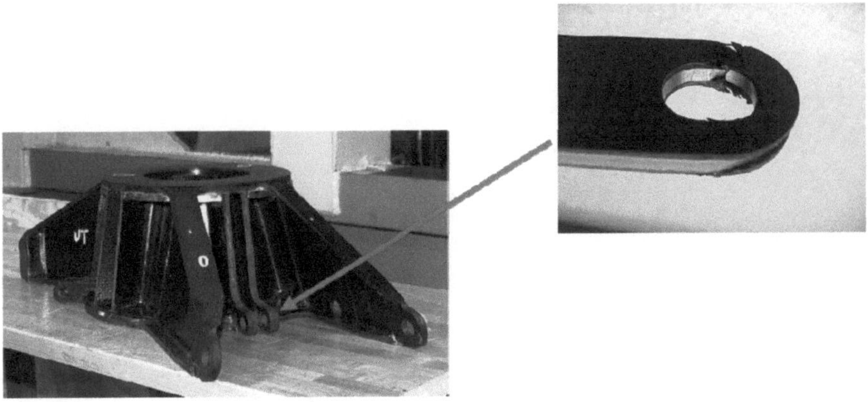

Fig. 15.7 Composite support fitting.

by the team as being the "right solution." The new configuration allowed the module to drop in from above onto an extended shelf on the structure.

The Department of Defense Advanced Research Projects Agency (DARPA) later initiated a program to develop a composite version of the titanium transmission support fitting (Fig. 15.7). The fitting was manufactured using a resin transfer molding process. It incorporated titanium shims in the servo lugs to maintain small clearances around this joint. The part was lighter and had a much lower cost than the baseline titanium fitting. The intent was to incorporate the composite fitting during subsequent production.

Chapter 16

The Hydraulic System

Bruce Kay

The hydraulic power system consisted of three fully independent subsystems (Fig. 16.1). Systems one and two exclusively powered the flight control servos and hydraulic permanent magnet generators (for flight-critical electrical power), and were identical in configuration and function. System three provided backup to systems one and two if either or both failed, and powered both systems during ground maintenance and preflight checkout. System

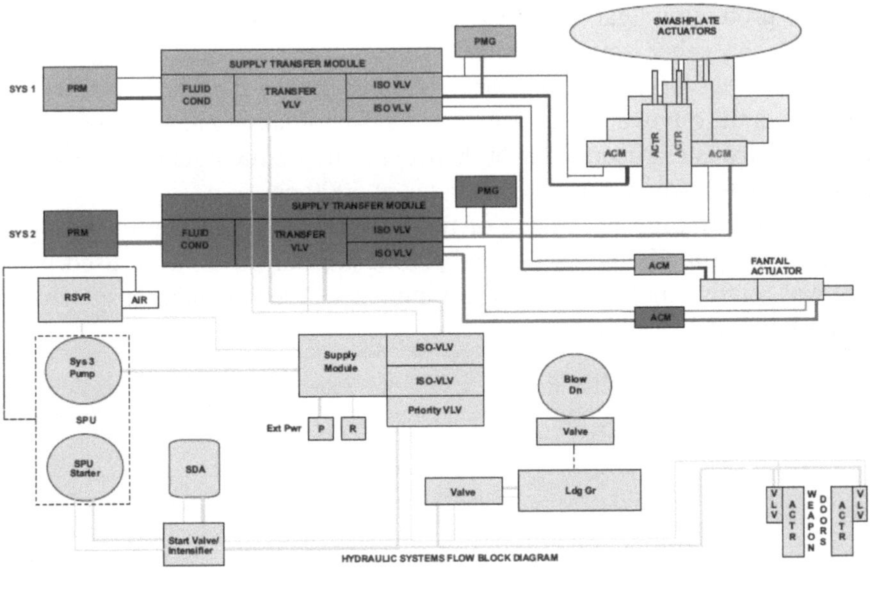

Fig. 16.1 Hydraulic system schematic diagram.

Fig. 16.2 Main rotor servo.

three also provided hydraulic power for utility functions, specifically landing gear operation and weapons bay actuation.

Systems one and two pumps were driven by the main rotor gearbox to supply hydraulic power whenever the main rotor was turning. System three was driven by the secondary power unit (SPU) gearbox. Therefore, powered operation of flight controls was assured on the ground and in all flight regimes, including autorotation. A hand pump could recharge the SPU hydraulic start accumulator in the event of a failed start or discharge due to prolonged inactivity.

The main rotor servos (Fig. 16.2) had dual redundant piston actuators that provided protection against ballistic threats. If one actuator was jammed or damaged, the other actuator could still control the rotor. Compensating hydraulic valves were used to balance forces between the dual actuators.

The hydraulic power system was a 3000-psi-class system; however, the SPU start subsystem operated at 5000 psi via a hydraulic intensifier. The landing gear blow-down system also operated at 5000 psi.

All rigid plumbing was made of titanium tubing and titanium fittings to minimize weight. Kevlar or braided stainless steel hoses were used where there was relative motion between components. A leak detection system was also built into the system.

Chapter 17

The Electrical System

Frank Schrader

History of Electrical Power Systems

In the early days of minimal electronics, aircraft systems were powered by simple 28-V direct current (DC) batteries. As power needs increased with more sophisticated systems, generators were installed to provide continuous DC power. Then more efficient 115/220-V alternating current (AC) power generators were installed to satisfy larger loads, with converters to supply 28-V DC for existing battery-powered systems.

Most of the essential systems like communication radios and engine controls used battery power, because engine starts required instruments and radios, and auxiliary power units were rarely installed on small aircraft. Eventually, military standards were created for most subsystems; MIL-STD-704 was created to standardize the power quality generated by the electrical power system and provided to military subsystems.

Until the 1980s, the standard power systems were 28-V DC and 115/220-V, 400-cycle three-phase AC systems. MIL-STD-704 defined acceptable "continuous" and "transient" power requirements; for instance, the 400-cycle waveform was allowed to vary no more than 5% in frequency during normal operation and had to control harmonic frequencies to minimize electronic interference.

As mission subsystem complexity increased, redundant power generators were installed, and they were wired through backup relays to provide critical power in the event of a single power generator failure. MIL-STD-704 also covered the transient nature of relay switching between power sources and allowed up to 50 milliseconds for "no power" and up to 7 seconds for "backup power outage."

The most weight-efficient AC generators were direct drive systems off the engines or transmission. The sources were geared down to 400 cycles per second and were designed for proper quality when the rotor rpm was 100%. Typically, they would come on line at 85% rpm and were allowed to vary transient voltage and frequency tolerances until steady state was reached.

The Comanche Approach

The Comanche electrical power system was architecturally similar to previously fielded systems. It contained multiple primary power generators for failure tolerance, multiple transformers for secondary power levels needed by specific equipment, and many relays for power isolation between the generators and for power sharing if generators failed. It had a battery to start the aircraft, an external power receptacle for ground maintenance without running aircraft engines, and extra power sources for the fly-by-wire flight control system to provide ultra-reliability.

Unlike earlier designs, however, Comanche was a 270-V DC system. Given the proven reliability of the traditional power systems, why would Comanche change to 270-V DC? The first and most important reason was weight; second was computer intolerance to transient power outages; and third was rotor speed flexibility.

Comanche was not the first aircraft to apply 270-V DC. Both of the other advanced Air Force and Navy stealth aircraft also chose to employ this approach, but that was not obvious to many of us on the Comanche program at that time.

The Benefits

Weight

The first weight savings were due to driving the generators at higher speed, because we no longer needed gearbox-reduced 400-cycle AC power. They could be designed for optimum power conversion with a direct feed off the transmission. Additionally, we decided to tap the cooling oil reserves of both the secondary power unit (SPU) and transmission to maximize internal heat transfer and not rely on extensive heavy ambient cooling fins. Third, we immediately converted all power to 270-V DC at the generator, reducing the amount that had to be converted to 28-V DC at the old 115-V AC/28-V DC transformers. Fourth, we insisted that most of the installed subsystems use 270-V DC directly, which forced most manufacturers to change their input power supplies to use 270-V DC SEM-E form factor cards instead of 115-V AC power supplies. This was weight-neutral for most of the suppliers, but an overall benefit to the 270-V system.

Lastly, we reduced wire gauges, because the higher voltage required less current for the same power transfer. And because the DC power created less eddy currents, we were able to use the aircraft as a ground for many subsystems, further reducing wire weight.

COMPUTER TOLERANCE

What do most computers do when exposed to a short power outage? They reboot. Because we had a centralized, rather than dispersed, data processing architecture, a long reboot effort would disable many critical displays and processors. There are two solutions to this. The first and heavy solution is to build extra battery capacity into every subsystem to sustain the equipment over the expected transient power outages. The second solution was to eliminate power transients. We chose the latter.

The primary reason all backup power relays "break before make" is that the 115-V AC power generators are not phase-synchronized, and the voltage differential between "ganged" generators could cause large current surges and unknown damage into a shorted line. But with two DC lines, there are fewer imbalances and minimal current flow between two "operating" generators. There are two conditions that might cause a backup relay to operate. One is a failed generator resulting in zero power; the other is a shorted feed line that overpowers the generator and drops it offline. We "make before break" with the 270-V relays and design them to withstand enough surges to determine if the other generator is failed or feeding into a short. If feeding into a short, the relay resets and disables that bus; however, if the generator fails, it restores power to the bus without excessive transients.

ROTOR FLEXIBILITY

Rotor speed variations allowed a wide degree of aircraft performance tuning for special conditions, such as air density compensation, aggressive maneuvering, or acoustic signature variation. By having no AC power frequency tied to rotor speed, the 270-V DC system allowed a greater rotor speed range than other helicopters.

OTHER DESIGN INNOVATIONS

Apart from the use of 270-V DC, emerging (in the 1970s) solid-state devices allowed replacement of electromechanical relays and circuit breakers. Solid-state light-emitting diodes (LEDs) furthered the use of fiber optic transmission lines for high speed data, which saved weight compared to copper wires and increased the capacity of the lines, negating parallel wire runs to achieve the same data rate.

These solid-state devices also provided the avenue for smarter diagnostics throughout the aircraft.

THE BASIC DESIGN

The elements of the electrical power system were:

- 270-V generators
- A 28-V DC battery
- DC-DC transformers for additional 28-V DC
- Solid-state circuit breakers to route the power to individual subsystems and provide protection to the aircraft in case of subsystem or wire failures
- Solid-state power relays to provide maximum backup power in the event of generator failures
- An external power receptacle for power cart connection

Every element, including the battery, became a new design.

After numerous trade studies and integrated product development team inputs (particularly from the diagnostics and flight controls groups), a robust, redundant, but lightweight architecture evolved. The triplex fly-by-wire flight controls required a triply redundant generator system, and the singular secondary power unit drove two additional hydraulically driven power sources dedicated to the flight control system. The circuit breakers were housed with the generator controllers in three units called power system controllers (PSCs). The contract for the development of the three generators and three PSCs was awarded to a Sundstrand Aerospace division in Lima, Ohio.

The final design elements and their suppliers were (see Fig. 17.1):

• Two main gear box 270-V DC generators (15,000 rpm)	Sundstrand
• One SPU 270-V DC generator (26,000 rpm)	Sundstrand
• Three power system controllers	Sundstrand
• Two hydraulic permanent magnet generators (HPMGs)	Hamilton Standard
• One 28-V lithium ion battery	Supplier competition
• Two 270–28-V DC transformers	Sundstrand (within PSCs)
• Seven solid-state power relays	Sundstrand
• One external power connector	(probably Ampex)

The power system controllers were the brains of the system (Fig. 17.2). Each provided local control of their respective generator and provided redundancy management and load shedding/load sharing capabilities. All the smart circuit breakers were also housed in the PSCs. At an early development stage, the two separate 270-V DC–28-V DC converters were also resized and carried in the PSCs. Until that time, we had three identical PSCs, which aided logistics but required careful balancing of circuit breakers (multiple on each SEM-E module) to minimize part number modules and retain commonality among PSCs. Eventually, with the elimination of intermediate-level maintenance and the ability to remove SEM-E modules in the field, PSC commonality was not an issue. We still did maintain only two configurations of the PSC.

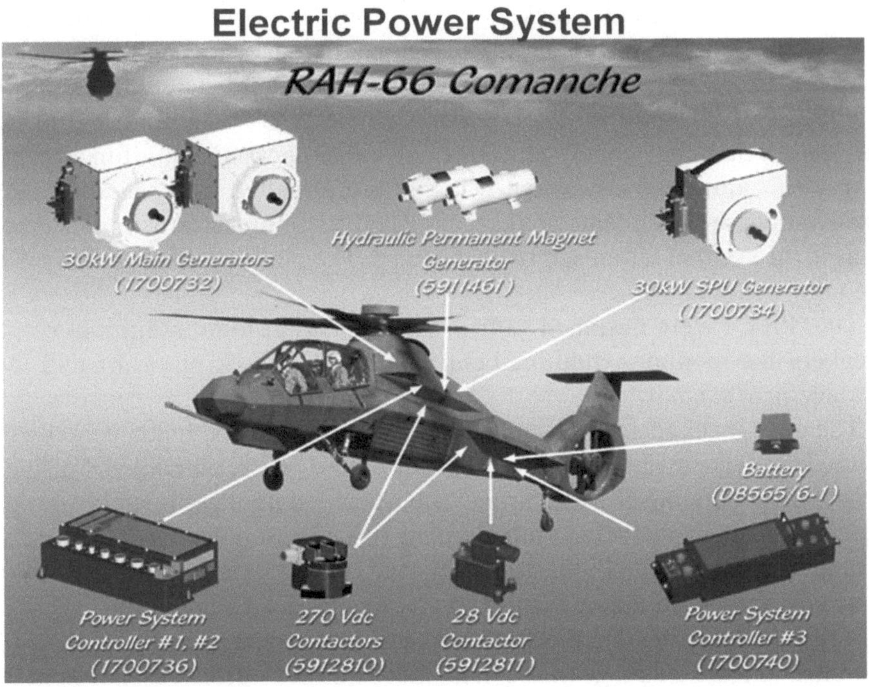

Fig. 17.1 The Electric Power System.

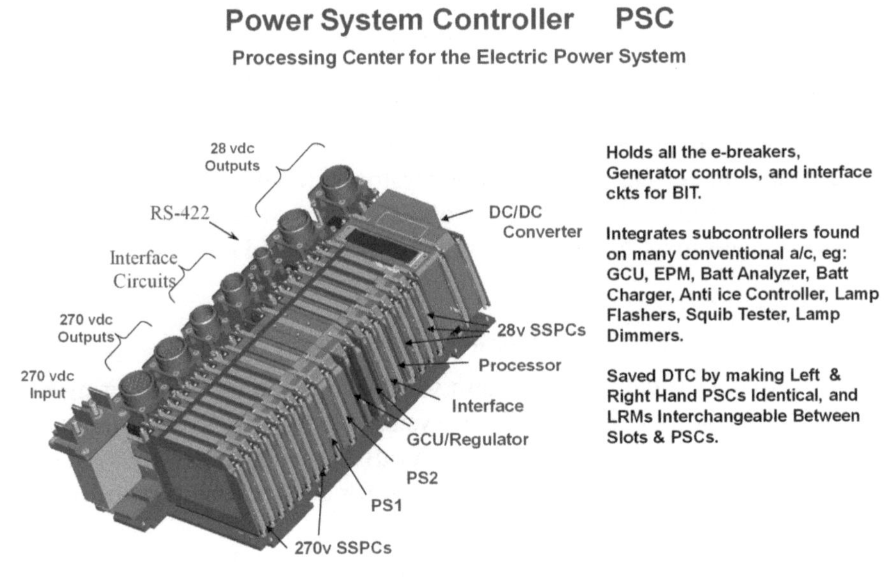

Fig. 17.2 Power System Controller.

The traditional circuit breakers were located in the cockpit overhead for easy access/visibility to the pilots in case one tripped or was pulled to disable a circuit. The Comanche replacements were called solid-state power controllers (SSPCs) and were remotely located in the PSCs. The pilots could still manipulate them and check status through either cockpit's multifunction displays. The additional advantage was that preformatted load shedding logic (mission phase dependent) could be enacted to optimize remaining subsystems in the event of failures or battle damage. The pilots could always override the logic as circumstances changed.

The SSPCs were designed to trip 10 times faster than traditional circuit breakers and monitor actual load currents for diagnostic reporting and load management reasons.

The three main generators were all 30-kW machines and fundamentally the same. They were maintenance-free, 12-pole, three-phase, oil-cooled, brushless generators, operating at 24,000 rpm. Their power output per pound of weight was four times that of standard high maintenance air-cooled brush generators operating at 6000 rpm. All the generators had small gear boxes with oil ports to mate with the transmission and SPU and a frangible shaft to eliminate generator seizures from affecting the drivers. The two main generators were vertically mounted on the main gearbox, and the SPU generator was horizontally mounted with a different gear reduction (and a different part number). The SPU provided power at 85% startup rpm, and the mains kicked in at 92% of rotor speed. In flight, the rotors could dip to 75% during autorotation, and the generators would still produce 50% of their rated power for 5 seconds. They could also be overloaded by 33% for 5 seconds without damage.

In the fly-by-wire system, the flight controls needed power to control the rotor as soon as the rotor started. Normally this would be provided by the SPU, which provided sufficient power to start all mission systems as well. As backup, the flight controls could be operated on battery for a short time, or an external power cart could be used.

The hydraulic permanent magnet generators (HPMGs) were provided for additional reliability. The flight controls' electric power reliability requirement was greater than the three generator systems could provide.

DEVELOPMENT HURDLES

Designing the electrical power system was truly a systems engineering task. MIL-STD-704 did not address 270-V systems, and every component was a new design. Comanche avionics and aircraft subsystems mostly needed to accept 270-V power. Exceptions were made for a limited number of off-the-shelf systems operating from 28 V.

The electrical integrated product development teams (IPDT) lead was Dan Dellaripa, who was an excellent manager and motivator. The systems design manager was Mike Bradford, whose skill at balancing engineering solutions with safety was critical to the design. Tony Sedowski ably led the implementation engineering crew.

Of course, the overriding tradeoff variable was weight. The constraints were to package everything for two-level maintenance and maximum life or lowest, design-to-cost, DTC. Thus, the packaging group strove to put all related electronics into the triple power system controllers. This included generator controls, smart circuit breakers, the 28-V DC supplies, and all interface circuits. Two-level maintenance drove the circuits into SEM-E module formats, which was no easy task for some heavy heat generators, like the 28-V DC converter. Additionally, the SEM-E connectors were all high density arrangements (300–600 pins). We used flexible flat-ribbon wiring within the PSC backplane to provide the multiple solder connections to each module.

The actual PSC box was really an enclosure designed to permanently stay in the aircraft, while individual failed cards were replaced. Special emphasis was placed on isolating functions on the modules to ensure that fault detection could be isolated to a single module to meet the Comanche diagnostic requirements. The concept never was verified with the PSC diagnostics, except analytically, because we never had a module failure in Dem/Val!

The Dem/Val generators also performed well, but we did have some frangible drive spline failures—believed to be due to the difficulty in defining transient startup torques. We eliminated the problem by either beefing up the spline or adjusting the starting time sequence.

The make-before-break relays were a new innovation, preventing transients and arcing. The solid-state nature of the relays added to the overall diagnostic capability of the Comanche. The relays had to hold up under a dead short for a short period to verify whether a generator had failed or had tripped due to a short. So that's 30 kW of power flowing through it. I remember Tony Sedowski's cheer when we passed the qualification test.

Tony was not as happy with the HPMG qualification. The problem lay in the weight constraints, which minimized the metal containers. We failed quite a few overpressure tests (4500 psi for the nominal 3000-psi hydraulics)—1000 psi more than the standard helicopters—before a successful total redesign. Naturally, all our vendors were struggling with the unknowns of higher hydraulic pressures and higher 270 voltages.

Sikorsky also took the lead for supplying 270-V power supplies (in SEM-E format) to any avionic design house that had to redesign its units for Comanche two-level maintenance. This included air vehicle interface control system (AVICS), flight computers, helmet integrated display and sighting system (HIDDS), mission computer clusters, and Target Acquisition Designation System/Pilot Night Vision System (TADS/PNVS). We competitively picked a single vendor, but it struggled with arcing between traces on multilayer boards carrying the 270 V. There were quite a few design iterations before proper isolation could be achieved for the rigorous qualification tests (which were more stringent than typical military qualifications).

When we embarked on engineering, manufacturing, and development (EMD), we ran into several problems due to the long delay between Dem/Val and EMD production lines. The Sundstrand (now Hamilton-Sundstrand) generators were originally hand wound by "little old ladies who had since retired," and the replacement staff didn't have the same techniques. So the production line struggled to meet the original specifications.

AIRCRAFT WIRING

Sikorsky's expertise in computer-designed wiring harnesses was well known and eliminated the need for physical mockups of the wiring before committing to production harnesses. In the 1980s, this was considered state of the art.

However, Comanche pushed the technology envelope in wiring in several areas: connectors, wire gauge reduction, and fiber optics. Battlefield repair was another technology reach.

For weight reduction, we reduced wire gauge below the previous Mil Spec recommendations of 22 gauge. This reduced many wires to single strands and required new tools to attach the wires to connector pins. We also had to consider new strain relief and clampihg methods for the lighter wiring bundles. We also committed to composite materials in the aircraft connectors before they were universally accepted. Sikorsky EMI engineers had to approve new coatings to enable shielding effectiveness.

FIBER OPTICS DEVELOPMENT

At the outset of Comanche, there were limited applications of fiber optics in lieu of traditional wiring. The only use we were aware of was a Collins dedicated control path between an HF transmitter-receiver and power amplifier, and our thought was that the application was only for electromagnetic interference reduction. Comanche made extensive use of fiber optics in high-speed data buses and digital video buses, where data speed requirements from the central data processors exceeded wire bus (MIL-STD-1553) frequency limits. Sikorsky developed an in-house expert who taught the rest of the companies how to install and maintain fiber optics in an Army environment. This engineer was so respected that he later moved to NAVAIR to spread his knowledge.

The most significant development was that of a fiber optic field repair kit. At that time, all fiber optic terminations were polished in "clean room" environments; Sikorsky was able to demonstrate that battle damage field repairs were feasible.

Another significant new development was fiber optic couplers. These devices allowed the transmitted light information to be split and sent to multiple receiving systems. This allowed for redundancy in the displays and central processing. The design was achieved by putting multiple glass strands into a bundle and then melting them to achieve multiple paths from the singular input to the outputs. Again, Comanche was at the forefront of these technology innovations.

Overall, the electrical power system was a great technical achievement. It was amazing to see how our electrical system reliability-driven design paid off in the end.

Chapter 18

Reliability and Maintainability

Steve Horling

Reliability, availability, and maintainability (RAM) was always of paramount importance from the initial envisioning of LHX. RAM's roots were imbedded in the weapons systems cornerstones of combat sustainability and affordability.

In order to be combat sustainable, a weapons system must be available to launch and must be able to complete the mission. Both of these criteria required LHX to be fully or at least partially mission capable to be combat sustainable. As a result, LHX RAM focused on the percentage of time the weapon system's mission essential functions (MEFs) are up and running and available for use. This focus was a radical departure from traditional RAM at the time, because availability was typically determined using mean time between failures (MTBF) and mean time to repair (MTTR) parameters. Neither MTBF nor MTTR focused on MEFs. The MEF RAM characteristics are primary inputs to combat sustainability models, which in part determine the force structure or fleet size. Consequently, MEF-focused reliability is critical to the robustness, and MEF-focused maintainability is critical to, in part, the ability to rapidly restore lost or degraded MEFs.

In seeking congressional approval for the program, RAM played an underlying early central role in the process. Wisely, from the onset of LHX, the U.S. Army recognized that it takes more than just lofty RAM parameters. A plan for success was needed that incorporated lessons learned. Previous weapon systems had also planned for RAM success and incorporated lessons learned with mixed outcomes. Why would LHX be significantly different?

THE DIFFERENCE IN THE LHX PROGRAM

By replacing traditional MTBF and MTTR analytics with MEF-focused requirements, the aircraft design was driven towards meeting the desired combat sustainability.

OPERATIONAL VS INHERENT RAM

MTBF and mean time between aborts (MTBA) were based on inherent reliability, which includes only select equipment hardware failures and, as a result, underestimates the total unscheduled maintenance demand. LHX characteristics would be operational parameters that included inherent, induced, and no-defect maintenance. Induced includes maintenance generated by external influences such as maintainer breakage. No-defect includes maintenance where no fault was found. Maintenance generated due to software anomalies was also included as operational. Simply stated, any time the aircraft was down for unscheduled maintenance for whatever reason was considered operational. This approach captures what actually occurs in the field, which was lost using inherent RAM parameters.

MAINTENANCE NOT FAILURES

Combat sustainability and support costs are driven in part by maintenance. The MTBF and MTBA parameters counted failures, not maintenance. As a result, new RAM parameters were introduced for LHX: mean time between essential maintenance actions (MTBEMA) and mean time between mission-affecting failures (MTBMAF). These parameters were operational and were based on maintenance demand. In addition, the traditional MTTR parameter was redefined as mean time to repair essential (MTTRe), which only included maintenance for essential maintenance actions.

TWO VS THREE MAINTENANCE LEVELS

The existing U.S. Army maintenance system at the time consisted of three maintenance levels: aviation unit maintenance (AVUM), aviation intermediate maintenance (AVIM), and depot. The AVUM level provides on-aircraft unscheduled and scheduled maintenance. On-aircraft maintenance tasks that require higher skill levels are provided by AVIM along with repair of select line replaceable units (LRUs). The intermediate level is generally in deployed mobile facilities that house diagnostic equipment, tools, spare parts, and repair materials. In order to increase mobility by reducing the logistics tail, and to reduce support costs, LHX would move to a two-level maintenance system. AVIM would be eliminated, and AVIM maintenance tasks would have to be either eliminated or absorbed by AVUM or depot.

Design Influence

An age-old question is how to translate the alphabet soup of RAM parameters into actual design. RAM engineers have traditionally contributed to the design process with design guides, board leaning, tradeoff analysis, and reports detailing the level of specification achievement. But LHX provided a new opportunity to move RAM deeper into the design process, into the integrated product development teams. As such, RAM engineering was now partnered in a collaborative environment up front in the real-time design process. The new mantra was "design influence" in lieu of a more traditionally viewed role of design analysis.

RAM was focusing on the ability of the weapon system to meet the primary sustainability and affordability goals. By the mid-1980s, U.S. Army Training and Doctrine Command (TRADOC) released the first draft of the LHX RAM rationale report that began to solidify for contractors the definitions of the new operational RAM parameters. The release of the detailed LHX RAM scoring criteria document finally resolved any outstanding contractor operational parameter definitional unknowns.

LHX Requirements

LHX/Comanche had a plethora of RAM requirements, the foremost of which were MTBEMA and MTTRe. These parameters were considered the most important because they are the two variables used to determine operational availability. The third availability-determining parameter, administrative and logistic delay time (ALDT), was specified as a given fixed value in the Performance Weapons System Specification (PWSS).

The following is an overview of the PWSS RAM requirements. All parameters are "operational" unless otherwise specified:

- *Mean time between essential maintenance actions (MTBEMA):* Reflects unscheduled maintenance actions for the loss or degradation of one or more MEFs, regardless of when the loss occurs and independent of the mission at hand. Therefore, the weapon system would be either not mission capable or partially mission capable. Independent of the mission at hand means that if an MEF associated with mission *B*, for example, is lost or degraded while on mission *A*, it's still an EMA. EMAs are a subset of all unscheduled maintenance actions (UMAs). The Dem/Val PWSS required LHX to demonstrate a lower 90% confidence MTBEMA of not less than 4.5 flight hours by the conclusion of the two-year RAM validation period following initial operational capability (IOC).
- *Mean time between mission-affecting failures (MTBMAF):* Reflects unscheduled maintenance actions for the loss or degradation of one or more MEFs occurring during the mission and independent of the mission

at hand. Failures resulting in a loss of redundancy are not MAFs as long as there was no degradation or loss of the MEF. MAFs are a subset of EMAs. The Dem/Val PWSS required LHX to demonstrate a lower 90% confidence MTBMAF of not less than 8.5 flight hours by the conclusion of the two-year RAM validation period following IOC.

- *Mean time to repair, essential (MTTRe):* The MTTRe parameter is similar to the traditional MTTR parameter, which represents user-level elapsed unscheduled maintenance time. However, MTTRe includes only essential maintenance actions; higher-level rollups are a weighted average based on EMA rate. The Dem/Val PWSS required LHX to demonstrate a MTTRe of not more than 0.86 hours by the conclusion of the two-year RAM validation period following IOC.
- *Maintenance ratio (MR):* The MR is also similar to past traditional maintenance man-hour per flight hour requirements and is defined as the sum of direct scheduled and unscheduled maintenance man-hours at a specific level of repair divided by the corresponding sum of flight hours (DMMH/FH). The Dem/Val PWSS required LHX to demonstrate a user-level MR value of not more than 2.6 DMMH/FH to a 90% confidence level by the conclusion of the two-year RAM validation period following IOC.

In addition to the quantitative RAM requirements listed, the Dem/Val LHX PWSS contained an extensive array of additional RAM requirement:

- Mean time between unscheduled maintenance actions
- Flight safety reliability
- Mean time between removals (rotating/dynamic components only)
- Time to first removal (rotating/dynamic components only)
- Mean time between removals (specified components)
- Self-deployment probability
- Training equipment and PSE/nonair vehicle equipment RAM
- Mean time to repair
- Mean time to repair depot level
- Maximum time to repair
- Engine access time
- Maintainability RAM with MOPP/Arctic gear
- Battle damage assessment and repair RAM
- Availability
- Formal RAM demonstrations at EUTE and IOTE (select parameters only)

RAM DESIGN

The RAM mantra instilled by the Comanche program manager was “design influence” via the IPTs. The quantitative PWSS RAM requirements were of little direct use to the design engineer. Rather, through a collaborative effort within the IPDT structure, the RAM engineer provided guidance, evaluated alternate designs, and contributed to tradeoffs in order to provide the design feedback necessary to achieve specification compliance.

The RAM influence resulted in the designs described in the earlier chapters. It's important here to describe how they also greatly influence aircraft maintainability.

AIRFRAME

Reducing weapon system downtime due to maintenance increases the system's availability and reduces maintenance costs, both of which were top LHX goals. Early in the program, access to components for replacement and/or repair was of prime importance. Ease of access criteria were established early in the program. This provided design guidance based on the anticipated frequency of access to any given aircraft area and/or equipment. Areas requiring the most frequent access were granted the easiest access, and those infrequently accessed were low on the accessibility priority list.

Traditional monocoque airframe construction did not lend itself to lightweight penetrations in the airframe skin needed to meet the maintainability access goal of walking up to the aircraft, opening a quick release hinged access door, and easily replacing the bad component. The Comanche's keel beam airframe design provided a lower weight solution for airframe penetrations; as a result, external, hinged, quick release access doors, some of which doubled as work platforms, became a reality. Figures 18.1 to 18.4 depict Comanche's extraordinary access features.

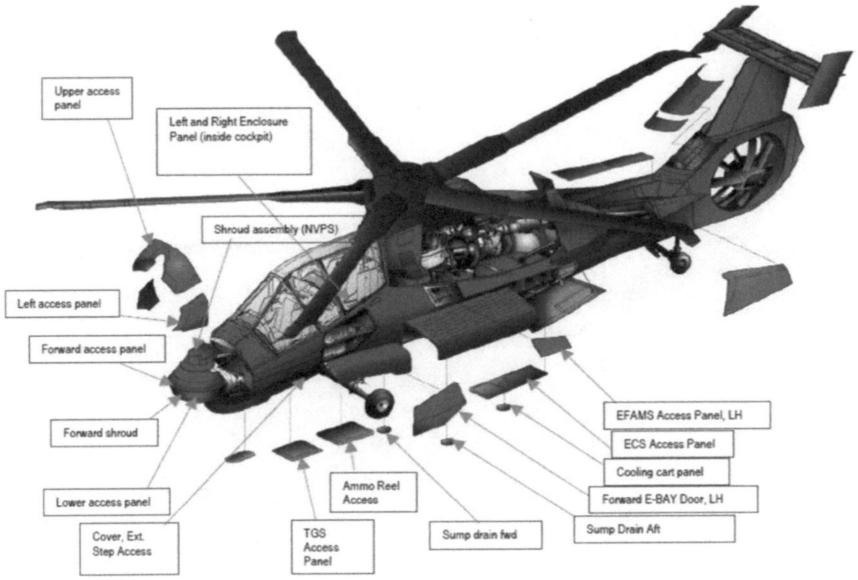

Fig. 18.1 Comanche exterior maintenance access panels, view 1.

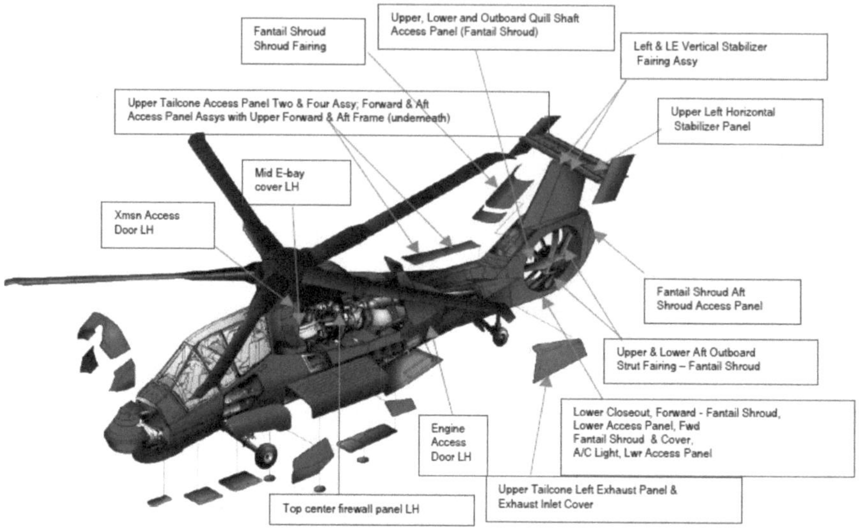

Fig. 18.2 Comanche exterior maintenance access panels, view 2.

The weapons bay door panels doubled as work platforms to access the main gearbox, main rotor head, and engines. These door panels could be easily unlatched from the weapons door structure and raised to various positions by maintainers and ground crew for platform use. The MEP bay doors when open also served as maintenance platforms to access secondary power unit (SPU)

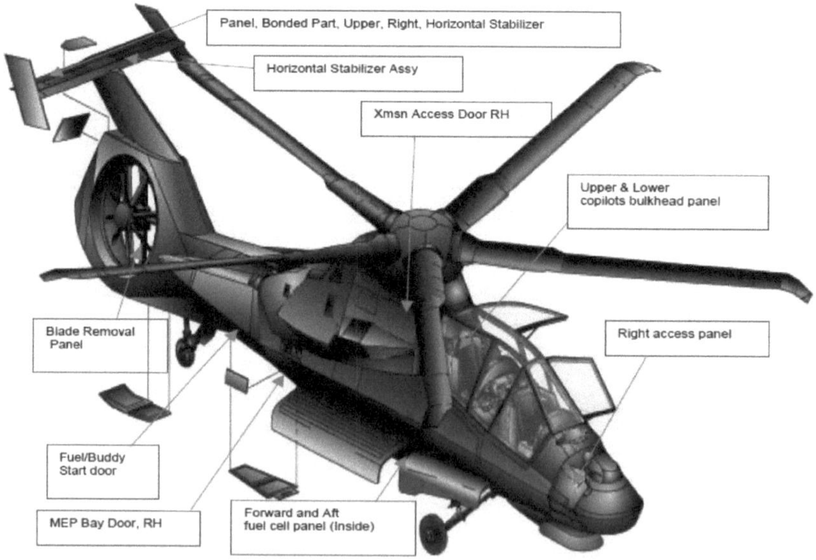

Fig. 18.3 Comanche exterior maintenance access panels, view 3.

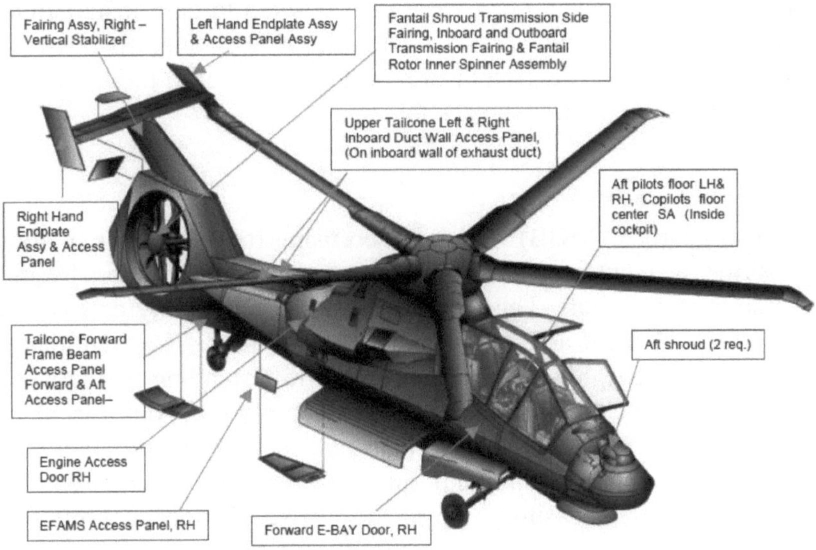

Fig. 18.4 Comanche exterior maintenance access panels, view 4.

and environmental control system (ECS) compartments. A tailcone work platform was provided by a walkway extending from the empennage forward to the main rotor cowl.

AVUM-level composite material repair kits were being developed to facilitate routine and battle damage repair. Adhesives with rapid cure times and/or those that could be easily accelerated were to be included in the kits. Compact portable support equipment was also envisioned such as heated vacuum bags to support rapid repair.

In addition, a limited number of common in-inventory standard tools were defined early in the LHX concept phase to provide design guidance for AVUM-level fastener selection.

LANDING GEAR

Landing gear components including fastening hardware and servicing ports were easily accessible with the gear extended. Sensors embedded within the system provided status and health to the flight crew and fault detection/isolation capability to the maintenance crew. In addition, sight indicators were provided for pressurization along with airframe jack pads for landing gear elevation during maintenance.

FLIGHT CONTROLS

All flight control hardware was designed as line replaceable units with Murphy-proofed design to prohibit incorrect installation along with electrical connector keying. Service loops were incorporated to ease installation and avoid induced damage, and were routed to minimize chafing. Extensive diagnostics were built into the system to report status, health, failure detection, and fault isolation to the line replaceable unit.

THE DRIVE SYSTEM

The main gearbox (MGB) was accessed using the weapons bay doors in open position as maintenance platforms to access the MGB doors located on either side of the aircraft. The two-stage oil filter assembly consisted of a disposable 3-μ element and a 75-μ cleanable element. The oil filters were accessible from the right side of the aircraft while standing on the weapons bay door maintenance platform. The filter assembly had both a local visual impending bypass indication and an electronic cockpit-reported bypass indication. Sensors monitored input and out pressures of the filter assembly. Oil pressure drops were measured across the heat exchanger as well as heat exchange oil temperature. A remote oil level sensor and oil level sight gauge were provided. The MGB had an inductive debris monitor and three magnetic plugs to facilitate evaluation of system status. Two additional strainer magnetic plug chip detectors monitored the return flow from the oil-cooled generators. A fourth chip detector monitored the return flow from the main rotor bearings. The sensor information was integrated within the computer clusters to report system status, health, and fault detection/isolation to the LRU level.

The tail rotor drive shaft consisted of four sections (Fig. 18.5) with sections 1, 2, and 3 being interchangeable. The fourth section was a quill shaft

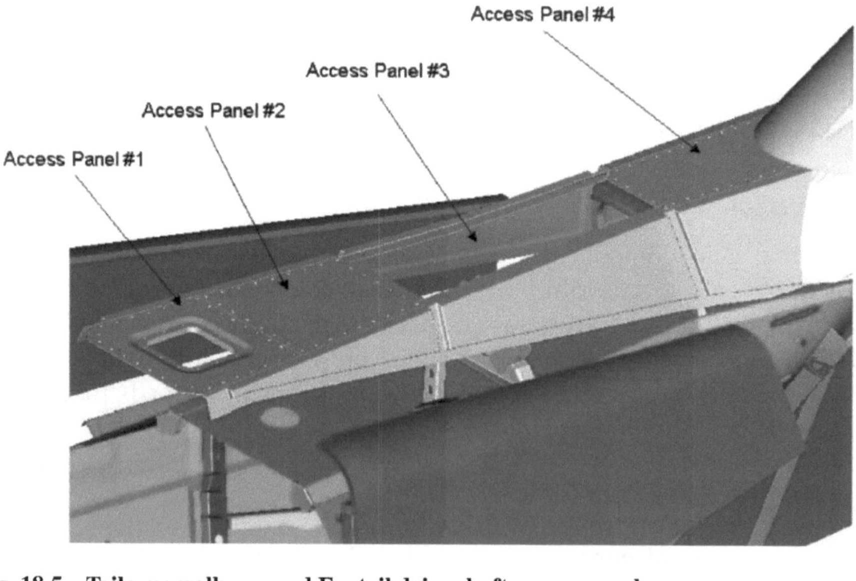

Fig. 18.5 Tailcone walkway and Fantail drive shaft access panels.

connecting the Fantail gearbox. The shafts were supported by elastomeric mounts attached to the airframe via bolt and nut plates to facilitate single tool interchange.

The Fantail gearbox was splash lubricated with sensors that provided debris monitoring as well as high-temperature and low-pressure indications to the flight/maintenance crew. The oil drain plug and the debris monitor were located along the underside of the gearbox. Each sensor could be individually removed and replaced on-aircraft at AVUM. A locally mounted sight gauge provided visual indication of oil level and did not require removal of any fairings or panels for viewing.

T802 ENGINES AND ENGINE INTERFACE/INSTALLATION

Access to the engines was accomplished through large cowls on either side of the aircraft. The maintainer could access the cowls by standing on the weapons bay door work platforms. When unlatched, the engine doors rotated in the horizontal plane aft on radius arms. Quick release pins were provided on the engine door radius arms to allow for door removal if desired. The engine compartment floor was rated for walking loads to allow for maintainer footing. The engine support tube design included enlarged gimbal slots for disconnecting the input drive shaft from the MGB. Also, a "steady rest" positioning ability to secure the engine during MGB removal, air inlet removal, or other maintenance actions dictated a change in engine position. Removable firewalls were provided to enhance the access of surrounding equipment. A removable baffle box and baffle box access panels provided the means to perform maintenance on the exhaust cone and nozzles. Engine sensors and controls were interfaced with the aircraft's on-board integrated diagnostics system to provide status and health monitoring to the flight crew, and to provide to the maintainers automatic fault detection/isolation to engine LRUs.

SECONDARY POWER UNIT

The SPU was located on the right side of the aircraft aft of the T802 engine and inboard of the engine exhaust manifold. The SPU was accessed by opening and swinging the engine doors aft on the radius arms (Fig. 18.6). Quick release pins were provided to disconnect the radius arm(s) to further rotate or remove the door to provide enhanced SPU access. Removable firewalls and walking loads beneath the engine exhaust S-duct enhanced maintainer access. The SPU could also be accessed by using the tailcone work platforms. Like the T802 engines, the SPU was of modular design, and removal of the turbine was seldom required. SPU sensors and controls were interfaced with the aircraft's on-board integrated diagnostic system, providing capabilities similar to the T802 engines.

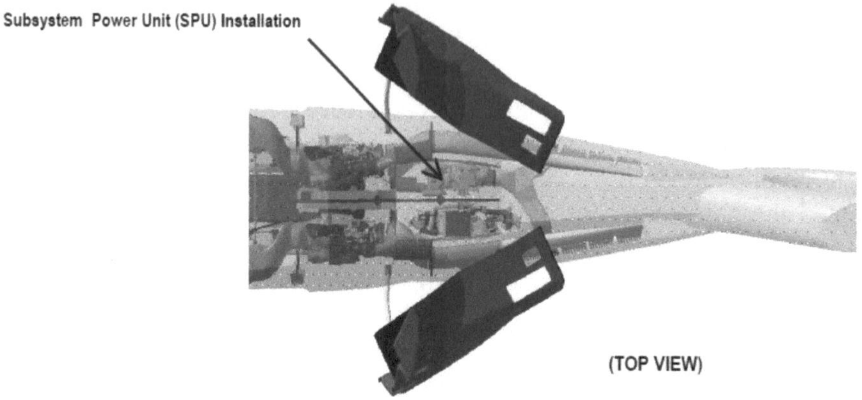

Fig. 18.6 SPU with access doors in open position.

ROTOR SYSTEM

The main rotor hub was accessed by standing on the MGB deck and reaching over or under the main rotor pylon. An engine inlet stepstool (EISS) was developed to facilitate access to rotor head components. The EISS was placed over the engine air inlet and provided the maintainer with additional height for maintenance actions at the limits of the fifth percentile maintainer's reach. The EISS was utilized for centrally located rotor hub components such as the slip ring, standpipe, and mast nuts. The majority of maintenance tasks could be accomplished without the use of the EISS. The main rotor blades, torque tube, and flexbeams could be removed without the EISS. The on-board integrated diagnostic system supported automated blade track and balance, which was accomplished in part by bending thermoplastic trim tabs. Balance was accomplished by adjusting weights located under the snubber bearing bolts and under the tip caps.

Fantail blade tracking requirements were eliminated due to the Fantail's rigid rotor concept and the close-tolerance fiberglass construction. Balance weights were located in the blades' spar transition area with balance set as part of the manufacturing process. Design and close tolerance manufacturing enabled blade replacement without the need for rebalancing. Verification of Fantail blade replacement required only a maintenance operational check and not a maintenance test flight.

ENVIRONMENTAL CONTROL AND NUCLEAR, BIOLOGICAL, AND CHEMICAL PROTECTION SYSTEM

The ECS/NBC system was located on the left side of the aircraft opposite the SPU aft of the left T802 engine. Local access features were similar to those described for the SPU including integrated diagnostic capabilities.

ELECTRONICS PACKAGING

Traditional helicopter maintenance of electronics consisted of repair by black box replacement. This repair method was acceptable in a three-level maintenance system where the removed LRU was sent to AVIM for diagnosis and replacement of the faulty circuit card. The circuit card would be sent to depot for piece part repair. This was not a viable solution for LHX in a two-level maintenance system, because the higher cost black box would have to be sent to depot for repair. Sending black boxes to depot was a turnaround time increase from the existing AVIM concept and subsequently increased the spares pipeline HPMG and stocking support cost. Obviously, sending black boxes to depot for repair was not in keeping with LHX's affordability and/or sustainability goals. Tradeoffs showed the two-level maintenance requirement could be met in part by replacing the circuit cards in lieu of black boxes at AVUM. Under this concept, lower cost circuit cards would be sent directly from AVUM to depot; however, circuit cards were traditionally designed for handling only in a controlled environment such as provided by AVIM or depot, not AVUM. As a result, LHX's electronics would be packaged as SEM-E modules in electronic racks. The SEM-Es were circuit card assemblies with covers to protect the delicate electronics during handing, electrostatic discharge, environment, and so forth. Integrated on-board diagnostics would isolate to the faulty SEM-E module, and the module would be accessed through the rack's hinged, quick release access doors. Entire racks could be replaced similar to the black box concept, if needed. The racks were strategically separated in the airframe to support combat sustainability. These AVUM replaceable SEM-E modules were designated as line replaceable modules in Comanche's two-level maintenance concept.

ELECTRICAL SYSTEM

The two main generators were located on the top of the MGB and could be quickly removed via the MGB doors by loosening V-band clamps. The SPU-mounted generator attachment was the same and could be accessed by opening the engine cowl door. Oil from the generators returned to the MGB by gravity, and therefore was of little issue during removal. The Hydraulic Permenent Magnet Generators (HPMG's) were located beneath the MGB and were also access via the MGB doors.

The two electrical power system controllers (PSCs) were located in the right and left mid-ebays. The PSCs were packaged in racks of LRMs. The on-board integrated diagnostic system would automatically fault detect and isolate to the LRM. After opening the PSC rack, the maintainer would replace the faulty LRM that was retained by toolless cam-operated wedge locks. All components of the electrical system were easily removed for maintenance. On-board diagnostics provided a high degree of automatic fault isolation to all electrical system LRUs and LRMs.

Using the SPU, SPU generator, line contactors, bus tie contactors, and PSCs, ground maintenance checkout power could be provided to any aircraft electrical load. If the SPU was not available, a ground power hookup beneath the left-hand MEP shelf was available. Diagnostic information was accessible via the crew station MFDs or the Comanche's portable maintenance aid (PMA). The PMA is a tablet-type device. It was tethered to the aircraft via a hard-wire 1553 data-bus interface and flew with the aircraft in the left-hand MEP bay. An additional 1553 data port was located in one of the main landing gear wheel wells. Maintainers could access maintenance procedures, maintenance records, and automated fault detection/isolation tools via the PMA. The PMA would also be used to fault isolate wiring failures by providing automated fault-trees to the maintainer.

HYDRAULIC SYSTEM

The two pump reservoir modules (PRMs) were located on the top of the MGB adjacent to the generators. Access was similar to the main electrical generators. Using the SPU, system 3 starter pump, and system 1 and 2 transfer and isolation valves within the supply transfer module assemblies (STMAs), ground maintenance checkout of hydraulic power could be provided to any aircraft hydraulic component. If the SPU was not available, a hydraulic ground cart could provide power via pressure and return ground quick disconnects on the system 3 supply module located behind the left-hand MGB access door. Also, if the SPU was unavailable, the weapons doors could be extended and retracted using the hand pump located beneath the right-hand MEP shelf. The hand pump could also be used to manually recharge or "top off" the SPU start accumulator. Hydraulic quick disconnects were strategically located throughout the system to eliminate bleeding and possible contamination of critical hydraulic paths. The on-board integrated diagnostic system and PMA provided similar capabilities to those described for the electrical system.

MISSION EQUIPMENT PACKAGE

The mission computer clusters (MCCs) consisted of two racks, one located in the right-hand MEP bay and the other in the left-hand MEP bay. The racks were fully populated with line replaceable modules (LRMs) (see Fig. 18.7). In the event of a malfunction, the on-board integrated diagnostic system would automatically fault isolate to the LRM level with PMA aid if desired. The MCCs also contained a separate test and maintenance bus that provided a communications path to perform software/hardware test and maintenance functions. The populated MCC racks could also be replaced as a single LRU in the event of catastrophic damage.

The AVIC and ICNIA racks, fully populated with LRMs, were also located within the right- and left-hand MEP bays adjacent to the MCCs. The HIDSS

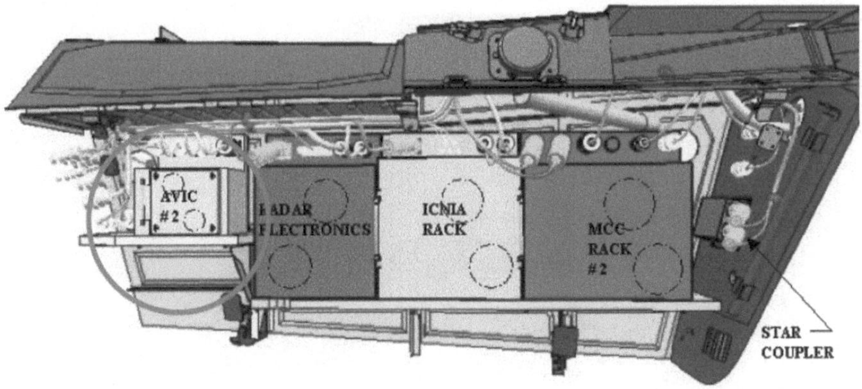

Fig. 18.7 Typical MEP bay packaging.

Fig. 18.8 Mission computer cluster rack.

expanded display electronics unit (EDEU) was also fully packaged with LRMs. The AVIC, ICNIA, and EDEU fully populated rack could also be individually replaced as LRUs (see Figs. 18.8 and 18.9).

The cockpit displays were generally only display heads with video processing conducted in the MCCs. The displays, along with other cockpit controls, were packaged as LRUs. The remaining MEP hardware was designed as LRU replacements with the exception of the EOSS. The EOSS was designed to be maintained at AVUM without removal of the complete EOSS from the aircraft. The EOSS was of modular design. The on-board integrated diagnostic

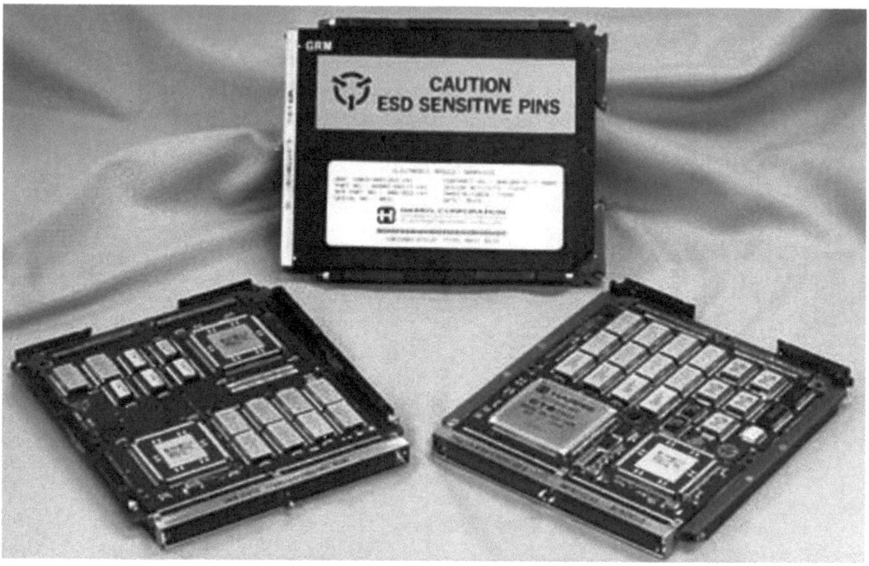

Fig. 18.9 Typical line replaceable module.

system provided system status and health to the flight crew and fault detection/isolation to the maintainers. Repair was accomplished by on-board replacement of the defective EOSS module.

ARMAMENT

The turreted gun system consisted of a number of assemblies that were LRUs: Gatling gun, turret, electronic control unit (ECU), load control unit (LCU), reel assembly, and chutes. Some of the subassemblies and/or components on these LRUs could be serviced/replaced on board the aircraft. Several required more than one person to remove and replace them. The reel required two or three maintainers along with a cradle, cables, and "come-along" to raise and lower it into position. The turret assembly required two maintainers and a spotter to raise and secure the unit to the bottom of the aircraft using attachable carrying handles. Removal of some components, such as the elevation drive gearbox and elevation resolver, required removal of the turret assembly. As with other systems, on-board diagnostics provided status to the flight crew and fault detection/isolation to the maintainers. In addition to other maintenance tasks previously described, the PMA also tracked life-limited parts such as barrels. Barrel life, along with all other limited life component lives, was automatically calculated and tracked at the aircraft level and communicated to the PMA.

The integrated retractable aircraft munitions subsystem (I-RAMS) hardware was easily accessible with the weapons bay doors open. The weapons bay door panels could be raised beyond the horizontal work platform position to permit look-down inspection or maintenance of weapons, rails, or the armament interface unit (AIU). Several interlocks were incorporated to ensure safe I-RAMS operation during maintenance. The weapons bay doors retracted slowly when the aircraft was on the ground to prevent injury should the doors be inadvertently commanded closed. Weapons bay doors were inhibited from retraction unless the weapons door panels were latched in the closed position. Similarly, for example, if a fuel cell was mounted to the enhanced fuel/armament management subsystem (E-FAMS) wings, an interlock inhibited extension of the I-RAMS.

RAM EPILOGUE

Military helicopter systems had repeatedly failed to meet their initial RAM quantitative requirements. The expectation was that the systems would demonstrate the mature requirements at the initial operational test and evaluation (IOT&E) or at least by the first production unit produced. These requirements were at best impractical and unachievable. There are a number of reasons for this:

- *Too-small product sample size:* Only a limited number of prototype helicopters were manufactured during the product development phase. The sample size was too small to contain all the dominant failure modes necessary for reliability development.
- *Insufficient testing:* Generally there were insufficient test hours in the development program to precipitate the dominant failure modes necessary for reliability development.
- *Nonrepresentative equipment:* The helicopter systems under test were not completely reliability representative of the production configuration. There were two basic reasons for this: (1) there was insufficient lead time available within the development program to incorporate all the correctives necessary to demonstration mature reliability, and (2) at the beginning of the initial production program, equipment was redesigned for a variety of reasons. As a result, new dominant failure modes were introduced into the production units. Consequently, new production units generally exhibited a backwards slip in reliability over their predecessors.

Comanche implemented a plan to achieve RAM maturity two years post initial operational capability (IOC). There was much scrutiny of this approach when it was first introduced, but the laws of physics and common sense prevailed, and it was adopted. This plan allowed the sample size to be increased to include fielded production aircraft, which also addressed the nonrepresentative equipment issue. The "two years post-IOC" would provide sufficient operational flight data to identify dominant failure modes. However, the lack of sufficient lead time to incorporate fixes still existed. To address this issue, Comanche implemented two methods of RAM compliance achievement: demonstrated and assessed. Demonstrated compliance was based on collected aircraft data and therefore included the benefits of those RAM corrective actions incorporated. Assessed compliance was based on the demonstrated value but also included the benefits of all unincorporated corrective actions with approved production lot commitments.

As of the termination, Comanche was on-plan for RAM success. There were a number of indicators of RAM program success:

- *RAM design influence:* At termination, Comanche's design was largely complete with EMD prototypes in fabrication. The impressive volume of detailed RAM features evident in design was a testament to the Comanche IPDT process.
- *Failure reporting, analysis, and corrective action system (FRACAS):* The FRACAS collected maintenance data from flight test aircraft. Every maintenance action was subject to root cause failure analysis and recommended corrective action. At termination, more than 1100 reliability corrective actions were implemented.
- *Maintainability corrective action system (MCAS):* The MCAS collected data from maintainability development tests and demonstrations. Each anomaly identified was subjected to root cause analysis and recommended corrective action. At termination, more than 350 maintainability corrective actions were implemented.
- *100% FRACAS and MCAS corrective actions implementation:* All FRACAS and MCAS approved corrective actions with effectivities of aircraft no. 3 and subs were incorporated on released drawings.
- *Reliability growth:* Maintenance data resulting from the flight test program were continuously analyzed to trend actual progress towards quantitative reliability. At termination, these reliability growth trends indicated that Comanche's primary reliability parameters were on-plan.
- *Maintainability demonstration:* A formal demonstration was conducted using Dem/Val aircraft to measure MTTRe and MTTR. The demonstration was conducted by the U.S. Army using Army maintainers. Comanche successfully passed the demonstration.

Comanche's program management was unique and successful. The keys were ownership and appreciation. Program leads essentially were empowered with total responsibility for their discipline: proposals, technical, estimates, cost, schedule, performance, negotiations—everything. Leads ran their own shop, true ownership. If it was going well, one had satisfaction in a job well done. If not, one only had to look in the mirror. If you needed support from others, you got it. It wasn't management oversight that made it happen, it was a culture. It's what was done—acting as a team, supporting others. There is a subtle difference between doing what's expected and "culture." This atmosphere encouraged the best individual performance, which was widely appreciated. The appreciation was genuine. There were no large-scale incentive programs to hand out gold stars for a job well done. These programs weren't needed. Daily, the appreciation of one's efforts from coworkers and superiors was felt and reciprocated. As an example, after returning home from giving a program-critical two-level maintenance briefing at the Pentagon, the RAM lead was welcomed by a standing ovation from supportability coworkers.

Chapter 19

THE FLIGHT TEST PROGRAM

RUS STILES
COMANCHE CHIEF PILOT

INTRODUCTION: DESIGN TRANSITION TO FLIGHT TEST

After the LHX design studies were completed and two demonstration/validation (Dem/Val) aircraft were constructed, the Comanche was poised to enter the testing phase. Maximum strength and fatigue properties of flight-critical components were tested to ensure parts would stand up to the loads expected during flight test. Ultimately, the transition from ground test to the flight program was facilitated by a ground test vehicle.

Although many programs construct a full-scale aircraft to validate aircraft systems prior to flight, the Comanche program was uniquely positioned to utilize a propulsion system test bed (PSTB; see Fig. 19.1). The drivetrain and rotor, electrical, hydraulic, and fly-by-wire (FBW) systems were mounted to

Fig. 19.1 The Comanche PSTB in operation, circa September 1995.

a support frame, eliminating the need to construct the full aircraft structure and cockpit. Control of the rotor was integrated within the FBW system and operated from an external blockhouse.

PSTB rehearsal of the most demanding flight profiles demonstrated the aircraft was ready for flight test. A few weeks later, the first prototype aircraft completed its initial flight.

The Comanche program had a number of unique features that would present challenges to flight test. Sikorsky had no real experience in joining with another company in flight testing a brand-new, fully integrated aircraft. Fortunately, Sikorsky and Boeing were reasonably well aligned in their flight testing philosophies, but the approach to flight test personnel assignments differed dramatically between the two companies. Sikorsky flight test was organized as a separate engineering unit, with dedicated flight test engineers. It was aligned with design engineering, with groups assigned to structures, propulsion, handling qualities and flight controls, and avionics. Sikorsky flight test personnel responsibilities included collecting and analyzing data, determining specification compliance and operational suitability, and risk management. Sikorsky flight test engineers also assisted the design group in making changes to the design.

Boeing flight test personnel were generalists, and the company drew from its design organization to support flight testing.

There were numerous discussions about the location of the flight testing. The Boeing flight test center was at the Wilmington, Delaware, airport, a public facility; Sikorsky facilities were at a private airport in West Palm Beach, Florida. At the time of the 1990 proposal and the 1991 best and final offer (BAFO), all the testing was to be conducted in West Palm Beach. However, soon after BAFO, an alternative plan was proposed: to conduct initial flight controls development on aircraft 2 and mission equipment package (MEP) development on aircraft 3, at Wilmington. This allowed the flight controls development to be close to the flight controls system integration rig (FCSIR) and simulator in Philadelphia. The MEP testing would be close to the full systems integration lab and the software development team in Philadelphia.

On January 1992 this became the new plan; however, this was when the program consisted of six test aircraft. When the contract was later modified to only two flight test aircraft, the plan reverted to conducting all the testing at West Palm Beach. This created a drain on Boeing engineering personnel resources, because they had to split their time between Philadelphia and West Palm Beach.

The program office allocated responsibility for completion of flight test and system development along the lines of the design responsibilities. Sikorsky had the major portions of flight test: structures and propulsion, which included engines; low observable exhaust; environmental control and filtration systems; and aircraft performance and handling qualities. The major portions of Boeing flight test allocations included flight controls and mission equipment.

The Comanche was the first major Army aviation procurement program to utilize the combined test team (CTT) concept. In addition to the Boeing Sikorsky contractor team, the Army contributed participants to the CTT, including personnel from the flight test center, program office, and Training and Doctrine Command (TRADOC). This was supposed to reduce the overall amount of flight testing by combining contractor and military test objectives. The CTT was also intended to provide early feedback on the operational suitability of the aircraft. The Army flight test center provided an experienced military test pilot who participated in daily flight activities, and the program office provided a liaison to facilitate day-to-day coordination.

Quoting Army CTT pilot Chief Warrant Officer (CWO) 4 John Armbrust [1]:

> [With the previous system] once the Army finally got its hands on the machine [that won the competition], it was pretty far down the development road. Then the government testing process was usually everything that had been done by the contractor. It added years and added cost. … The combined test team concept gets Army input from the operational and flight perspective from day one. … I think we have substantially reduced the chance of the aircraft going into production with design deficiencies—and saved time and money.

Despite the challenges and considering the unconventional nature of the Comanche design, the flight test program was quite successful. There were no major incidents, and the basic aircraft met or exceeded all operational objectives. Forward flight speeds in excess of 200 kt were achieved, and the aircraft exhibited astonishing capabilities in sideward and rearward flight. The Comanche could easily reach speeds in excess of 60 kt sideways and 80 kt rearward, while maintaining high levels of maneuverability. This represented a generational leap forward in mission capability; at low speeds, the relative direction of the wind would now be hardly a consideration. This differed dramatically from the legacy aircraft the Comanche was to replace, which had very limited low-speed performance capabilities that had introduced the term *loss of tail rotor effectiveness (LTE)* into the training vocabulary of Army aviation. With Comanche, LTE was a thing of the past.

The aircraft was extremely agile, yet it was easy to control, thanks to the fly-by-wire control system, which enabled customization of the handling characteristics. Although the Army resisted providing a flight clearance to permit aerobatics, the "split S" (half-roll, half-loop) was demonstrated on several occasions (Fig. 19.2). The rationale provided by the test team was that all-attitude control capability had to be demonstrated to validate the algorithms used by the control system to calculate vehicles rates and attitudes. In reality, the test team really wanted to demonstrate the high degree of agility and controllability provided by the bearingless rotor and the FBW system.

Fig. 19.2 "Split S" maneuver.

TECHNICAL CHALLENGES FACED EARLY AND OVERCOME

The Comanche prototype aircraft were so unique, from both a hardware and software standpoint, that the flight test team broke new ground in virtually every aspect of flight testing. The composite materials in flight-critical areas of the aircraft structure had properties that were new and untested. There was considerable uncertainty in how flight loads would be transmitted through the structure to and from critical components such as the rotor hub, transmission, and drivetrain. This led to the inclusion of extensive instrumentation, all of which had to be closely monitored until the loads characteristics were better understood.

The FBW system detected pilot inputs at the control sticks, and then software embedded in the flight control computers determined what command was sent to the actuators. Whereas with a conventional mechanical control system the pilot has full control of the size, duration, and quickness of control inputs, the FBW system added static and dynamic shaping to improve aircraft response. This prevented traditional control buildup techniques. Data had to be collected early in the program to characterize the size and sharpness of commands transmitted to flight control actuators and validate the simulator models, while avoiding structural limits. In many cases, the pilots had to manually limit control input size and quickness for buildup; this was particularly challenging with the small displacement sidearm controllers.

The low observable qualities of the aircraft largely dictated the outer mold lines, which left internal systems very tightly packed. Essentially, there was little room for internal components to "breathe," making extricating heat difficult. Very close watch had to be kept on the thermal health of computer processors, control valves, hydraulics, and the like. The only other known stealth aircraft at the time, the F-117 (see Fig. 19.3), reportedly had similar issues, but the programs were so compartmentalized, little could be learned from the other.

Fig. 19.3 The Air Force F-117 Nighthawk and Army RAH-66 Comanche shared contemporary approaches to low observable design.

Because of the highly integrated nature of the avionics, the failure of a computer processor potentially had a major impact on the functionality of the aircraft. Although most systems were redundant, single failures were taken seriously. Until the cause of a single failure was understood, it had to be assumed the backup system could be affected in a similar way.

In many ways, the Comanche was essentially a transition in technologies. Few in the aerospace industry had any experience with low observable designs and the effect on performance. Fly-by-wire flight controls were not common on either military or commercial aircraft, and had never before been implemented in a helicopter. Whereas other aircraft had achieved high levels of maneuverability, the Comanche, through the FBW-controlled bearingless main rotor and Fantail, achieved superior agility. FBW promised to eliminate much of the complex pilot workload challenges associated with taming high-agility rotors, by getting rid of cross coupling, nonlinear response, and airspeed-dependent sensitivity, all while providing intrinsic limit avoidance. The FBW also provided unparalleled open loop (hands off) stability.

Glass cockpits with digital and graphical displays were just starting to appear in the civil aviation environment, but had not yet reached the military cockpit. Comanche designers pioneered the use of multifunction cockpit displays and automation for the military helicopter mission. Thus, the flight test team was often required to invent new test techniques to test, evaluate, and develop the design into an operationally viable system. During the years of YRAH-66 flight test, advanced cockpit technologies began to proliferate on other platforms, with inevitable improvements as technology matured. Many of these improvements were cosmetic, but this led some Comanche observers to the frustrating impression of impending obsolescence in what was really the most advanced vehicle of its time.

Then there were the numerous Comanche program restructures. This often led to temporary termination of flight testing to meet funding profiles. Of the eight years of flight testing, four of them achieved less than 60 flight hours during the entire year. Months of down-time were not uncommon, and flight test personnel often had to be assigned to other programs. Although these assignments were intended to be temporary, the reality was many people never returned. When the aircraft was on flight status, the flight rates were very respectable with few maintenance issues, but the low number of annual flight test hours gave the erroneous impression of poor reliability.

Some of the starts and stops were due to the fact that the program only funded two prototypes. Only the first aircraft incorporated full structural instrumentation, so any modification that would require verification of the flight envelope had to be done on aircraft 1. Aircraft 2 was intended to be the mission equipment aircraft, but in reality aircraft 2 had little more mission capability than aircraft 1. But without instrumentation, it couldn't fly until aircraft 1 was ready. So, during modification periods, both aircraft often sat on the ground.

Despite these challenges, the flight test group was a highly motivated team focused on safe development as rapidly as possible. Boeing and Sikorsky personnel worked as a coordinated unit; it was impossible to tell which company contributed any given person.

Boeing was responsible for the FBW system and had recent experience with a difficult FBW development on the V-22 Osprey tiltrotor. Learning Osprey lessons, Boeing built a full-up "iron bird" laboratory that accurately replicated all the Comanche hardware. This flight control lab had the ability to introduce faults such as hydraulic leaks, broken wires, and the like, allowing end-to-end verification of software functionality. This gave engineers confidence that automated fault reaction would be correct.

Sikorsky utilized drivetrain test stands to verify the adequacy of Comanche lubrication systems and load capacity. An end-to-end electrical lab was also constructed at the Sikorsky facility in Stratford, Connecticut.

The Propulsion System Test Bed (PSTB)

As stated earlier, the PSTB was the ground test vehicle to prove the aircraft was ready for flight. The job of the PSTB was to verify that the drivetrain could sustain the power levels needed for flight, the FBW and full authority digital engine controls operated predictably, and ancillary systems supported aircraft operation in a reliable and robust manner. Completion of a preflight acceptance test profile was a necessary prerequisite to receiving a clearance to fly.

There were operational advantages to the PSTB configuration. Unlike a traditional ground test vehicle, where pilots occupy the aircraft cockpit and flight test engineers monitor telemetry in a separate building, the Comanche test team operated from a single room. A pilot seat was surrounded by workstations for structures, propulsion, flight controls, and systems (electrical, hydraulic, fuel, etc.). If there was a question about a system anomaly, the whole team could discuss and resolve the issue face-to-face. The pilots were especially pleased to operate in an air-conditioned environment, especially during the summer in South Florida. This close team arrangement set a very positive tone for the flight test development.

The remote control arrangement, along with a large capacity underground fuel tank, allowed the PSTB to be operated virtually continuously; when a team member needed rest they simply left their workstation (or cockpit seat), and a replacement took their place.

Although the PSTB was an excellent vehicle for shakedown of critical drive systems, there were new technologies embedded in the air vehicle that could not be included on the PSTB. Probably the most significant was the Comanche mission equipment package (MEP), which managed many of the aircraft systems. This included orderly start and shutdown of interdependent support systems. Generators, for example, require adequate rotational speed to provide specified power. In a conventional aircraft, the pilot manually turns generator switches on once the proper operating rpm is reached. On the Comanche, this function was allocated to the automated system manager embedded in the MEP.

The MEP was also the source of data for the pilot to determine system health. Failure modes were announced "by exception." This differed from conventional aircraft, which provided a dedicated panel listing all the possible failures; the pilot monitored the panel for failure annunciations and mentally filtered these annunciations by the operating environment. For example, in a conventional aircraft, if the engines were not running, oil pressure would be shown as low, and pilots had to determine this was normal due to the engine off configuration. On Comanche, these annunciations were filtered automatically. In the low engine oil pressure example, the MEP watched for engine operation before declaring an abnormality. Even though this functionality was simulated in the laboratory, understandably it often did not initially operate correctly in the real-world environment.

The MEP was not yet developed at the time of PSTB operations, so the Sikorsky instrumentation group constructed an avionics suite that mirrored what the MEP would do, but without much of the automation envisioned for the aircraft. This so-called "iMEP" allowed the pilot to monitor drivetrain health such as pressures and temperatures, control subsystems such as generators, and monitor systems for faults.

The inability to shake down the MEP along with other systems would later present problems for the aircraft.

The first PSTB run commenced in June 1995. The goal was to conclude PSTB preflight qualification runs in time for first flight of the aircraft by the end of the year. Although for the most part the Comanche systems operated as designed, on 3 July 1995, the main gearbox experienced a major failure. During a high-power event, a portion of one of the engine drive gears fractured, separated from the web, and exited the side of the gearbox (see Figs. 19.4 and 19.5). A rapid loss of oil pressure announced the failure, and operations were terminated. Suddenly, the scheduled goal of a first flight date in fall of 1995 was in serious jeopardy.

An investigation revealed that the gear in question was susceptible to resonance at very high power levels. The fix was to add a coating to the gear to act as a dampening agent. Detailed design and implementation of this change was predicted to take several months. After careful analysis, it was shown that the drivetrain could, even without the fix, provide enough power with

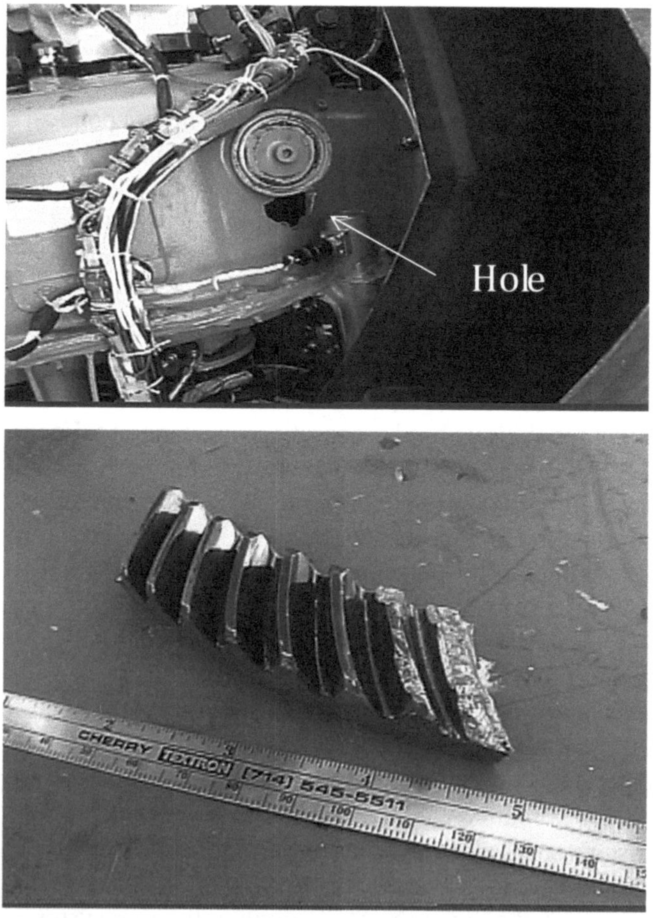

Fig. 19.4 Main gearbox failure, July 1995.

Fig. 19.5 The PSTB uncovered a gear problem that threatened to delay the first Comanche flight.

adequate margin and conservatism for limited initial flight testing. PSTB operations resumed on the unmodified gearbox with limitations in place but with the goal of proving that first flight power requirements could be achieved safely.

On 7 December 1995, the PSTB commenced the final 50-hour run to satisfy the preflight acceptance requirement. At the same time, preparation of aircraft 1 (Tail No 94-0327) for flight was underway.

PREPARATION FOR FIRST FLIGHT

A series of ground run tests was required on the prototype aircraft before it could be cleared for flight. Although the PSTB verified a robust drivetrain and control system, the systems also had to operate correctly on the aircraft itself. This was where the MEP, which controlled the support systems (generators, environmental control, fault detection, and annunciation), came into play.

Problems plagued the mission equipment package. Early on, the MEP had a tendency to erroneously indicate component failures when they didn't exist. More importantly, it was simply unreliable. Initially, the MEP would not start up automatically, and as a workaround, a support vehicle had to be attached to the aircraft via a data umbilical. After the aircraft was operating on generator power, engineers initialized the MEP from outside the aircraft. The MEP also had a tendency to—without warning—stop operating, causing the cockpit displays to go black. Eventually, the pilots were provided with a pushbutton in the cockpit to command the MEP to reset (akin to "Ctrl-Alt-Delete" on a personal computer). But this led to additional problems.

Whenever the MEP was restarted, the secondary power unit would mysteriously fail. This was an ominous preview of what happens when software is put in full control of aircraft systems. It turned out that during a MEP reset, the system would initiate a built-in test (BIT) of the fire warning and extinguishing microprocessor. During this check, the fire warning lights would illuminate.

What was not understood at the time was that the software engineers had interpreted a somewhat nebulous requirement for "automated response to failures" in such a way as to autonomously command an SPU shutdown anytime a fire was detected. The mechanism to determine the presence of fire was illumination of the fire lights, which coincidentally came on during BIT. This explained the mysterious SPU shutdowns. A software change, which took several months to implement, removed this function. Although this type of problem was largely due to the fact that Comanche was one of the first highly integrated computer-controlled aircraft, driven by requirements that were not well understood at a systems level, it was typical of the types of issues that had to be discovered and corrected on the flight line. The pilots demanded more control, which led to additional software changes.

Aircraft ground checkout was slow and tedious. Relatively small problems—overheating of components, oil and hydraulic leaks, relatively minor component failures—could stop progress for days at a time. The FBW system worked almost flawlessly, but the rest of the aircraft systems also had to be operating properly before a first flight release could be granted. Having a program office representative on site streamlined this process.

COMANCHE FINALLY FLIES

There was substantial political pressure to fly the Comanche by the end of 1995. The program had been delayed so many times that some felt pushing first flight into 1996 would be seen as delay of another "year" and might be the breaking point leading to program cancellation. This perception—being at the precipice of cancellation—was pervasive throughout the flight test community; it seemed that the flight test team was continually defending the program through one demonstration or another.

The program and company leadership were remarkably supportive of the test team. The program had assigned a Boeing engineer, Clarence (Hutch) Hutchinson, as lead for development in West Palm Beach. This was initially met with skepticism in Florida—how could a Boeing engineer from Philadelphia know how business was done at the Sikorsky Development Center? Hutch had considerable experience, having been chief of test at Boeing Helicopters, and brought so much development knowledge to the team that he was quickly accepted and admired as a valued leader.

As Comanche chief program test pilot, I had an experience with Sikorsky leadership that clearly demonstrated safety remained the highest company priority. After several frustrating delays, it was apparent that first flight might not happen in 1995. One day, it was announced that the corporate jet was inbound with high-level Sikorsky executives from Connecticut. Most of the test team thought they were probably coming down to pressure us to get the flight done before the end of the year. I soon learned one of the visitors was Eugene Buckley, president of Sikorsky Aircraft. After he arrived, Buckley asked to talk to me personally.

We were conducting a ground run at the time, but Buckley said he would wait in the hangar. After we shut down, I walked over with several people to meet with him, but was told that the meeting was to be private. I was understandably nervous—having come from a military background, a private meeting with the commanding officer was rarely a good thing. I walked into the otherwise empty hangar and there were three people there to greet me: Buckley and two high-level Sikorsky executives. After a short greeting, Buckley looked at me and said, "I wanted to come down here to tell you in person—that aircraft does not fly until you say it is ready to fly, is that clear?" The other two executives said nothing. A minute later, all three left the hangar, got on the company jet, and departed back to Connecticut.

I wondered, why fly all the way down to Florida just to say that? It didn't take long for me to figure it out. This visit was all about communicating the importance of safety to the executives that accompanied the president. It was an example of leadership one rarely encountered in a civilian or military environment. It made me want to achieve that first flight date far more than any political pressure.

As the Comanche team was getting close to final preparations for first flight in late December, small issues created frustrating delays. Relatively minor problems, such as the failure of a pneumatic fitting, prevented systems needed for flight from operating normally. Nevertheless, the team kept forging ahead, finding and solving problems. After one day off for Christmas, ground runs continued with hopes that a first flight in 1995 was still achievable. But it was not to be.

On December 30, the failure of a hydraulic instrumentation transducer flooded the engine deck with hydraulic fluid. Hydraulic fluid is very corrosive and flammable, so it had to be removed completely. The engine deck was very crowded; components had to be removed to get to some of the surfaces where the fluid pooled. Knowing it could not be done in a day, the team faced the difficult admission that first flight was not going to be completed in 1995.

Upon return from the New Year's break, it took only two additional days to ready the aircraft for flight. The YRAH-66 flew for the first time on 4 January 1996 (Fig. 19.6).

Because the aircraft was equipped with the original main transmission design that failed on the PSTB, it was cleared for only one flight. In the first flight configuration, the published maximum power limit was only 6% above the power required to hover. As a result, first flight was limited to low speed

Fig. 19.6 The "official" first flight photo, signed by pilots Rus Stiles and Bob Gradle.

Fig. 19.7 The 15 January 1996 cover of *Aviation Week & Space Technology* announced the first flight of the YRAH-66.

flight in all directions, some hovering turns, and takeoffs and landings. Despite these limitations, much was learned, and the predictable handling qualities proved the value of simulation.

The FBW system performed as predicted. Despite initially failing to make the transition to flight mode due to a stuck weight-on-wheels switch (a scenario that was practiced in the simulator), the bearingless main rotor was stable and easy to control. Subsequent takeoffs and landings properly transitioned. The aircraft was remarkably smooth—main rotor track and balance were eerily precise with virtually no tuning required. As might have been expected, the MEP failed prior to completing the flight, but the team stubbornly worked through the reset and completed the planned test card. The aircraft was under power for 2 hours, and the flight duration was 34 minutes.

One of the traditions expected from a first flight of this magnitude was to be on the cover of *Aviation Week & Space Technology* (Fig. 19.7), a magazine widely read by the aerospace industry. It happened, but the caption "Comanche Finally Makes First Flight" was certainly a negative comment and quite disappointing to the test team that worked so hard to make the flight happen.

Quoting me after the flight, the *AvWeek* author said [2]: "Stiles told debriefers he was surprised by low vibration levels of the prototype's five-bladed, high-technology composite main rotor, which he said were on par with the levels of production-quality rotorcraft. 'I would be very proud to deliver it just this way' he said."

Boeing Sikorsky program director Jim Morris mentioned how this first flight was a boost for the contractors. "This was the culmination of 10 years of work for the contractors and their employees who have endured the delays [due to the funding stretch-outs]."

EXPLORING THE ENVELOPE

The second flight did not take place until 24 August 1996, after the main transmission had been modified and PSTB validation completed. Subsequent progress was rapid; less than a month later, on flight 8, CWO4 John Armbrust, the Army test pilot assigned to the CTT, flew the aircraft. Incorporating a military test pilot this early in testing was unprecedented and generated a lot of enthusiasm for the program. The following are comments by CWO4 Armbrust after his flight:

> The objective of our flight was to gather handling qualities data at a hover and at forward flight to 80 knots, in turning flight, and during partial power descents. In all these flight regimes, the helicopter exhibited surprisingly strong roll and pitch control power. Only in a high performance fighter aircraft have I felt as much roll control power. At a hover, the yaw rate feedback control loop felt very nearly like a "heading hold" feature. This in spite of the fact that the level of flight control augmentation flown

> (mission primary flight control system) is essentially a degraded mode, and would not normally be seen by the aviator unless multiple failures in the FCS had occurred. The aircraft did display some sideslip-to-pitch coupling during flight that requires the attention of the developers, and there are a few other areas which require "tweaks" and should be worked as development continues. The flight was an hour long, but seemed only minutes.
>
> I have been working with the program since 1991, and have been on site here at WPB since July of 1995. This flight, while a greatly rewarding experience for me, is proof of the Army's commitment to the Combined Test Team approach to RDT&E.

The next six Comanche flights verified the low-speed flight envelope and collected frequency response data for characterizing the FBW software. This was an important step in ensuring stability and optimizing the response to pilot inputs. It also gave the structures engineers an idea of the magnitude of critical loads. These data would be the first step to activating the higher order modes of the flight control system, which were designed to provide low workload and high-level mission capability.

There were four levels of FBW functionality, each dependent on increasing levels of sensor feedback to provide higher levels of stability. The mode flown initially during flight test was known as Mission Primary Flight Control System (PFCS). Mission PFCS incorporated stability augmentation only in the yaw axis. It was supported by quadruple yaw rate sources and thus was highly fault tolerant.

The next higher level of augmentation was known as Automatic Flight Control System (AFCS). The AFCS provided rate command/attitude hold capability and was most like conventional aircraft. The AFCS provided uncoupled rate response at all speeds, was tolerant to weight and center of gravity changes, and incorporated some passive load limiting.

The highest level of augmentation was known as Velocity Stabilization (VelStab). The VelStab mode provided an attitude command/velocity hold capability that was well-suited to low-speed, low-altitude flight in degraded visual conditions. The pilot could select the VelStab mode with the press of a button on the upfront flight control panel or by depressing the cyclic trim switch. When VelStab was first selected on, the control system commanded the aircraft to wings level and captured airspeed. This effectively provided an automated "unusual attitude recovery" mode. In conventional aircraft, pilots received training to address spatial disorientation, but in the Comanche this would have been replaced with a procedure to simply activate VelStab. The aircraft offered many opportunities to reduce training requirements due to the advanced control system, although this was not widely acknowledged.

The fourth level of control allowed flight with no augmentation whatsoever and was known as Core PFCS. In the Core PFCS mode, pilot commands were shaped via software algorithms to remove much of the nonlinearity associated with raw "stick-to-head" control of the conventional helicopter but required no feedback data. The pilot could not select Core PFCS, because it was considered a fly-home mode should all the aircraft state sensors be lost—a highly unlikely scenario. Eventually, the flight envelope was demonstrated in Core PFCS, albeit with high pilot effort. The capability to fly the aircraft without any sensors was unique to the YRH-66. Most military FBW aircraft require some stability augmentation for controllability, and if lost, the aircraft is no longer flightworthy. (Crashes of an F-22 fighter and B-2 bomber have been attributed to loss of vehicle state sensors.)

On flight 12, on 30 September 1996, the first VIP flight demonstration was conducted for Assistant Secretary of the Army for Research, Development and Acquisition Gilbert Decker, who visited West Palm Beach for a program status review.

After the VIP demonstration flight, the aircraft was put into modification status for two months. Issues had surfaced with rotor stability and tail shake. A low-frequency rotor mode, known as regressive lag mode (RLM), had surfaced, and pylon wake impingement on the tail created concerns for ride quality and loads. Changes were made to the main rotor dampers, engine drive system, and horizontal tail.

Flight 13 took place on 21 November 1996. No issues were noted with the modifications, so the envelope development continued. The focus was on rotor stability, expansion of the autorotation envelope, and investigations into the tail shake issues. Progress was steady, and the second flight demonstration took place on flight 18 for Senator Daniel Coats (R, Indiana).

It should be noted that almost all of the flight demonstrations were short (10-minute) segments of a regular data flight, usually after data collection was complete for the day. This took some creative orchestration, because the VIP had to be "entertained" (with program briefings or visits to the telemetry room) while the aircraft was conducting the flight test. Data collection had to be terminated with enough fuel to complete the demonstration, and the airfield had to be closed while the Comanche completed its maneuvers. The test pilots had to "shift gears" from data collection to marketing mode. The flight demonstrations displayed the aircraft through nearly its full flight envelope, at quite low altitudes, and this took considerable piloting concentration. In the end, the team got pretty good at pulling these off and would admit it was rewarding to have the opportunity to show what the aircraft was capable of doing.

On flight 19, the automatic flight control system (AFCS) mode was first engaged. Flight control workload was suddenly and dramatically reduced, and the advantage of FBW became instantly clear. Unlike virtually every conventional helicopter, control response was uncoupled—a command in one axis resulted in response in only that axis, with no off-axis coordination required.

Trim was completely automatic; all the pilot had to do was let go of the controller, and the aircraft remained in the current attitude. Although the wisdom of the twist grip control for yaw was routinely questioned during the design phase, in the AFCS mode, there was no reason for the pilot to make any yaw inputs except at low speed when the aircraft was turned without changing bank angle. The so-called "snap turn," a heading change using only the yaw axis, was accomplished by simply twisting the control grip in the direction of the turn without making other axis inputs—at speeds exceeding 100 kt indicated airspeed (KIAS). This was an impressive capability, albeit with dubious operational applications.

The next flight took the Comanche to 140 KIAS. Achieving nearly its maximum forward speed in level flight in only 20 flights was quite an achievement for a prototype aircraft of unconventional design. Some tuning of the horizontal tail and expansion of power available resulted in an eventual maximum level flight speed over 170 KIAS.

The next few months, now into the spring of 1997, consisted of normal development activities: envelope clearance, system checkouts, and incredibly boring cooling surveys, which consisted of long duration, steady-flight conditions to confirm internal components were able to expel heat throughout the full flight envelope.

The first low observable survey was conducted during this period: a preliminary evaluation of the aircraft acoustic signature.

The Sikorsky Development Flight Center proved to be well worth the cost. It was quite remote, so no public (read "press") observation was possible. The large campus included a surveyed acoustic range, slope landing facility, long runway for low speed envelope development, and wide-open airspace over unpopulated areas. Florida earned its motto, the Sunshine State: a miniscule number of test flights were canceled due to adverse weather.

By the end of this flight test period, in March 1997, the YRAH-66 had flown to 10,000 ft, pulled a 1.7-*g* load factor, and executed a 120-kt autorotation, all in the AFCS mode of flight.

The backup engine governor was demonstrated, along with Core PFCS, the "last-chance" flight control mode. At least two identical failures had to occur before these modes would activate, an extremely unlikely occurrence during normal operations.

SHAKES AND CHANGES

At the end of March 1997, the aircraft went down for another series of modifications. Changes were made to the horizontal tail and mission systems based on data collected during the previous flight test period. The next flight would be 5 months later. Some on the test team were thankful for this break, particularly the pilots. The air conditioning system was not performing well, and this allowed testers to forego flying the aircraft over the hottest months of the year.

During the down periods, considerable time was spent in the simulators continuing to develop the software for the flight controls and mission systems. Significant effort was expended to provide a realistic tactical environment in the simulator, including representation of threat systems, terrain, and adverse weather. The crew interface included high-fidelity hardware to replicate the aircraft and incorporated some of the more advanced systems, such as the helmet-mounted display, that had yet to reach flight test. The intent was to create an environment where a non–test pilot could evaluate the mission systems, and dozens of operational pilots participated in several "full mission" simulation trials.

The three flying months September through November 1997 focused on completion of the Comanche envelope development and continued investigation into systems issues. Maximum speed in level flight, high rotor speed autorotations, maximum load factor, maximum sideslip, and maximum yaw rate were all achieved. Sideslip capabilities, largely due to the Fantail design, were well over twice those of conventional helicopters. The hover yaw rate was also noteworthy, exceeding a stabilized 60 deg/s in less than 180 deg of rotation. Stopping the aircraft from this high yaw rate resulted in some load issues for the Fantail transmission, and a solution using flight control software was sought. This was the first of several issues borne from the extreme capability of the aircraft that were solved using the power of the full authority FBW system, something that would be unheard of when considering a conventional control system.

A pesky handling qualities issue that resulted in a low-frequency "tail wag" was noted during this test period. A lot of data were collected in the hopes that the flight control system would solve the problem. Unfortunately, for this issue the FBW system could not eliminate the problem. The eventual solution—addition of horizontal tail endplates—took a while for the program to accept.

A short modification period took place from November 1997 to February 1998. The year 1998 would prove to be a very productive one for Comanche flight test. The variable rotor speed modes—one of which automatically changed the operating speed from 100% to 107% rotor speed (Nr) to enhance maneuverability at higher speeds, and another that provided pilot-commanded low Nr to reduce the acoustic signature of the aircraft—were activated and the envelope cleared for unrestricted use. Maximum level flight speed was 177 kt. A dive speed of 196 KIAS was achieved, and level flight in the rearward direction reached 70 kt.

The aircraft was presented to the national media in 1998. Several press days were held, each accompanied by a flight demonstration. The Discovery channel did a full episode dedicated to the aircraft. In addition, congressional delegations and high level Army officers also viewed aircraft flight demonstrations. The unique capabilities and unconventional appearance of the Comanche intrigued both those familiar with helicopters and the general public. Chapter 26, Marketing, has more detail on these activities.

As the aircraft envelope opened up, more and more aggressive maneuvers were conducted. The regressive lag mode (RLM), the low-frequency rotor instability observed early in the program and addressed by a damper change, once again surfaced at higher load factors. Under some circumstances (largely dependent on bank angle and load factor), RLM would result in uncommanded pitch oscillations. This was due, in part, because the control laws mandated aft pitch input with high bank angles to provide the feel of positive maneuver stability, otherwise known as "stick force per *g*." As the aircraft began oscillating, the pilot's arm reacted to the airframe movement and caused unintended commands, thus increasing the oscillations. The first-generation controller lacked sufficient mechanical damping to resist rapid arm movement. Relaxing bank angle or load factor eliminated the oscillations, because the pilot held less force on the stick and because the rotor generated lower loads. But because the aircraft required high *g* capability for air-to-air and close air support maneuvers, it was clear that a complete solution to this issue would be required. A detailed discussion of the regressive lag mode and its resolution is contained in Chapter 6, Handling Qualities.

Tail shake continued to be an issue. Design engineers employed a Cray supercomputer to model the flow around the main rotor pylon to the tail in an effort to determine the source of the turbulent flow. Flight test resorted to an old but proven method of "tufting" the aircraft to help engineering visualize air flow over the aircraft (Fig. 19.8). This, combined with the computer analysis, resulted in a redesign effort for the pylon cowlings. Because of the low observable requirements, this entailed a combination of aerodynamic and radar cross-section attributes.

Fig. 19.8 The "tufted" Comanche.

MODIFICATIONS FOCUS ON AERODYNAMIC FIXES

The aircraft was put into another modification period in January 1999. The most visible change was the new main rotor pylon shown in Fig. 19.9. This, combined with the addition of asymmetric gurney flaps on the horizontal tail, reduced the tail shake to an acceptable level from a ride quality and structural standpoint.

Directional stability remained an issue. Once again, there were two fundamental paths to the solution. The first made software modifications to the flight control laws, and the second introduced hardware changes. Although several innovative methods for anticipating and correcting the directional oscillations with the flight controls were attempted, the fact was that the issue (and solution) were rooted in aerodynamic hardware.

The Comanche vertical tail had been optimized so that in cruise flight, Fantail pitch was near neutral. This reduced the power consumed by the Fantail, thus providing more power to the main rotor for achieving high speed and cruise efficiency. However, unlike exposed tail rotors, when a directional perturbation was encountered, the Fantail rotor contributed little aerodynamically to correct it. Buried deep inside the duct, the Fantail rotor disc did not see a change in inflow until there was a relatively large change in the fuselage angle relative to the direction of flight.

At the same time, a relatively large change in Fantail pitch, from neutral, was required before any aircraft response was achieved. Due to the null zone, the aircraft would respond only after a large change in Fantail pitch had been commanded, which then resulted in higher-than-desired yaw rate. The aircraft state sensors would detect the undesired yaw rate and immediately command a reduction in Fantail pitch, which would put the Fantail back into the null zone with a yaw rate building in the opposite direction. The corrective command started the erroneous response going the other way. This back-and-forth yaw command resulted in a yaw oscillation labeled a "tail wag."

Fig. 19.9 The modified main rotor pylon introduced in 1999 reduced Comanche tail shake.

Fig. 19.10 Leading and trailing edge vertical tail slats were tried on the Comanche to counter persistent "tail wag."

A number of experiments were conducted to better understand the aerodynamics of the empennage. Leading and trailing edge slats were rapidly prototyped and flown to see if an aerodynamic solution could be created to address the fundamental lack of directional stability (Fig. 19.10). The slats did not show enough promise to pursue further.

A Gurney flap (Fig. 19.11) was added to the vertical tail to change the cruise pitch of the Fantail from neutral to a positive value, thus eliminating the null zone in yaw response. Although this worked reasonably well, it had two very negative consequences. First, the Fantail was now sapping power intended for the main rotor in cruise flight, reducing maximum forward

Fig. 19.11 The vertical tail Gurney flap addressed tail wag but imposed power penalties.

Fig. 19.12 An interim endplate configuration helped achieve a "tail wag" solution.

speed and cruise efficiency. Second, it introduced a serious problem in the event of a Fantail failure. Loss of thrust from the Fantail resulted in an uncontrollable yaw rate, even during autorotation. The Gurney flap solution was scrapped.

Ultimately, aerodynamic endplates were required to increase the total area of the vertical tail and provide enough inherent restoring moment to provide adequate directional stability in forward flight (Figs. 19.12 and 19.13). Several iterations were flown; not surprisingly, the final solution was associated with the largest additional vertical tail area. By carefully locating the position of these endplates, parasitic drag, which had the potential for increasing power required in sideward flight, was minimized. Even a small improvement in RCS was realized, because the endplates blanked view of the horizontal/vertical tail hinge point from certain angles.

SYSTEMS TESTING

The ability to expose operational pilots to the Comanche design early in the development process had its drawbacks. Operational pilots assigned to Training and Doctrine Command (TRADOC), the primary requirements developer, came from either the OH-58D Scout or AH-64 Apache gunship, both with technology pedigrees from the 1970s and earlier. The Comanche was to provide a great improvement in the helicopter crew interface, thanks to generational improvements in technology. Operational pilots, however, were devoted to their familiar platforms and, although open to brand new technology such as the digital map, resisted evolutionary changes to their legacy systems such as helmet displays. TRADOC pilots resisted even when

Fig. 19.13 The final Comanche endplate configuration fixed tail wag with acceptable Fantail power consumption and improved RCS.

Comanche concepts were introduced in the simulator with favorable results, but the design was different from what they knew.

The combined test team concept was supposed to address this by providing test-pilot-qualified Army pilots to the program. These pilots had been exposed to new technologies during their test pilot school training and follow-on research and development assignments. Combining that expertise with their operational experience was the best way to ensure new concepts could be presented, but only accepted if they met operational requirements. These Army test pilots were an inherent part of the CTT. But, the Army TRADOC operational pilots seemed to mistrust their test pilot counterparts as being "in bed" with the contractor. Thankfully, though, the Army test pilots provided insightful, articulate operational input to the program office.

In its September 1998 issue, *Army Magazine* had a cover story on CWO4 Armbrust, the first Army test pilot assigned to the program [3]. It detailed his biography and how he became an Army test pilot.

About Comanche, he mentioned how Army rules had to be modified to allow him to participate in envelope expansion flights, not just observe them after the contractor pilots had reached specific points. By September 1998 he had participated in about 25% of the flight time on the aircraft.

A comment on workload [3]:

> In the OH-58 A/C/D, pilots had to spend 70 percent of the time flying the aircraft, just keeping the greasy side down, and the other 30 percent being soldiers—doing the mission that they were out there to do. ... Now it looks as if we can reverse those numbers with the Comanche, with pilots devoting 30 percent of their time and effort flying the aircraft and the rest to executing the battle.

He also explained a problem with being an Army test pilot assigned full time to the program. People wanted to know his views on the abilities of the aircraft and how the program was going. He liked doing this, but always made sure he was not overselling the aircraft. He worked daily with the contractor test team and thought of them all as friends, but did not want to be seen as a Comanche PR guy. He had to remain objective.

One of his quotes accurately expressed the feelings of many on the program [3]:

> He [Armbrust] said that it [progress] had been especially difficult because until now the Comanche test program had only one aircraft. The original plan was to have five to seven prototypes at the start of the program. Having more than one aircraft allows parallel testing. ... It is serial with one aircraft. Every time a new widget needs to be added, it stops flight testing. If we had more aircraft, we could continue. We have been restricted to one aircraft and then hear complaints about how long tests take.

On 30 March 1999, the second prototype Comanche aircraft, tail number 95-0001, made its maiden flight. Aircraft 2 was intended to be primarily a systems development platform. Very little of the mission equipment functionality was ready, however, so it spent most of its time conducting surveys.

Initial infrared signature and acoustic assessments were conducted over the course of 1999. Although the detailed results were classified, data indicated that the aircraft would meet its operational requirements. One test mated the RAH-66 Comanche side-by-side with a UH-60L Black Hawk, which incorporated a state-of-the-art hover infrared exhaust suppressor. The raw photos showed a startling difference in the aircraft signatures (see Fig. 19.14, white indicates heat).

Photos of both YRAH-66 aircraft flying together provided an indication of program progress (Fig. 19.15). Marketing continued to be an important element of the Comanche program, though it seemed like the program was constantly selling itself. This was just one disturbing indication of the lack of total customer commitment at this time.

As 1999 was coming to a close, a major update to the flight control software was provided. This mainly addressed issues with turn coordination and high torque loads during hover turns.

Fig. 19.14 Infrared photo shows comparative signatures of the YRAH-66 (left) and UH-60L (right) hovering over a runway. (Hot areas are white.)

Fig. 19.15 The second Dem/Val aircraft introduced the more powerful T800-LHT-801 engines planned for production Comanche.

At the same time, a weakness of the highly integrated and very complex nature of the MEP became evident in a simple cooling fan. The MEP was programmed so that if the primary cooling fan indicated "failed" during startup, the MEP would not operate. There was such a high failure rate of this fan that the Comanche program ran out of test assets and could only operate one of the two prototype aircraft. The irony was that there was nothing actually wrong with the fan; the speed sensor that reported the health of the fan was dysfunctional. No consideration was made for the ambient temperature (a failed-fan signal would have prevented MEP startup even in the Arctic), and the pilot had no override of the function. Ultimately, although the desire for high levels of automated control might have reduced pilot workload, a degree of manual control would have been more appropriate for a military aircraft where availability could mean the difference between life and death.

SYSTEMS VALIDATION TESTING ADVANCES

Aircraft 2 entered a modification period that included additional instrumentation, so it could conduct more strenuous development testing. Aircraft 1 resumed flight test duties in January 2000 and focused on handling qualities improvements. Several iterations of the vertical tail endplates provided data for a final solution. Considerable experimentation with the yaw control laws addressed the high torque loads during hover turns, and a quite innovative solution was finally found.

Slope landings were conducted at an on-site facility built especially for Comanche. There were concerns about the control laws that commanded the transition from air mode to ground mode. The aircraft had to be able to land not only on level ground, where the left wheel touched first, but also on sloped surfaces where any wheel might make initial ground contact. Weight-on-wheels sensors were required for each gear leg. Furthermore, unlike on level ground, the aircraft on slopes had to remain easily controllable during the deliberate transition of touchdown of the remaining wheels. In air mode, the control laws attempted to maintain the trim attitude. But in ground mode, the attitude had to be ignored to allow the pilot to touch down on a slope of any direction.

The solution was to terminate air mode upon the initial ground contact, freeze the rotor trim at that point, and allow the pilot to cushion the touchdown of the remaining wheels. The process sounds simple, but it entailed designing weight-on-wheels switches with the proper sensitivity, desensitizing the control laws in ground mode, and allowing for a slow transition to neutral control position once all gear were on the ground. The entire solution was exacerbated by the high moment generation capability of the bearingless main rotor—which created the ability for substantial pitch and roll control on the ground—and high structural loads.

Foreseeing this issue with takeoff and landing control, the companies conducted a joint flight test on the Shadow aircraft in early 1992—in time to make inputs to the Comanche design. (See Chapter 8, Flight Controls.) Ultimately, this early risk reduction flight testing paid off. Virtually no issues were found with the takeoff and landing control laws, and the aircraft exceeded all of the specification requirements with virtually no flight control changes.

The only issue discovered during the landing tests was the long, relatively soft main landing gear struts (for crash survivability) compressed substantially during side slope landings, causing the aircraft deck angle to continue down slope as collective pitch was reduced to achieve full weight on wheels (Fig. 19.16). This was extremely uncomfortable for the pilots, and in some conditions resulted in close proximity of blade tips to the ground. The maximum slope angle requirement was 12 deg of side slope, and this was achieved. But at one point, almost 20 deg of deck angle was observed. The soft struts also affected rolling landings, so a new design was undertaken.

Fig. 19.16 Slope landing tests validated a combination of hardware and software changes made to adapt control laws to landing on inclines.

During earlier performance assessments, an issue was encountered during high sideward flight speeds. The problem was inherent in the design of the infrared suppressed exhaust (Fig. 19.17).

Fig. 19.17 The Comanche infrared exhaust-suppressing manifold created a series of venturis to dilute and shape hot exhaust.

Fig. 19.18 An overhead view shows Comanche tail boom slots that admitted diluting air for the exhaust suppressor.

As air entered the slot on top of the tail boom (Fig. 19.18), the venturis accelerated it across the exhaust. The large volume of air being pumped through the exhaust manifold cooled the exiting air sufficiently to reduce the infrared signature. In hover, the downwash of the main rotor helped add energy to the air flowing into the tail boom slots, adding to the efficiency of the infrared suppressor. As sideward flight speeds increased, however, the tail boom inlets eventually stalled, which prevented air from entering the suppressor cavity. The result was a rapid increase in the temperature of the air exiting the suppressor. During one flight, the lower tail boom exceeded temperature limits, which resulted in damage to the composite structure.

One of the propulsion flight test engineers had an idea to address the problem: install an air dam on top of the tail boom to prevent the inlet slot from stalling in sideward flight. A prototype air dam was installed on the tail boom to validate the theory (see Fig. 19.19). Obviously this was not an elegant or production-quality solution, but it did permit a quick flight evaluation.

Data indicated the theory had merit, so the doors surrounding the infrared suppressors were modified so that the slot was lowered below the top of the tail boom. This did not affect the operation of the suppressor in hover, but it again prevented the inlet from stalling. In June 2000, several flights were conducted to collect data on the effectiveness of the modification, both on aircraft performance and on the infrared suppressor. This provided design engineering the data needed to develop a permanent solution, which included a combination of exhaust modifications, changes to tail boom materials, and aerodynamic shaping.

Later in the test program, some additional mission systems were incorporated. The night vision pilotage system accumulated several hours of flight testing and demonstrated the potential of the combined systems. Other

Fig. 19.19 Comanche tail boom air dam.

Comanche mission systems tested included the digital map and the advanced diagnostic system.

One of the program changes that would have a major effect on flight test was the addition of the mast-mounted radar (Fig. 19.20). The Comanche radar was a smaller but more advanced version of the LONGBOW radar developed for the AH-64 Apache. Although the Comanche main rotor mast was, like that of the Apache, designed to accommodate a sensor, some changes were required to install a derotation unit and special fairing to interface with the radar. This activity was conducted on Comanche No. 2. First flight with the radar fairing was conducted on 10 October 2000. A series of envelope expansion flights concluded with full envelope clearance with a dummy radar unit

Fig. 19.20 Comanche tested with a mast-mounted radar fairing.

Fig. 19.21 A Hellfire missile mounted on the internal weapons bay door of the Comanche.

(aerodynamically and structurally representative, but without the electronics installed) by the end of the year.

Late 2000 and early 2001 focused on verifying the structural integrity of the various changes to the empennage and main rotor hub, in concert with the radar aero dummy. To optimize handling qualities, several iterations of the horizontal tail and vertical tail endplate incidence were flown. As the helicopter configuration stabilized, flight test expanded to include potential weapons systems such as the Hellfire missile. To verify the concept of internal weapon storage, a Hellfire missile was mounted on the center rail of a weapons bay door and deployed and stowed in flight throughout the flight envelope (Fig. 19.21). No problems were encountered.

HANDLING QUALITIES FOCUS

At this stage of the test program, it was appropriate to commence formal evaluations of the aircraft handling qualities. The Comanche was the first new helicopter to be subject to the ADS-33 handling qualities specification. This specification evaluated short flight scenarios, known as "mission task elements," which had very precise performance requirements (Fig. 19.22). Pilots flew the scenarios as aggressively as possible while attempting to meet the required task parameters.

One scenario required an aggressive acceleration and deceleration. From a hover, the aircraft had to reach maximum acceleration within 1.5 seconds, achieve 50 kt, and then decelerate to a hover over a precise point. The pilot had to maintain altitude below 50 ft above ground, lateral track within 10 ft, and heading within 10 deg. During deceleration, the helicopter had to achieve

Fig. 19.22 Comanche No. 2 conducted ADS-33 mission task elements including (left) aggressive acceleration and deceleration to (right) a precision hover.

at least 30-deg nose-up attitude. Other maneuvers were similar but focused on different aspects of helicopter handling.

Trained test pilots conducted the maneuvers mainly because they had the skills required to rate the workload associated with a particular level of performance. Needless to say, these were not maneuvers one would attempt on an immature prototype helicopter. The fact that they were conducted on YRAH-66 No. 2 in mid-2001 was an indication of the progress the Comanche team had made in aircraft development.

Aircraft 2 was put into a modification period in June 2001 and would not fly again until May 2002.

A major milestone was met on 1 June 2001, when the upgraded T800-LHT-801 engine was flown on aircraft 1. The remainder of 2001 focused on the increased power with respect to structures, performance, and handling qualities. In addition, a new anhedral blade tip, designed to improve performance and reduce acoustic signature, was flown (Fig. 19.23). This was similar to the blade tip designed for the improved UH-60 Black Hawk rotor blade and proved to be essentially trouble free, even during a high speed dive where the tip inadvertently exceeded Mach 1.

The aircraft was designed to accept external wings for mounting missiles and rockets when low RCS was not required. To assess the operational capability in this configuration, a set of external wings was installed (Fig. 19.24). The flight test wings were weighted to replicate the maximum load anticipated during an external stores mission.

A very successful 2001 closed out with two days of marketing support, where an IMAX movie of the aircraft was filmed, clips of which are still included in many aviation documentaries.

In January 2002, aircraft 1 completed evaluation of a new version of blade dampers, with promising results. In addition, new multistage landing gear struts, which improved stiffness during slope landings, were installed, and the slope tests were repeated with good results and much more comfortable deck angles.

Fig. 19.23 Anhedral blade tips.

Aircraft 1 was then put into another modification period to update all its systems to the latest configuration. Little did we know, the aircraft would never fly again.

In May 2002, the second prototype emerged from its modification period with additional mission system capability. Intensive systems development was planned, but due to delays in software deliveries, not much of the mission system was available until late in the year. One accomplishment of significance was the flight of the second-generation night pilotage system (Fig. 19.25). This included an advanced infrared imaging system, wide field-of-view helmet display (Fig. 19.26), and all-aspect helmet tracking system.

Fig. 19.24 Aircraft 1 tested with external wings.

Fig. 19.25 Aircraft with the Comanche night vision pilotage system, including the multisensor nose turret.

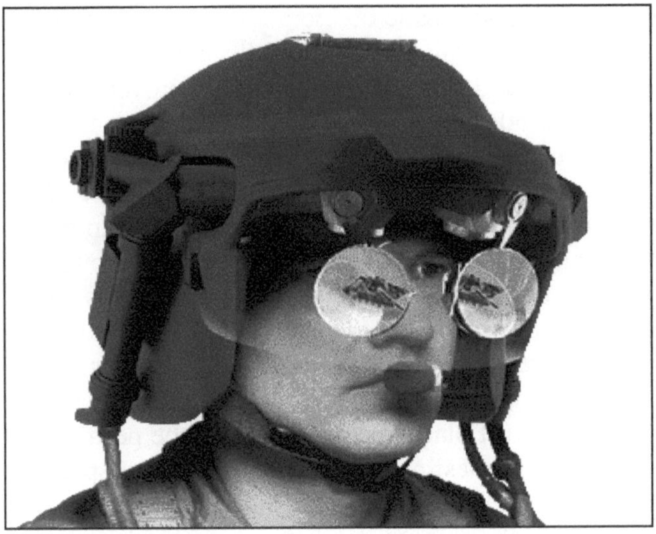

Fig. 19.26 Wide field-of-view helmet display.

The first flight took place in June 2002. Many automated image optimization functions were available for evaluation, and the system showed real promise. After several flights totaling approximately 10 hours of flight test, the system was removed. At this time there was considerable controversy within the Army as to the requirement for inclusion of image intensifiers in the night pilotage turret; this uncertainty led to putting further flight test on hold.

Shakedown of basic mission system functionality, including improvements to the caution alerting and system status displays, encompassed several flights. Results met expectations, although the system continued to suffer from sudden in-flight shutdowns. There was never a safety issue, because a backup display with data sourced from the flight control computers provided safety of flight information and system status. The MEP reset switch, which would restore the cockpit displays, also worked in most situations.

Although the helmet integrated display and sighting system (HIDSS) demonstrated a dramatic improvement over the legacy system, the helmet tracking system in flight test never truly achieved the kind of fast, jitter-free response that would have put the complaints to bed. Evaluators felt the system would have been worth fielding, but required ongoing improvement. Innovative design changes were being developed in the lab, but never had the opportunity to get to flight test. Interestingly, the contact analog approach later migrated to other helmet display systems in fixed-wing fighter aircraft, and the helmet trackers indeed received the same improvements being envisioned for Comanche.

The remainder of flight testing in 2002 focused on handling qualities improvements. Noteworthy was the engagement of higher orders of flight control functionality, including automatic deceleration to hover, hover position hold, and altitude hold. Although this type of capability existed on legacy models, the FBW system demonstrated an order of magnitude improvement in precision and robust tolerance to winds and turbulence, while retaining, through the unique trim controller, the capability for full pilot control with the modes engaged.

Comanche system maturation continued as the flight test program progressed into 2003. Additional modifications were made to the main landing gear struts to enhance the quality of landings. The struts continued to provide nice cushioning during normal landings, but improvements were made in the stiffness of the struts as weight settled on the gear. During higher-speed rolling landings, this improved stiffness also improved aircraft stability during roll out, improving the pilot's ability to maintain heading while slowing to a stop.

The Comanche's novel blade-mounted pitot probe permitted the accurate calculation of airspeed, sideslip, and wind, even at lower speeds. This system is described in detail in Chapter 3. Flight testing of the system indicated a high level of confidence in the technology; however, as with most new pitot systems, the blade-mounted solution required considerable calibration. In particular, the system accuracy varied with blade angle of attack. Fortunately, the data to determine angle of attack were easily acquired, and corrections were programmed into the software.

The Comanche aircraft was faster than the chase helicopter, so a fixed-wing aircraft was needed to complete the air data calibration at higher speeds (Fig. 19.27). The U.S. Army provided a T-34 from the flight test center at Fort Rucker, Alabama, modified with a calibrated air data system.

Fig. 19.27 The T-34 calibrated airspeed support aircraft chasing the Comanche during high-speed air data calibration.

Early in 2003, the Army finally acquiesced to allowing a non–test pilot to fly the Comanche. TRADOC had formed a team of experienced aviators to "get the word out" regarding the capabilities of the aircraft and decided that the spokesperson for this group should have first-hand experience with the aircraft. Consequently, CW5 Steve Saunders became the first operational Army aviator to fly the YRAH-66. As expected, he had nothing but accolades about the ease of handling, performance, and maneuverability of the aircraft.

The Army continued its marketing effort with an unusual event. There was a popular TV show at the time titled *American Chopper*. Each episode was devoted to the building of a themed motorcycle, and the Army commissioned the build of a Comanche version. At the completion of the build, the TV show crew traveled to West Palm Beach with the Comanche motorcycle to film with the aircraft (Fig. 19.28). It was a nice break for a lot of people who worked so hard to develop the aircraft. It was no secret that many of the mechanics who worked on Comanche were bikers and were especially thrilled to have the TV show on site.

Other Comanche systems that continued development during 2003 included the night pilotage system and helmet display, improvements in moving map and flight plan software, flight control software, and anti-ice systems. In addition, thermal and acoustic surveys were conducted as the aircraft approached full maturity. Additional handling qualities evaluations were conducted with positive results. Comanche was becoming an operationally capable aircraft.

Fig. 19.28 The Comanche made a marketing appearance on the *American Chopper* TV show.

Late in 2003, the Army requested a high-ranking general officer be given a demonstration flight in the aircraft. This was a very positive indication that the Army was gaining full confidence in the aircraft, and hopefully the program in general. On 3 December 2003, Lieutenant General Richard Cody, Army deputy chief of staff, flew the aircraft. The full flight envelope was demonstrated, and the general took the controls for much of the flight. His postflight comments were overwhelmingly positive, marking the aircraft as the future of Army Aviation. It's a little hard to believe that the general was unaware that in less than two months, the program would be canceled.

Ironically, the final two flights of the Comanche, in February 2004, were devoted to a survey of the infrared signature of the aircraft against actual threat systems. Although the results remain classified, comments from the operators of the threat systems confirmed that Comanche would easily meet the operational requirements. Yet one of the reasons given for program cancellation was that it would not meet infrared signature requirements—one of the continuing mysteries of the Comanche program.

REFERENCES

[1] Sikorsky Flight Test Report, Flight 9 Summary, September 20, 1996

[2] James T. McKenn, "First Flight Boosts Comanche Program," *Aviation Week and Space Technology*, 15 January 1996, pp. 44–45.

[3] Dennis Steele, "Comanche Pilot—Living the Dream," *Army Magazine*, September 1998, pp. 24–31.

Chapter 20

WEIGHT

BILL HARPER AND ART LINDEN

Aircraft weight was perhaps the most critical technical parameter on the Comanche program. During the conceptual stage, a gross weight of 8000 lb was thought to be appropriate. This grew as the concept evolved and additional requirements were added. The Army eventually specified an empty weight of 7500 lb, believing that by limiting aircraft weight to that value it could prevent requirements growth and limit people from adding bells and whistles.

Bill Harper relates how this evolved:

> In February 1984 Charles C. Crawford published his report on *Scrubbing Requirements to reduce LHX Size/Cost to Meet Affordability Objectives Specified by the Undersecretary of the Army.*
>
> Charlie started with the AH-1S Cobra aircraft with full weapons load and sufficient fuel to fly the SCAT mission profile. This resulted in a 9910 lb. takeoff gross weight (TOGW). Technology improvements were incorporated to reduce the size and weight of the aircraft and the power required. Advanced technology engines were installed, ...composite technology was assumed for the airframe ... the number of missiles and ammunition for the turreted gun was reduced but air-to-air stinger missiles were added. Adding the full mission equipment package ... [brought] the aircraft TOGW [to] 9215 lbs., with an empty weight of 6500 lbs.
>
> This was his dual crew LHX SCAT configuration. He also did a utility configuration, single pilot configurations, and tilt rotor configurations. A side note, the dual crew tilt rotor LHX SCAT configuration empty weight was 1444 lbs. heavier than the helicopter configuration, and the dual pilot helicopter was 603 lbs. heavier than the single pilot helicopter.

> Armed with this information, in 1985 Mr. [Robert] Hubbard [Army deputy program manager,] informed [Army program manager] Gen. Andresen that the study did not include the survivability features required for the LHX, and this is when Andresen dictated the 7500 lbs. empty weight requirement.
>
> . . . In almost every case where a component weight was reduced, a successful tech base program was identified or started to substantiate the weight reduction assumed. While ACAP included signature reduction requirements, they were satisfied with shaping and parasitic radar absorbing material to reduce radar signature. Since there remained a risk with the weight for signature reduction, a full-scale model test activity was added to the Dem/Val contract to help verify the weight needed. The Advanced Rotorcraft Technology Integration (ARTI) contract was to help verify the MEP weight.
>
> The 1989 Boeing Sikorsky weight empty estimate was 7489 lbs.

A Note of Confusion

Starting the program off in 1990, the Army admitted that it was not sure that all of its detailed requirements could be achieved within the 7500-lb weight limit. The competing contractors were directed to "tailor" the requirements, meaning that we should delete whatever was necessary to get to the 7500-lb limit. We could use whatever design and simulation tools we had to determine what to leave out.

This meant that the two competihg teams would be submitting two different designs.

The Optimization Process

Our plan was to use Dem/Val as a tool by which to conduct the tradeoffs, establish the full-scale development aircraft specifications, and optimize the aircraft requirements to produce the best combat solution within the weight limit.

This set the stage that everything had to earn its way onto the LHX platform. The optimization process defined an aircraft that maximized the operational effectiveness, survivability, and suitability of the weapon system within the weight and cost constraints. The results of completed trades, design analyses, tests, demonstrations, and operational analyses studies were all utilized in the optimization process.

The process used linear programming. The objective function was to maximize military worth, and the two main constraints were the weight and cost limits. The first step in the optimization process was to identify the candidate options to be considered. These could be design options such as type of auxiliary power unit and gun type; performance attributes such as yaw rate, load factor, and targeting range; or survivability levels such as crashworthiness and ballistic tolerance. The candidate features or options were then attributed to quantify the differences between the options.

This process was first used to support the definition of the preliminary design in July 1989. The next application was to support definition of the best technical approach (BTA) for the Dem/Val proposal. This process was eventually adopted by the Army for the MS II analysis of alternatives (AoA) evaluation.

Boeing Sikorsky's "optimized" 1991 proposed design included almost all of the Army's required features; however, it also did the following:

- Reduced the crashworthiness level from 42 to 38 ft/s
- Reduced the structural limit load factor
- Reduced the fuel tank ballistic protection from 14.5 mm self-sealing to 12.7 mm
- Tailored the main gearbox rating
- Delayed for a future preplanned product improvement:
  - Radar
  - Cockpit fragmentation barrier
  - Leg armor

The competing SuperTeam design had a different list of tailored requirements, somewhat more extensive than ours.

Army Restructure Number One

After our original proposal was submitted, but before we were put on contract, the Army made a major change in the development approach. Originally the program was to be full-scale development (FSD) of the full weapons system. The new revised approach kept us in the Demonstration/Validation phase with the FSD decision put off to 1994. The causes for this were a concern that the technology might not be ready for the ambitious schedule combined with a reduction in available Army funding.

As a result of this, we now had prototype aircraft that were not the configuration we would have in production. This would cause extensive problems in the future because we now had two designs—one for the experimental prototypes and a second for the promised production design.

It also meant that we put off many weight-reduction ideas to the later FSD program.

The Design Process

As the design process began, each design team was given its requirements for system weight. These were developed from an extensive weight estimation process and were clearly as tight as we dared to meet the overall weight requirement. Then, to give the designers an additional challenge, each was

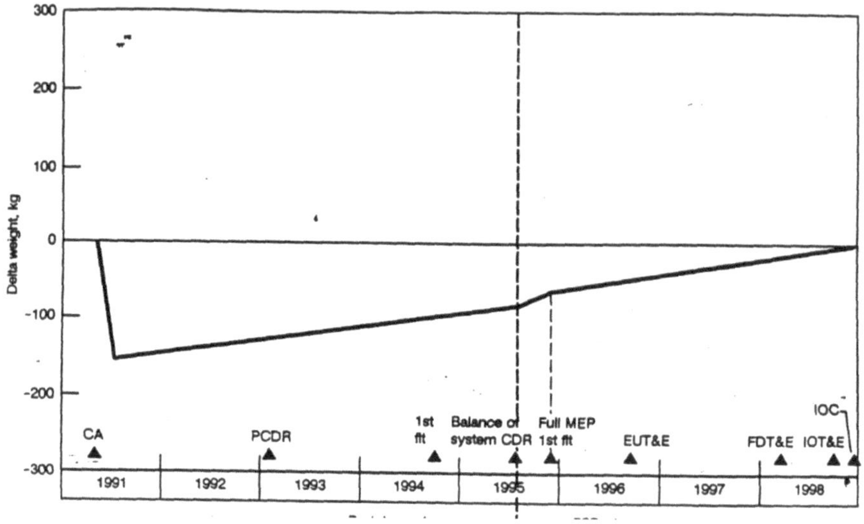

Fig. 20.1 Full aircraft planned value profile.

given a planned value profile (PVP). (See Fig. 20.1.) These PVPs were developed recognizing that weight almost always grows during the design process, so we had to start at a lower value if we were to end at the required point. All these subsystem PVPs added up to the full aircraft PVP.

Years later, in retrospect, our chief of design, Bruce Kay, had these thoughts on the process:

> Weight control/management was a constant battle that lasted through the entire duration of the program. The designers believed that this situation occurred because of overly optimistic, unrealistic parametric weight allocations. From the outset, the Army always briefed LHX as a 7500 pound, $7.5M helicopter. There were extreme competitive pressures to meet this objective that led to optimistic weight predictions. The weights engineers were uncompromising and ideological. They firmly believed that aircraft weight was primarily determined by the allocations distributed and managed by the weights department. If left to the discretions of the designers and structures analysts it would always be heavier. They also believed that parametric weight predictions are more accurate than bottoms up preliminary analyses. These tenets are true within the envelope from which the parametrics were extracted. Unfortunately the parametrics used for LHX were derived from large utility and cargo helicopters, and, in the view of the designers, did not adequately address the unique features associated with small low RCS attack helicopters.
>
> So, from the outset, weight control was a struggle. To make matters worse there was tremendous procrastination over any weight growth. Weight growth had to be signed off by the Chief of Design and/or the Chief Engineer. During this process, some growth was mitigated, but the

> remaining growth was carried in a weight growth risk category for extended periods until all avenues of avoidance were eliminated. On the other hand there were continuous Weight Improvement Programs (WIPS) that were grasped with minimal substantiation. Even when rejected as unfeasible by the IPDTs, the WIPS were carried almost forever. They never died, and constantly re-surfaced for re-examination. Working toward difficult weight allocations cost a lot of time and money, and ultimately led to several weight target increases during numerous program restructures.

ADDITIONAL PROGRAM RESTRUCTURES

Throughout the 13-year program, a major condition that affected aircraft weight was the continual program restructures. These occurred generally every two or three years and were usually due to changes—reductions—in the Army's available funding. Each program restructure required us to rebaseline the aircraft weight. We reviewed the progress that had been made up to that point and updated the empty weight estimate. Usually it went up, due to some optimistic original assumptions that had not been achieved.

Also impactful was the Army's opportunity to add requirements into this new baseline. Internal Army reviews led to more people advocating for new features, as well as other technical experts using the opportunity to update their requirements for the latest new technical breakthrough. The advocates strongly believed that these additions were necessary to keep the program sold and were therefore included as we moved forward.

What started out as a 7500-lb empty weight aircraft with a 10,088-lb mission takeoff weight, by 2004 had grown to a 9609-lb empty weight aircraft with a 12,838-lb mission weight.

Keeping the required aircraft performance levels as the weight grew became a major issue and eventually included two growth engines and an increase in main rotor diameter.

The second program restructure, in 1993, increased the maximum allowable empty weight by 450 lb, to 7950 lb. This allowable weigh increase was done with a "Design Flexibility" clause. The Army's Tom House showed that in addition to accommodating a weight empty growth to 7950 lb, we could also carry all the mission and aircraft survivability equipment (ASE) kits (total of 274 lb) and still achieve the 500 ft/min vertical rate of climb, providing we used nearly 100% engine Maximum Rated Power (MRP) (instead of 95%). He also indicated that to carry LONGBOW (which adds 570 lb), the engine must be grown 7% or so to achieve the 500 ft/min climb.

So we added the features required by the Army, primarily the LONGBOW radar. Other additions were a high frequency radio, infrared (IR) and radiofrequency (RF) jammers, and additional armor. Again, factions in the Army believed these additions were required to keep the program sold. A 12% increase in engine power was then required to maintain aircraft performance.

A third restructure, the "Streamlined" program in 1994, put integrated communications, navigation, and identification avionics (ICNIA) into the program and added a fifth-percentile female requirement, the integrated mission support system (IMSS), air bags, use of beryllium aluminum, and an image-intensified charged coupled device TV camera. It eliminated the gun and electro-optical target acquisition and designation system (EOTADS) and delayed LONGBOW development.

Weight status at this time was as follows, as reported by *Aviation Week & Space Technology*, 1 May 1995 [1]:

> Aircraft weight also is on target. The current Comanche has a baseline weight, set by the contractors, of 3,535 kg. (7,779.2 lb.), and weight estimates are 'slightly exceeding this figure.' The aircraft, however, is below its specification weight of 3,626 kg. (7,977.2 lbs.).

The next restructure (the fourth), in May 1995, was the result of serious threats to the future of the program. Program funding was reduced by $2 billion, and the Office of the Secretary of Defense (OSD) was considering canceling the program. This restructure, called the early operational capability (EOC), moved the LONGBOW to (Pre-planned Product Improvement) and added SADA (Standard Advanced Dewar Assembly). Link 16, SATCOM (Satellite Communications), and LO (Low Observable) kits. By now the aircraft weight empty had grown to 8265 lb. This incorporated both the new Army requirements and updated weights for numerous aircraft components that did not hit their weight targets.

By 1996, aircraft 1 was flying, and a weight empty check was appropriate. A point paper developed by Darrell Harrison and Mike Ritchey of the Army program office, dated 11 July 1996, was sent to Deputy Chief of Staff, Operations (DCSOPS) in response to their views on the program:

> ...the Comanche weight empty for production was specified at 7500 pounds. Disregarding instrumentation, the first prototype aircraft does weigh 8780 pounds. But this prototype was never planned to meet the production weight goal. As with most aircraft programs, we have very definite plans to reduce this weight during the development process.... As part of the current restructuring of the program, we have also decided to put our major production specification emphasis on aircraft performance, and not weight. Specifically, the latest specification has 500 feet per minute vertical rate of climb as an absolute requirement. This is certainly more important to the user of the aircraft than empty weight. And the Army has budgeted the money in the development program to achieve this performance requirements.

With the production design now at 8265 pounds, *Aviation Week & Space Technology*, in its 21 October 1996 issue, quoted Art Linden, the Boeing Sikorsky program director, as saying that "8,800 lbs. [is] seen as the top limit for a production aircraft to maintain performance" [2].

Moving forward a few more years, in its 27 September 1998 issue, *Helicopter News* said: "The Comanche program manager [that again would be Art Linden] continues to downplay the aircraft's ballooning weight. 'The bad news is that Comanche's weight has increased by 900 pounds...but the good news is that Comanche's weight has only increased by 900 pounds,' he added" [3].

Inside the Army added a further (and alarming) update in its 14 December 1998 issue [4]:

> According to the [CAIG] group "The contractor's production weight goal is ten percent below the current aircraft weight."
>
> "The reduction will require a nearly complete redesign and the use of exotic materials, some of which have not yet been used on aircraft, to achieve weight requirements" the memo states.
>
> Comanche proponents firmly disagree. . . . the helicopter's weight now includes calibration and other testing gear that will not be present on the fielded aircraft.

And from *Flight International*, 26 May 1999 [5]:

> RAH-66 internal fuel capacity is unchanged, but mission fuel weight has been increased by 180kg (400lb) to 870kg [1918lb].
>
> . . . "The aircraft could always take that weight," says Daryl [sic] Harrison, US Army deputy programme manager, noting that the first prototype is operating at over 5,900 kg [13,010 lb].
>
> . . . The US Army at the same time has increased the RAH-66's design target weight empty to 4,060 kg [8952 lb] and raised its threshold not-to-exceed weight to 4,220 kg [9305 lb].

By 20 September 1999, in an article by Stanley Kandebo, "GAO Hits Comanche Development Plan," *Aviation Week & Space Technology* reported that [6]:

> The report also resurrects the specter of weight growth, saying the Army now plans to modify the Comanche rotor to provide more lift capability as a hedge against weight.
>
> The Army says the change in rotor diameter from 39 ft. to 40 ft. is true. "As we started negotiating the EMD contract, it became obvious we could spend significant amounts of money to eliminate every ounce of weight from the aircraft that was possible. To save money on the weight reduction, we decided earlier this year to increase lift and to grow the main rotor diameter by a foot." [Army Program Manager BG Joe] Bergantz said. He added that the Comanche was originally designed with clearances for a 40-ft.-dia. main rotor.

The problem remained that the prototype aircraft did not reflect the weight required for the production design. Production designs that were not funded included:

- Steel environmental control system (ECS) bleed-air ducts
- Noncommon chassis for electrical power system controllers
- Steel propulsion system components
- External main gearbox lubrication tubes
- Fluid-elastic dampers
- Firewall material
- Nuclear, biological, and chemical (NBC) compliant material
- Composite material density
- Composite gearbox housing
- Radar cross-section (RCS) compliant material
- Various other prototype design solutions

These remained major concerns in the transition to EMD.

AND YET ANOTHER RESTRUCTURE

Soon after we entered the EMD program, we got into another restructure. This extended the program schedule by 7 years and introduced the aircraft in phases. The intent was to reduce technical and cost risk, which both the Army and Boeing Sikorsky program management teams thought was too high.

The aircraft would be introduced in three blocks, building up to the final production design. The first block went back to the idea of delivering just the reconnaissance configuration and would also be use for initial training. The second block updated the computers and added Link-16 and unmanned aerial vehicle (UAV) control, and the third block added the full mission equipment package (MEP), external stores (EFAMS), air-to-air missiles, and microclimate cooling. This third element also included major weight reduction initiatives.

The restructure proposal started with the then-current weight empty at 10,219 lb and the primary mission gross weight (PMGW) at 13,338 lb. (The 10,219 lb empty weight caught some in the government a little off guard because Boeing Sikorsky had recently signed a contract implying/stating the aircraft could be brought in under 9600 lb.)

A weight statement from that time gets close to this value and details where the growth came from:

With the block development approach, each block had a different weight empty goal. Although functionality and capability increased with each block, there were weight reduction plans to reduce weight empty back to 9609 lb.

Block I would have an empty weight to 9925 lb and a mission weight of 13,010 lb. Block II added more capability and also included weight reduction

Original weight empty	7500 lb
LONGBOW radar provisions	64
Add combat kits	149
Other new requirements	319
Additional 2000/02 requirement additions	705
Prototype development weight growth	421
Growth rotor	439
Growth engine (801)	52
801+70 SHP engine	60
Weight reductions lost	277
Contingency for weight growth	200
Total weight empty	10,186 lb

to bring the empty weight to 9609 lb and the mission takeoff weight to 12,449 lb. Block III got to the production design and kept the empty weight at 9609 lb while adding in the full MEP and other final items. To do this, a further 817-lb weight empty reduction was optimistically projected.

The Boeing Sikorsky design team soon realized that we could not on our own create a closed design solution at 9609 lb that was acceptable to the Army. Something had to change; systems/capabilities removed, maneuver spectrum reduced, change in rotor system—something.

There were 500 lb of parasitic radar cross section (RCS) reduction features on the aircraft. One option might have been to define these as kits and reduce the maneuver envelope when the kit items were on the aircraft. We could then use in-ground-effect hover for the first segment of the mission when the aircraft was in the LO configuration. We might go back to single-store stations on the EFAMs or go to a 50-caliber gun. The bottom line was that we had to ask the Army to reduce the users' expectations.

The final weight reduction of 817 lb for the production design included 207 lb from mission tailoring and 610 lb of aircraft design savings.

REQUIREMENTS CHANGES VS DEVELOPMENT GROWTH

A great debate among those involved with the Comanche program was how much of the weight growth was due to the Army adding requirements, and how much was due to development issues with the original design? And were there other considerations?

Going back to 1993, we were visited by a representative from OSD who ran all the government R&D facilities. With him were all the directors of the labs. Their task was to coordinate their activities to support the technology needs of the program. We used optimistic weight assumptions based on their projections. As Army funding was reduced, the lab support disappeared, along with our optimistic weight projections.

As for growing Army requirements, there were a total of 85+ user operational requirements document (ORD) changes from 1991 to 2001. These included:

- Radar warning receiver (RWR)
- Radar frequency interferometer (RFI)
- Armor
- High frequency (HF) radio
- Radar cross section (RCS) combat kits
- Link 16
- SATCOM
- SADA
- Fifth-percentile female
- Integrated mission support system (IMSS)
- Air bags
- Image-intensified charge-coupled device TV camera (I^2 on nose)
- Internal weapons bay fuel tanks
- Two-axis seats for the fifth-percentile female
- Chemical detectors
- Two-station external fuel and armament management system (EFAMS)
- Increased gun rounds
- Growth T800 engines

From the contractor side we had many challenging component weight estimates that failed to materialize. In a number of cases we just had poor weight predictive methodology. And, of course, ground and flight testing led to design fixes that usually added weight.

By 2002, the empty weight had grown by 2719 lb to 10,219 lb from the original 7500 lb. It looks like 926 lb could be tied directly to aircraft development growth, 1362 lb to capability growth, and 431 lb to bookkeeping changes (i.e., moving some mission equipment from useful load to empty weight).

This conclusion is open to debate. As the weight grew, parts of the basic aircraft (such as the rotor and transmission) had to be redesigned for more power. Was this due to Army-directed additional requirements or to development growth in the basic design? It's all arguable. A general conclusion by the authors is that it should be considered as a 50/50 situation—50% of weight growth due to requirements increases and 50% due to development growth in the basic aircraft design.

The resulting production design was to have a mission takeoff weight of 12,836 lb.

A REAL-WORLD CHECK

When U.S. Army Lieutenant General Cody flew the aircraft in December 2003, it had a TOGW of 13,018 lb and, according to our chief program pilot

Rus Stiles, "…it watered his eyes." Rus also mentioned that we flew the aircraft a couple of times at ferry weights, which were in the neighborhood of 16,000 lb. "We had ballasted wings, and added ballast to the ammo bay and weapons bay doors. In all cases, the aircraft flew remarkably well in part due to the model following fly-by-wire system which, unlike conventional controls systems, compensated for changes in weight and CG. Even putting the radar above the rotor wasn't really noticeable."

APPENDIX A: A "WEIGHT KILL TEAM" EXAMPLE

During the life of the program we had many weight reduction exercises. As an example, here is one from 2001:

In late 2001, weight growth was again an issue. The 30 October 2001 EMD weight empty status was at 9717 lb. Various weight risks were projected to grow this to 10,006 lb. And following the planned value profile, it was estimated that this would grow further to 10,206 lb for the production design.

We put together a "weight kill team" to attack the issue. The focus was to explore ways of bringing the Block III weight empty back down to the goal and prevent additional weight growth during the remainder of the development program. This looked at a large number of weight reduction options, most of them rather small, but they all would add up.

We started with a comparison with other helicopters (Fig. 20A.1). It was concluded that at this weight, Comanche compared favorably with other helicopters and complied with the performance weapons system specification (PWSS).

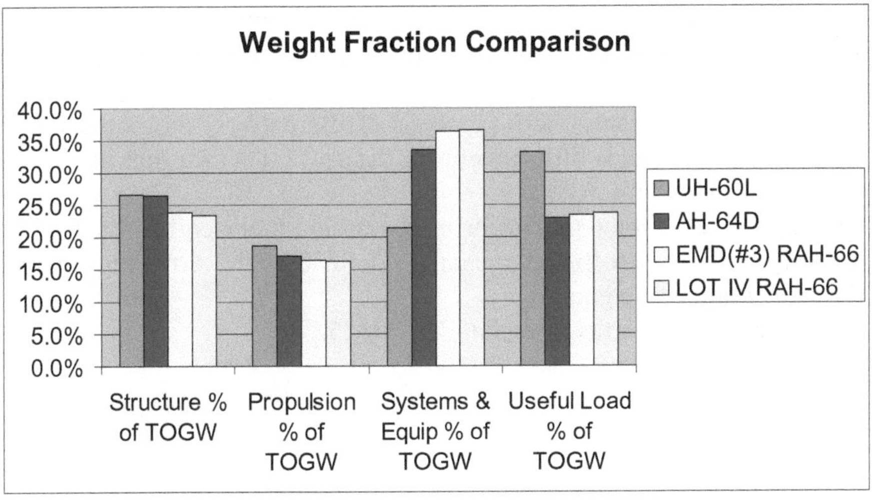

Fig. 20A.1 Weight fraction comparison.

ADDITIONAL WEIGHT REDUCTION OPPORTUNITIES

	Weight (lb)
B07 Delete MCC fault tolerance	-6.0
B12 EPLRS as Optional Equipment	-10.0
B12 HF (assumed conformal Antenna) as Optional Equipment	-22.5
B12 L16 except IFF/L16 PA as Optional Equipment	-7.2
B12 L16 + Alternate IFF PA as Optional Equipment	-3.6
B12 Delete HF, wire and antenna	-13.0
B12 Install SATCOM	+22.0
B09 Delete I2 on Nose, add I2 Helmet Camera (+5 lb)	-30.0
B09 Delete second PCD & wiring	-3.8
B09 LWR as Optional Equipment	-10.0
G02 Design for 320 rounds in lieu of 500	-10.0
G02 Restrict 20mm gun traverse in azimuth direction	-15.0
M04 Eliminate CSMU	-7.0
A10 OTTIS wiring and bracketry option	-7.6
A02 Take PMA off, and put log books on displays. CIK capability	-0.3
A10 Take PMA off, and put log books on displays. CIK capability	-1.3
Total Weight Reduction Options (less GFE)	-280.4

Fig. 20A.2 Additional weight reduction opportunities.

The weight kill team then found additional opportunities to reduce the weight further by 316 lb. This was 280 lb on the aircraft and 36 lb on the engines. The aircraft weight savings were as shown in Fig. 20A.2.

At this late date there was still a discussion of the need for the Comanche radar. A trade study considered putting the radar on only one third of the fleet, reducing the weight of the other two thirds of the aircraft (Figs. 20A.3 and 20A.4).

Note the last comment: "AOA [analysis of alternatives] shows that adding Radar to Comanche adds little benefit for the cost." This was not included in the final recommendations.

The analysis also noted that the primary mission fuel weight grew from the original 1588 lb to 1991 lb due to changes defined by the Army plus that due to flying at a higher weight.

The conclusions chart is shown in Fig. 20A.5.

COMANCHE RADAR OPTION

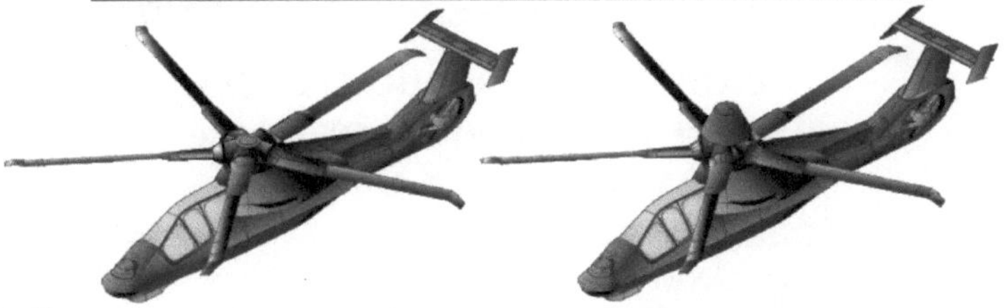

Description

Install A-Kit items on dedicated CR aircraft only. Save weight impact for 2/3 of fleet.

DTC Impact $-273,063

Kit Weight	
A-Kit	
Item	Weight Pounds
CR/ADS Standpipe	8.7
ADS Slipring	0
Standpipe Wire Harness	8
Lower Clamp Fitting	1.9
Center Support	17.6
Transmission to REU W	13
MEP Modules	13
Rotor Fairing w/cover	12
Fusion Modules	8
Total	82.2
B-Kit	
MMA/REU/Pedstal/Dero	267.4
Total	349.6

Fig. 20A.3 Comanche radar option.

COMANCHE RADAR OPTION

- **Why penalize Performance of 2/3rds of the Fleet by adding 82.2 lbs for possible downstream need?**
- **Why accept the added cost ($273K) to every aircraft?**
- **AOA shows that adding Radar to Comanche adds little benefit for the cost.**

AOA Result

Comanche with EOSS and ATDC does the Job And RADAR adds little for 82.2 lbs

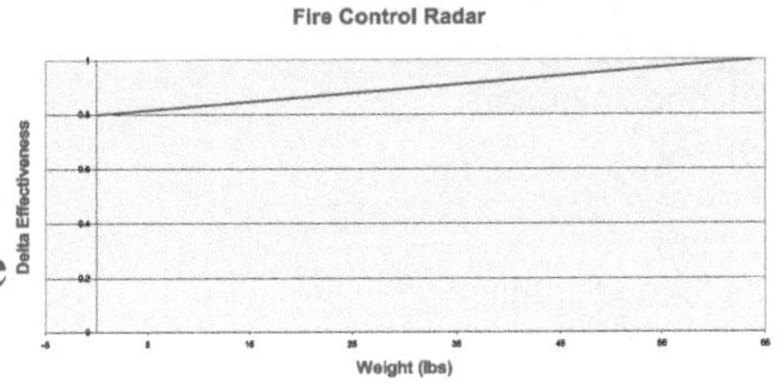

Fig. 20A.4 Comanche radar option.

OPPORTUNITIES FOUND SUMMARY
KILL TEAM RECOMMENDATION

- **WEIGHT EMPTY REDUCED BY 316.4 lbs.**
- **USEFUL LOAD SAVINGS OF 243.9 lbs.**
- **DTC SAVINGS OF $978,000**
- **NRE COST COVERED WITHIN THE RE-ALIGNED PROGRAM**
- **MINOR IMPACT ON PRIMARY MISSION EFFECTIVENESS**
- **INCORPORATED BY LOT IV OR EARLIER**

Fig. 20A.5 Opportunities found summary kill team recommendation.

Appendix B: Weight for RCS Reduction

There was always an interest in how much the radar signature reduction cost in weight. As of 2002, the weight of the RCS reduction approaches was as shown as follows.

	Empty Weight (in lb)	LO Kit (in lb)
Rotor		
Hub fairing		79.4
Blade		
Leading edge treatment	24.2	
Rotor body interference treatment	14.6	
Trailing edge treatment	?	
Additional tracking weights	7.1	
Torque tube		
Fairings		34.3
Removable		19.8
Inboard cuff fairing closeout		9.1
Bolt covers		12.4
Subtotal	45.9 lb	155.0 lb
Fuselage		
Conductive seals on MLG, doors, etc.	14.6	
Main rotor pylon treatment	42.6	
Forward and mid fuselage RBI treatment	66.7	
Gun fairing		5.0
Exhaust treatment, including baffles	13.5	
Forward fuselage lower chine		7.5
Upper tailcone RBI treatment	19.2	
Lower tailcone treatment	1.6	
Fantail shroud	42.6	
Vertical tail	2.3	
Horizontal tail	5.0	
Fantail blades and spinner	0.5	
Canopy transparencies	16.5	
Subtotal	225.1 lb	12.5 lb
Total	271.0 lb	167 lb

REFERENCE

[1] "Comanche System Made of New Alloy," *Aviation Week & Space Technology*, 1 May 1995, p. 65.
[2] David A. Fulghun, "Tests Confirm Low-Observability of Comanche Scout Helicopter," *Aviation Week & Space Technology*, 21 October 1996, pp. 26–27.
[3] *Helicopter News*, 27 September 1998.
[4] Top Pentagon Cost "Analysis Warn Comanche Is Underfunded, Overweight," *Inside the Army*, 14 December 1998.
[5] *Flight International*, 26 May 1999.
[6] Kandebo, S., "GAO Hits Comanche Development Plan," *Aviation Week and Space Technology*, Vol. 151, No. 12, 20 September 1999.

Part Two

THE COMANCHE PROGRAM

Chapter 21

Early Program and Contract Activity

Arthur W. Linden

In addition to all the technical activity on the program, the Army went through many iterations on fitting its desired LHX/Comanche program into realistic levels of funding and determining what type of contracts were appropriate. Numerous plans were developed prior to the April 1991 selection of Boeing Sikorsky for the Dem/Val prototype contract.

In January 1983, an Army LHX Special Working Group (SWG) developed the framework for the program, and in May 1983, a Justification for a Major System New Start (JMSNS) was approved, authorizing initiation of concept exploration for the LHX.

In December 1983, the deputy secretary of defense signed a program budget decision memorandum endorsing the LHX program. A "streamlined acquisition" approach was defined, meaning that the first weapon system was to be designed and procured as a total system—air vehicle with integrated mission equipment, engines, training devices, and combat mission simulators. The plan was to field the total weapon system with organizational, doctrinal, leadership, technical, and training components. It included a family of scout and attack (SCAT) and utility helicopters with a high degree of commonality that could be effectively operated by a single pilot. The prime contractor would develop and produce the training systems. Only the engines would be government-furnished equipment (GFE).

In January 1984, the Army's first LHX program manager was assigned, a colonel. In August, he was replaced by Brigadier General Ronald Andreson. Also in 1984, the Army established the LHX Program Management Office at the Aviation Systems Command (AVSCOM) in St. Louis. In March 1985,

Brig. Gen. Andreson was assigned to work directly for the commanding general of AVSCOM.

It was next concluded that the program was too large to give to one company. At that time, the Army projected a total of 4929 aircraft in both the SCAT and utility configurations. It asked the four major helicopter companies in the United States to form two teams of two companies each. Sikorsky teamed with Boeing, and Bell teamed with McDonnell Douglas (formerly Hughes). It was envisioned that each team would combine to create its best competitive aircraft.

In August 1984, an LHX acquisition strategy was created that defined a competitive full-scale development (FSD) program. The plan was for an air vehicle fly-off of the two competitive aircraft with the winner entering production. Each team would fly five aircraft, and the government would hold a competitive test in 1992. This would lead to a downselect to one team. After downselect, the two teammates would compete with each other for the buys of various production lots.

In March 1985, a letter of agreement (LOA), describing the essential LHX characteristics, was signed by the Army Training and Doctrine Command (TRADOC) and the Army Materiel Command (AMC).

Also in 1985, risk reduction contracts were signed with the Boeing Sikorsky First Team and the McDonnell Bell SuperTeam. These were for design and testing of the aircraft's mission equipment including breadboard and brassboards of the mission computer, electro-optical target acquisition and designation system (EOTADS), pilotage system, and helmet-mounted display, plus airframe wind tunnel testing and simulation. Also included were plans to achieve reliability, availability, and maintainability (RAM)/integrated logistics support (ILS), Manpower and Personnal Integration (MANPRINT)/training, pilot production, and production competition goals. Risk reduction was intended to enhance the ability to meet production and operating cost guarantees when the FSD proposals were submitted.

Special emphasis was placed on supportability and total life-cycle costs. It was thought that these attributes had not received sufficient emphasis in the early design stages of previous programs, and that substantial gains were possible if these were considered early in the conceptual design phase.

These were firm fixed price contracts, with no government-funded facilitation.

In 1985, the Army also signed two firm fixed price contracts for FSD of the LHX T800 engines, one with an AVCO Lycoming/Pratt & Whitney team and the other with a Garrett/Allison (LHTEC) team. The engines were to be 1200 shp, with improved reliability and maintainability, two-level maintenance, and failure detection and isolation of field-replaceable items.

A Scaled-Back Approach

Later in 1985, another Army restructure separated the Milestone (MS) I and MS II decision points, which were originally combined. It also canceled several risk reduction contracts and stretched the program to 60 months, 2 years later than the original target.

Phase I would include detailed design and would be a firm fixed price contract. To substantially reduce costs, one team would be cut after Phase I. This eliminated the highly desirable fly-off between two competing designs. A significantly modified second draft of the request for proposals (RFP) was released, including classified annexes.

In September 1986, the Army released revised risk reduction contracts to the teams for the higher-than-medium-risk technologies that had been identified.

Scaled Back Even Further

In 1987, the FY88 Defense Appropriations Bill was released. It reduced the FY88 LHX budget request of $267 million to $70 million. It was then decided that the goal of developing both utility and recon/gunship configurations was probably too ambitious, and the utility configuration was dropped.

In April 1987, the Army formulated and approved the LHX Research, Development, Test and Evaluation (RDT&E) prototype competition strategy that transitioned the concept exploration efforts into a 56-month Demonstration/Validation (Dem/Val) phase beginning in FY88. It now included design and engineering, prototype fabrication and test, tradeoffs, and planning to define the best solution to the operational requirements. Engineering, manufacturing, and development (EMD) was projected for 26 months after MS II, with a fixed price contact in September 1992. The low-rate initial production (LRIP) contract would be in November 1994.

In April, the Department of the Army and the Office of the Secretary of Defense also conducted a Milestone I review. On 19 May 1987, the Secretary of Defense Decision Memorandum (SDDM) was issued for the LHX.

On July 30, the next draft Dem/Val RFP was released to the companies. The system specification was now for guidance only, because it was thought that not everything the Army desired could be included in a now-required 7500-lb weight-empty limit. The contractors could eliminate some requirements—at their own choosing—to meet the weight limit.

Other government activities in 1987 included the release of the operational requirements document (ORD), an LHX analysis of alternatives (AOA), and the completion of a cost and operational effectiveness analysis (COEA).

At that time, the Army showed a projection of how the LHX would change the mix of its reconnaissance/attack helicopter fleet (Fig. 21.1).

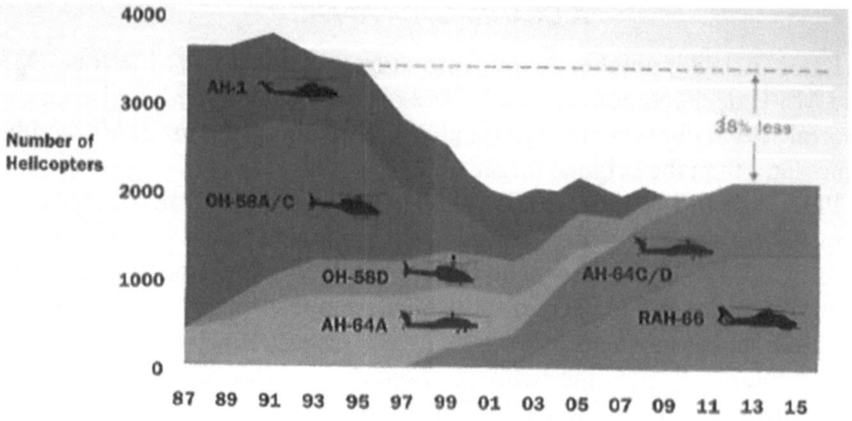

Fig. 21.1 Projected future mix of army helicopters.

AND ANOTHER CUTBACK

In January 1988, the program underwent an Army Aviation Modernization Defense Acquisition Board (DAB) review. This cited affordability issues and concluded that the program was "no longer a viable program." It would be refocused again, this time on an austere preliminary design and risk reduction (PDRR) effort with emphasis on reducing risks with the MEP. It transitioned concept exploration into an 18-month Dem/Val phase in FY89, with competition between the two contractor teams. It included MEP component demonstrations and MEP architecture specification definition. Breadboard and brassboard demonstrations would be conducted, along with other testing and simulations. On the air vehicle side, it included preliminary design and engineering, fabrication of aircraft mockups, and preliminary design of major systems such as the drivetrain and flight controls.

Evaluation of each team's LHX performance would be reviewed, and then there would be a downselect to one contractor team to perform a 69-month FSD phase. The FSD contract award was planned for December 1990, first flight for August 1993, low-rate initial production (LRIP) contract award for November 1994, and initial operational capability (IOC) for November 1996.

The total projected production quantity was now reduced to 2096.

In June 1988, the Dem/Val RFP was released to industry. On September 9, the contractor proposals were submitted. On September 29, both contractor teams received letter contracts to continue their programs.

A unique element of the acquisition strategy was the idea of international cooperative development. The Army was exploring international investments and technology transfer opportunities. Industry was to submit an international cooperation plan and potential partners with their RFP responses for review by the Source Selection Evaluation Board (SSEB). The benefits were standardization and interoperability, expanded production base, additional sources for components and an expanded vendor base, improved logistics support, and lower overall cost.

Eventually this idea was not pursued.

Considerable Boeing Sikorsky investment of independent research and development (IR&D) funds had been used up to this point to support the aircraft preliminary design activities. IR&D funded technology demonstrations included flying the Shadow and the Fantail demonstrator aircraft, testing the bearingless main rotor on a whirl stand, conducting other extensive wind tunnel tests, and building full-scale fuselage sections to demonstrate producibility. After the Dem/Val contract award, the Shadow and Fantail demonstrator aircraft programs continued with CR&D support.

Once we and the SuperTeam were under the Dem/Val contracts in 1988, we had little additional changes in the contractual nature of the program until we submitted our full-scale development proposal.

However, the Army was still changing things.

Chapter 22

Teaming With Boeing

William F. Paul
Then–Sikorsky President
With Art Linden

One of Sikorsky's most important decisions early in the Light Helicopter Experimental (LHX) program was the selection of a partner company to execute the LHX design, development, and production. The U.S. Army wanted a teaming arrangement that would reduce the four prospective bidders (Bell, Boeing, McDonnell Douglas, and Sikorsky) down to two competing teams. From the Army's standpoint, industry teaming would bring to the LHX more technical and managerial talent as well as greater financial resources to strengthen its chances of success.

In a January 2014 interview, William F. Paul, the president of Sikorsky Aircraft in 1985, related the process Sikorsky went through to get to teaming with Boeing. Bill went back in history and related that in the 1970s when the Navy had a requirement for the Light Airborne Multi-Purpose System (LAMPS III) program, IBM won the contract for systems integration while Sikorsky received a contract for the SH-60B air vehicle. When Bill went to the first meeting with IBM, the senior IBM person greeted him with, "Welcome to the team." This was a wake-up call; Sikorsky had always been an original equipment manufacturer (OEM). Here, Sikorsky was to be simply the supplier of the basic helicopter.

There was a lot of tension between IBM and Sikorsky. IBM was going to build a flight test facility at Owego, New York. Once it built that facility, it was going to try to convince the Navy to have Sikorsky simply deliver the Seahawk aircraft in a partially completed state, before they even flew, and IBM would complete them. "It was a $14-million aircraft and our portion was like $2.5 million."

Bill felt that it became obvious from a strategic perspective that Sikorsky could not let that type of contracting continue. "But we were still about two years away from [having that systems engineering capability]." The first thing he did was put together a team of people "…to develop the capability. …We got a number of guys thinking and putting it together as a kind of research program …". On the next [Inner Zone ASW helicopter] proposal,

> …we won the helicopter program but lost the systems integration contract, which was expected. However, both Seahawk programs did give Sikorsky valuable experience for the future.
>
> Then along comes the Australian [S-70B Role Adaptive Weapons System] program. It also required a systems approach. We bid on that program knowing that we would be losing money…. We lost money because we had to learn how to do a systems engineering program. You can either buy that capability or do it yourself. Either way, you lose money the first time you do it. So we did it, and we learned more about that ….

For LHX, "… that capability allowed us to manage that big program with another teammate with similar experience."

In the process of selecting the best team member, Sikorsky assessed its ability to design, develop, and produce the LHX, and determined that the integration of software-intensive cockpits, weapons systems, and defensive systems was the area where Sikorsky could significantly benefit by teaming with an acknowledged expert.

The two companies who were regarded as systems integration experts and who had expressed an interest in teaming with Sikorsky were Boeing and McDonnell Douglas. Bill and Bob Zincone, Sikorsky executive vice president at that time, visited both companies and heard "very effective" presentations on their capabilities. Sikorsky then put two teams together under Bob's leadership, one for each company, to get into the details and advantages of working as a team and how a teaming arrangement would work. Both companies were judged to have the right experience, adequate engineering manpower, and outstanding software development and systems integration laboratories. And they both had long lists of past successful programs.

Initially the teaming decision was swinging towards McDonnell Douglas. Bill commented: "What a powerful team Sikorsky and McDonnell would make. We were both winners of the previous [Army] contracts [for the Black Hawk and the Apache]."

However, Sikorsky selected Boeing as its team member for several reasons beyond systems engineering expertise. Boeing's heritage and product line were similar to those of Sikorsky, and Boeing was clearly skilled in all aspects of rotary wing engineering, manufacturing, and product support. Boeing and Sikorsky were both familiar with the Army technical, contracts, and logistics communities, and that familiarity would benefit the LHX program.

Fig. 22.1 The Boeing Sikorsky LHX team was made official at the 1985 Paris Air Show. From left: Boeing Helicopters President Joseph Mallen, UTC Chairman Harry J. Gray, Boeing Chairman T. A. Wilson, and Sikorsky President William F. Paul.

Additionally "big Boeing" (the fixed-wing side of the company) had a strong systems integration facility in Wichita, Kansas, which would be available for LHX. Lastly, physical proximity between Boeing in Philadelphia and Sikorsky in Connecticut would help communications and be an acceptable burden on employees assigned to either facility.

Sikorsky and Boeing signed a memorandum of understanding to bid as a competitive team on 3 June 1985, and formed the "First Team" for LHX. We formalized the agreement at the 1985 Paris Air Show, fig. 22.1. The Army would have one contract with the Boeing Sikorsky Joint Program Office, and that office would coordinate and direct the two companies. The senior position of LHX/Comanche program director would be rotated between the two companies every four years.

According to the January 1986 issue of *Sikorsky Technology* [1], "UTC Chairman Harry Gray hailed this partnership as 'an exciting challenge and commitment of the complementary resources and strengths of two great companies.'"

In that same issue of *Sikorsky Technology*, Sikorsky LHX Program Director Lou Cotton said [1]:

> Sikorsky and Boeing make a nice mesh. Sikorsky Aircraft has premier production know-how, high-speed aerodynamics skills, and advanced flight control and transmission experience. Boeing has strong systems

> integration capability developed over the years on fixed-wing planes. Boeing also has advanced composite capabilities. Our team brings the two strongest players to every major part of the LHX.... We'll field the best engineering team and go on to beat the competition by fielding the best high-tech, low-risk LHX.

Bell and McDonnell Douglas later formed their LHX "SuperTeam" in 1986.

The wining of the Demonstration/Validation and Engineering Manufacturing and Development contracts, and the outstanding flight capabilities demonstrated by the RAH-66 Comanche aircraft throughout development, confirmed that Boeing was the right choice as Sikorsky's team member.

Reference

[1] *Sikorsky Technology*, January 1986.

Chapter 23

THE FIRST TEAM

ARTHUR W. LINDEN

After Boeing and Sikorsky teamed to work together on LHX, and before Bell and McDonnell Douglas combined to form the SuperTeam, we set out to create the First Team, using industry leaders in all of the critical LHX technologies. By going first we had the ability to choose the strongest team.

The First Team at the time of the April 1991 proposal consisted of:

- *Boeing Helicopters:* Responsible for integrating the Comanche's avionics system, electronic flight controls, and mission equipment package (MEP), plus it was responsible for overall supportability management.
- *Sikorsky Aircraft:* Responsible for integrating the Comanche air vehicle and managing all simulation activities. We also would do the final aircraft assembly in Stratford and conduct the flight testing at Sikorsky's West Palm Beach, Florida, test facility.
- *Boeing Defense and Space Group, Electronics Systems Division, Seattle:* Responsible for developing the Comanche's flight control computers. Its flight control computer experience at that time included the ADOCS Light Hawk and Sea Lance and Boeing's SRAM II missiles; the V-22 Osprey; and Boeing's 737, 747, 757, and 767 jetliners.
- *CAE-LINK Corporation, Binghamton, New York:* Provided the integrated training systems. LINK's helicopter simulators included the McDonnell Douglas AH-64 Apache, Bell AH-1D Cobra and UH-1H Huey, Sikorsky UH-60, and Boeing CH-47D.
- *General Electric (GE) Armament Systems, Burlington, Vermont:* Developed the lightweight turreted, three-barrel Gatling gun. At that time, GE provided the turret system for the Bell AH-1 as well as other high-rate-of-fire weapons and ammunition handling systems.

- *Hamilton Standard, Windsor Locks, Connecticut:* Developed the display electronics unit (DEU) with Kaiser Electronics. It also would provide the air data system and the environmental control/collective protection system that guarded against environmental contaminants and also provided avionics and cockpit cooling.
- *Harris Corporation, Melbourne, Florida:* Provided the Comanche's three-dimensional digital map display, the fiber-optic high-speed data bus, and a fiber-optic sensor data distribution network. At that time, Harris had developed or was developing similar digital map displays for the General Dynamics F-16 and Lockheed F-117A and the F-22.
- *Kaiser Electronics, San Jose, California:* Developed the Comanche's helmet-integrated display-sighting system (HIDSS) with Hamilton Standard. Kaiser was the world's most experienced manufacturer of airborne displays and processors, with several thousand systems in use worldwide on more than a dozen combat aircraft designs. The company had integrated the helmet tracking system for the Shadow.
- *Martin Marietta, Orlando, Florida:* Responsible for developing Comanche's electro-optic sensor systems. The company had more than 20 years of experience in fielding electro-optical navigation and targeting systems, including the target acquisition designation system/night vision pilotage sensor (TADS/NVPS) for the AH-64 Apache, and the Air Force's low altitude navigation and targeting infrared system for night (LANTIRN). At that time it was also codeveloping the LONGBOW integrated mast-mounted fire control radar targeting system.
- *TRW Military Electronic and Avionics Division, San Diego, California:* Provided the RAH-66/F-22 common integrated communication, navigation, and identification avionics (ICNIA) system and aircraft survivability equipment (ASE). It was also codeveloper of the integrated electronic warfare system (INEWS).
- *Westinghouse Electronic Systems Group, Baltimore, Maryland:* Worked closely with Martin Marietta and Boeing to develop the Comanche's aided target detection/classification software. It was also codeveloping the ASE with TRW. Additionally, it was developing the very high speed integrated circuit (VHSIC) programmable signal processor and radar for the F-16, and codeveloping the LONGBOW and INEWS. These programs were directly applicable to the Comanche.

Germany's MBB was on the First Team at the time of the 1990 proposal. We had a teaming agreement with it to review our design based on its experience with bearingless rotors. At that time, it had is model BO-105 in production using a soft-inplane bearingless concept. It concurred with our approach and we felt it was appropriate for the LHX requirements. In addition, there were discussions regarding MBB designing and producing the torque tube.

Soon after proposal submittal, however, these discussions were terminated; the upcoming wind tunnel test of our rotor overtook events with MBB.

Light Helicopter Turbine Engine Company (LHTEC) was a teaming of General Motor's Allison Gas Turbine Division in Indianapolis, Indiana, and Allied Signal's Garrett Engine Division in Phoenix, Arizona, to provide Comanche's T801-LHT-800 turboshaft engines under a separate government contract.

In addition to the major First Team companies, we stressed in our program brochures that we had more than 1100 suppliers in 45 states. The only states missing were Alaska, Hawaii, Louisiana, Montana, and North Dakota.

Some of these team members would change as the program progressed. Soon after proposal submittal, Collins came onboard for the ICNIA, multifunction displays, and controls. Hamilton Standard took over responsibility for the flight control computers.

At the time of first flight, January 1996, some other First Team members had changed. Hughes Training was now in charge of integrated training. Lear Astronics was added for the flight control computer and controller grips. Litton provided inertial navigation sensors; MOOG provided the flight control actuators; Sundstrand provided the power generation, control, and distribution systems; and Williams International provided the secondary power unit. Martin Marietta was now Lockheed Martin. GE Armament was sold to General Dynamics Armament and Technical Products.

By the time of the 2000 EMD proposal, Boeing Defense and Space, Westinghouse, CAE-LINK, and Harris were no longer First Team members. BAE Systems was now providing the flight control computers and cockpit grips. Northrop Grumman was providing the data and signal processors and the aided target detection/classification system. Hamilton Standard was now Hamilton Sundstrand and was providing the electric power system, air vehicle interface control system (AVICS), rotary air data system (RADS), and environmental control system (ECS). A company called Longbow Ltd would provide the radar.

Chapter 24

New Management Approaches

Bill Harper, Bruce Kay, Art Linden and Rus Stiles

LHX was the first new helicopter program in 20 years, so both the Army and the companies had many new ideas, not just for the aircraft design, but also for program management. The complexity of the LHX would require extensive coordination among two major helicopter companies, the Army, the Department of Defense (DoD), and many diverse suppliers. Technical advances would have to be developed, integrated, and managed across suppliers as well as the Army and other DoD laboratories. The necessity of teaming with a competitive helicopter company would present a number of challenges.

Many felt that there were lessons to be learned from past programs. One of these was to proceed in a more structured development process that emphasized technology maturation and risk reduction efforts prior to entering the engineering, manufacturing, and development (EMD) phase where the production platform is designed and manufacturing approaches are established. It was thought that problems would be identified through prototyping and fixed early in the program when their cost and schedule impact would be minimal.

Another idea was to include "downstream" departments early in the program. On previous programs, under tight cost and schedule constraints, it was thought that not enough emphasis was placed on supportability and reducing manufacturing and operational costs. Engineering would design the aircraft with minimal input from other departments, then turn the design over to manufacturing engineering who also would work pretty much on its own, before turning it over to manufacturing. Tasks like maintenance and repair, reliability and maintainability, and production cost were introduced later in the program and had virtually no impact on the basic design. The Army wanted all of these disciplines involved at the beginning of the program to achieve a lower total life cycle cost and to make the aircraft easier to maintain and keep on flight status.

INTEGRATED PRODUCT DEVELOPMENT TEAMS

Integrated product development teams (IPDTs) were introduced at the beginning of the program. They included design, test, supportability, and manufacturing engineers, plus representatives from the manufacturing, quality, purchasing, and finance departments. The Army also provided active soldiers to provide a clear perspective from the user's standpoint.

The IPDT members worked in a matrix organization. They were assigned full time to the program yet retained their position in their functional organizations. All daily direction came from the program, with the functional department always available for technical guidance and additional help if required. The functional departments provided the necessary people to the program and reassigned them to other activities when they were finished with their LHX/Comanche assignment. Performance evaluations were done by the program and turned over to the functional chief. If necessary, a negotiation between the two parties would resolve any issues.

The aircraft was broken down into sections, with IPDTs responsible for complete ownership of that section from design through fabrication and test. Each IPDT would have a core staff and a leader supported by a colocated team of specialists to perform the necessary engineering, producibility, and supportability tasks.

When we initiated this, we truly believed we were revolutionizing the design process; we saw all the opportunities to reduce development time and cost. The Army and the two companies were very positive about the process. After some initial learning and confusion, there was never any question as to whether it was the right way to go.

SYSTEMS ENGINEERING

The LHX requirements were much more complex than those of previous programs. In addition to the conventional engineering requirements of performance, handling qualities, weights, and so forth, extensive requirements for the mission systems, design-to-cost, supportability, and even costs for the eventual retirement of the aircraft were part of the Comanche design requirements.

Management of the design was historically the role of the chief engineer, but as systems became more complex and more complicated technically, with more sophisticated technologies and software, management was relegated to a systems engineering (SE) organization, occasionally led by the chief engineer.

Sikorsky didn't have an SE organization when the LHX program launched. We did, however, make great strides in systems engineering with the Pilot–Vehicle Interface Mechanization Specification (PVIMS), which defined the pilot–vehicle interfaces needed to minimize workload and maximize mission success. The PVIMS contained mission walkthroughs, display formats and moding structures, control requirements, and implementation rules. The government provided the weapon system specifications (WSSs) and later the performance weapon system specification (PWSS). The PVIMS requirements and interfaces were allocated to the various segments (subsystems) through the weapon system design document (WSDD) through the use of whitepapers. This approach was unique to the Comanche program, at least at Sikorsky. The weapon system integration plan (WSIP) and test plan linked the segment verification requirements, processes, locations, and compliance requirements to the PWSS, PVIMS, and WSDD requirements. This logical sequence of activities and documents was critically important to ensuring that all the hardware and software requirements were defined and understood prior to the launch of the subsystem and component designs and software development.

To keep control of all these requirements, a system of product unique identifiers (PUIs) was developed. This flowed down all the top-level requirements to each component in the system. Every designer had a detailed list of what their part had to accomplish in every discipline. Only by using this system was it possible to create an overall design approach that met all the myriad requirements of the Army.

Combined Test Team Approach

In a vein similar to the IPDTs, and using an approach unprecedented in Army aircraft acquisition, it was decided that Army test pilots should fly the flight development aircraft alongside the contractor pilots. Initially the test pilot team consisted of one test pilot from Boeing, two from Sikorsky, and one from the Army.

An army pilot headquartered at the U.S. Army Aviation Technical Test Center at Fort Rucker, Alabama, Chief Warrant Officer (CWO) 4 John Armbrust, started working on the program in 1991. He was on-site at Sikorsky's West Palm Beach, Florida, test center starting in July 1995. He stated in the September 1998 issue of *Army Magazine* that when he reported to the team, the rules for Army flight tests "...essentially said that Army test pilots would not fly any test point or fly any part of the envelope that had not been previously experienced by a contractor test pilot. But everything here was envelope expansion" [1]. Thus, the rules had to be modified. His first Comanche flight was on 20 September 1996, the ninth flight of aircraft 1.

THE BOEING SIKORSKY JOINT PROGRAM OFFICE

We needed to create a management structure to integrate these two companies into an effective team. The solution was for each company to have a generally conventional program structure, but with two upper-level reporting organizations. The first was to their company senior management and the other was to the Joint Program Office (JPO).

The JPO represented the Boeing Sikorsky team's centralized high-level program decision-making element, facilitated through senior-level subject matter experts. This allowed the companies and subcontractor team members the freedom to focus and perform their contractual tasks. There was a JPO program director for each basic discipline: weapons system integration, supportability, finance, contracts, operations, marketing, and the combined test team, plus support staff as required (Fig. 24.1). The focus of the JPO was strategic planning, coordination with the Army, and resolution of critical issues between companies, suppliers, and the government. The JPO typically had about 40 people.

The company management organization provided the functional support and directed the individual Comanche program offices as needed to accomplish their specific tasks within budget and schedule.

EARNED VALUE MEASUREMENT SYSTEM

Because of cost and schedule problems with previous programs, the government instituted an improved cost/schedule control system called the Earned Value Measurement System (EVMS). The system managed and controlled workflow. Work was planned out down to "work packages," which typically encompassed about 160 hours of planned work. Although this

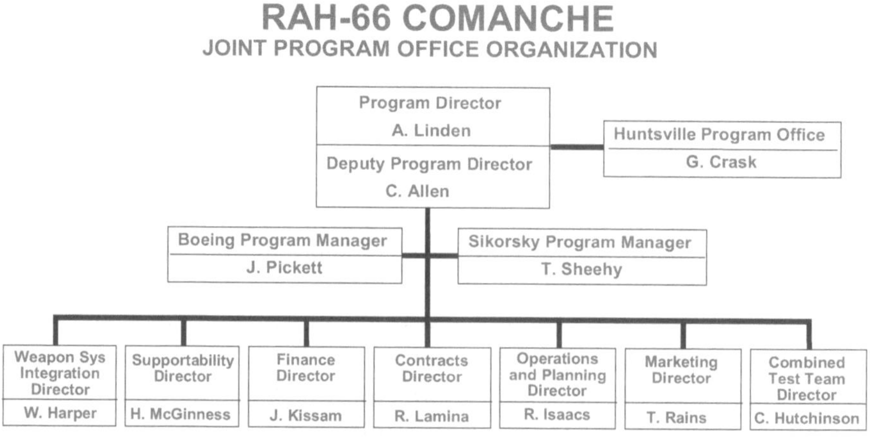

Fig. 24.1 Joint Program Office organization, 1999.

resulted in extensive program planning, it produced a plan that could easily be monitored and that gave the management team a clear early indication if any part of the program was going over budget or getting behind schedule. Sikorsky's system was called the Cost/Schedule Control System (CSCS) and was DOD certified. It measured performance against predefined budgets and schedules. For the IPDTs, work packages were broken down by part models, and then further broken down into smaller increments of approximately one man-month duration. In addition, task predecessors and successors were defined to show what had to be defined in order to start a task, and what would be produced in order to proceed to the next task. This entailed hundreds of pages of documentation. Changes to work packages could only be made prior to starting work. The system was statused on a monthly basis, and any variances exceeding a nominal tolerance had to be explained to senior company management, including Sikorsky president Gene Buckley, who took an active interest. With a variance, it was always smart to present a recovery plan to solve the problem.

HOW DID IT ALL WORK OUT?

IPDT PROCESS

The IPDT process was kind of like going back to the early days of aviation when everybody worked together as a team (Fig. 24.2).

The IPDT leader was the chairman of the team. His job was to seek consensus among the members. He would make sure that minority opinions were clearly and openly expressed. In the case of deadlocks, the IPDT leader would be the tiebreaker. The IPDT leader would govern the priorities of the team and would be a direct report to the program manager.

Fig. 24.2 Original Sikorsky aero engineering team with Igor Sikorsky, circa 1925.

Initially all this IPDT talk sounded like just another way of doing business as usual with a new name that was fashionable at the time. That was Chief of Design Bruce Kay's opinion when the new organization was announced. He felt that he would keep his old job and maintain the prestige and authority that went with that position.

However, as more details of the new organization were described, it was apparent that this was not meant to be a paper tiger. The candidates being considered as IPDT leaders were "heavyweights" reporting directly to the program manager and would not be mere figureheads. The IPDT team member positions were staffed with Sikorsky's A-team. With the preceding in mind, Bruce reversed his decision and was reassigned to be IPDT leader over fuselage IPDT and propulsion IPDT.

Participation from the Operations department was perhaps one of the team's strongest suits. The Operations department assigned full-time factory supervisors to the IPDT to make sure manufacturing lessons learned were incorporated into the design, manufacturing, and tooling plans. Its mode of operation was to take full-size 3-D layouts of the design, hang them on their office walls, stare at them, and envision just how they were going to put the aircraft together.

The departmental buy-in for the manufacturing labor allocations was very important. The perspectives and wisdom brought to the team from front-line manufacturing supervision was invaluable.

Keeping unit costs within budget was a requirement that had to be met. Accordingly a design-to-cost (DTC) organization was set up under the Operations department. Its responsibility was to allocate costs down to the part level and maintain a running status. This was analogous to the weight control methodology. In actual fact, DTC was unable to keep up with the design activities and merely book-kept allocations that added up to the total unit cost.

Nevertheless, for parts built in-house, intensive producibility reviews were conducted, with the factory signing up to build its parts for the allocated budgets.

A major portion of the helicopter cost was from purchased parts and material. Early supplier involvement in the process provided products with maximum value at the lowest possible cost. The key step in this process was early release of requests for information (RFIs) to potential suppliers and collaborative development of the final competitive procurement documents. This process allowed for greater exploitation of available technology, eliminated gold plating, and heightened competition. This was in stark contrast to past procurement policies that relied on the release of elaborately prepared requests for proposals (RFPs) that left little time to do anything more than merely select the low bidder or disqualify noncompliant proposals. At the end of the process, suppliers were signed up to fixed price contracts.

As a result of all this effort, the affordability requirements were accomplished through the hard work of the manufacturing engineers, factory representatives, and purchasing agents within the IPDTs, so at the top level, DTC requirements were achieved.

Continuous evaluation of supportability attributes was performed by the IPDT members. Prediction of maintenance requirements is typical of the analyses used to influence the design. These data were used in part to locate equipment; for example, components having the highest maintenance activity were placed in positions most amenable to frequent access. Outstanding supportability was cited by the Army as a major discriminator in the eventual evaluation of our Dem/Val proposal.

The Army also stationed a group of user representatives at Sikorsky's facility. These military representatives sat in on IPDT meetings and actively participated in design trades by presenting the user's perspective. By having the customer present as the design matured, potential problems were identified early in the process so that satisfactory solutions could be developed in a timely manner. One of these user representatives, former crew chief Sergeant Paul Sargent, or Sgt. Sargent, as he was called, was especially helpful. When he arrived in Connecticut he asked his commanding officer what his duties were to be. The major replied, "Look out for the soldier," and that is exactly what Sgt. Sargent did. He would go to any length to get his point across, escalating to whatever level was necessary to make sure the voice of the soldier was acknowledged.

Concurrent engineering did not happen by merely assigning multidisciplined specialists to a team and colocating them. Communication mechanisms had to be established and rigidly adhered to. The IPDTs did this through a series of standing meetings with fixed agendas. This was a radical departure from how business was normally conducted in the past. Some departments considered other departments their natural enemies and behaved like cats and dogs—"cats are not supposed to fraternize with dogs" and vice versa. Other participants felt their role was to represent their functional boss, and they therefore should not provide positions without their functional boss's concurrence. In other words, they were messengers. Some did not understand that they could influence the design by speaking up. Still others thought that the loudest voice would prevail (Fig. 24.3).

So, some level setting was in order to prevent chaos.

A considerable amount of training and coaching was required to get the IPDTs to run smoothly. They received training in Total Quality Management (TQM), structured problem solving, and conflict management. One of the most powerful training courses given to the team was conflict management. This course pointed out that conflict was healthy and a vital ingredient needed to develop a good, balanced design. The trick was to handle the conflict without offending participants and not making things personal. Techniques to

Fig. 24.3 The first IPDT meetings.

accomplish this were provided and practiced with role-playing. Things were a lot better after this, and the teams started to function proactively.

In an effort to constantly improve the IPDT process, periodic effectiveness surveys were conducted. The purpose of these studies was to solicit input on the effectiveness of the team as a whole and of its individual members. Self-assessments were used in conjunction with Pareto analyses to identify group obstacles and formulate improvements to bolster team weaknesses. Probing questions were asked of all members, and the results were analyzed.

Regularly scheduled technical meetings were used to present designs to the team. Any issues that were raised were documented in meeting minutes, and actions were assigned. For this process to work, it was imperative that all team members were actively involved. That meant meeting attendance was mandatory, as was completing assignments on schedule. This did not always happen, which meant that any absences ot delinquencies would hold up the team's progress and waste their time. Personal appeals by the team leader only provided mediocre results with some individuals. To incentivize these habitual slackers, it was decided that individual team member survey contribution results would be shared with the entire team.

Critiques were taken very seriously by team members because they came from colleagues and were shown to be statistically accurate. Once these issues were put on the table for discussion, they were dealt with head-on. Deficiencies were by and large corrected, and internal relationships were substantially strengthened after the survey. The survey also showed that virtually all respondents strongly felt that the IPDT process was a much better way to do business than previous design procedures.

Design activities also had to be coordinated with Boeing, and the JPO oversaw this integration. The JPO was aligned with the Army customer's Program Management Office (PMO), and the Army PMO had its own customers and clients. The PMO had to satisfy its user communities, headquartered at Fort Rucker, and relied on the Army Engineering Directorate (AED) for technical evaluations and qualification. So, all in all, there was a very formidable bureaucracy to manipulate to get things done.

A concern was that anyone within this labyrinth of organizations could raise or perpetuate issues that would require attention and disrupt the normal flow of work. And, of course, the program manager kept everyone busy by issuing directives and administering technical, schedule, and budget-related status reports. There were also very frequent fire drills for urgent presentations on topical issues and to host visiting dignitaries. In addition, the program seemed to be constantly engaged in preparing "what if" restructure proposals to address fluid funding profiles and other programmatic issues.

The IPDTs were subject to numerous levels of oversight and checks and balances. The traditional organizational structure was not entirely dismantled. Although IPDT members reported directly to their IPDT leaders, they also reported dotted line back to their functional departments as well as to functional managers on the program. These functional managers were part of the program manager's staff along with the IPDT leaders. The program technical managers included the chief engineer, operations manager, supportability manager, and quality manager. The chief engineer's direct reports included all the flight science disciplines such as aerodynamics, dynamics, weights, and loads, as well as radar signatures, survivability/vulnerability, and systems engineering groups.

In a manner similar to engineering, the operations director had a staff of direct reports for manufacturing, manufacturing engineering, purchasing, and design-to-cost. They would provide the checks and balances to the operations people working directly in the IPDTs. The operations director was Bill Mulvey. One of Bill's roving operatives was Lou Cimmino, who "managed by walking around." He used this technique very effectively. In the old days, this was called "walking the boards." In the digital age, it meant talking to the designers and looking at their part models on their computer screens. Lou was able to master this approach to detect and nip potential problems in the bud. It is unfortunate that the trend for design reviews has shifted to PowerPoint presentations, which are not as probing. Walking the boards might have been old fashioned, but it worked when performed by knowledgeable interrogators.

Nevertheless, decision making was almost always left to the IPDT. Other organizations and senior management could question the IPDT and ask that additional information be considered in their deliberations, but rarely were IPDT decisions or recommendations overridden. This power occurred naturally when the IPDT was functioning properly. A team staffed with competent representatives from all disciplines forced decisions to be made on a factual basis. Once the team substantiated a decision to its members and achieved buy-in, it was very difficult for outsiders to challenge it. As a rule, the IPDTs functioned as democracies where everyone was given the opportunity to voice their opinions, but in the end it was up to the IPDT leader to weigh the issues, formulate a decision, and move on.

Dr. Ken Rosen, Sikorsky engineering vice president, conducted weekly reviews every Friday afternoon for 12 years with a rotation of the IPDTs. Sikorsky president Gene Buckley had specifically told Ken that he was technically responsible for the program, and Ken took his responsibilities seriously. During the flight test phase he also conducted daily phone calls with the West Palm Beach (WPB) test team and took numerous trips to WPB for reviews. He also got heavily involved with our suppliers, including Boeing's subcontractors, working with both their engineers and company management.

One reason the process worked so well is that it had the full support of Sikorsky's president, Gene Buckley. With Gene on board and having weekly meetings with the Comanche team, everyone was focused on moving the program forward.

As an example of the result of this IPDT process, at the airframe critical design review, a four-day, multiple-session event, only 46 chits (questions or comments) were written against the airframe IPDT presentations. Usually these reviews draw hundreds of chits. This low number of chits was indicative of the good communications, buy-in, and overall integration that occurred in the events leading up to the critical design review.

As the program went along, it was agreed that the IPDT concept was an excellent idea. It led to an aircraft that required no major design changes once it entered flight test. The emphasis on supportability clearly helped us win the original contract. The inclusion of the manufacturing engineering, manufacturing, and purchasing departments early in the design led to an aircraft that was manufacturable, again without many changes required in the processes as we went along, as well as meeting the design-to-cost requirements.

Systems Engineering

The Systems Engineering group was the primary contact with the customer in terms of specification requirements. A major role was to assure spec compliance. To this end, it parsed the government specification line by line and flowed it down to the parties responsible for carrying out those requirements. They did this with the product unique identifiers (PUIs) for each sentence or partial sentence that cited a requirement. There were literally thousands of these PUIs. The IPDTs then had to take each PUI and describe how it was going to comply with the requirement. In the case where multiple parties were involved, it was necessary to coordinate with each party. This entailed a lot of paperwork and was looked upon with disdain. The IPDTs preferred to concentrate on designing and solving technical problems; however, the process did prevent requirements from falling through the cracks. Specifications often had to be modified or tailored, and the systems engineers had to substantiate any changes with a substantial amount of documentation.

The Combined Test Team

Due to the length of the test program, several Army pilots rotated through the Combined Test Team. All were highly qualified U.S. Navy Test Pilot School graduates with extensive operational experience. By combining operational knowledge with engineering savvy, they were able to provide valuable input into the potential mission suitability of the aircraft at an early stage. They advised the contractor flight test engineers on appropriate profiles to be added to test cards in addition to the traditional engineering maneuvers. This was particularly valuable when evaluating the handling qualities provided by the all-new fly-by-wire control systems.

Because the test program was conducted at a Sikorsky facility, Sikorsky procedures dominated daily activities. The Army participants enthusiastically embraced the Sikorsky risk management approach to flight testing. Although somewhat different from the procedures that might be found at an Army test site, they praised the discipline and thoroughness of test planning and execution. The results speak for themselves. There were no major incidents during the test program—every flight returned to home base safely.

Also on site for the flight test program were some Army pilots assigned to the Training and Doctrine Command (TRADOC), temporarily transferred from Fort Rucker to West Palm Beach. They were focused on the ability of the aircraft to meet operational requirements and had no engineering or test background. As a result, they were administratively banned from flying the aircraft. In addition, they often found it difficult to accept the fact that early in the test program there would be issues to be resolved, but their job was to look ahead and assess whether the aircraft showed potential for mission capability. Some were simply ill equipped to make such judgments. Although they might have been able to reconcile this by discussing their concerns with their Army test pilot counterparts, more often than not the operators felt the test pilots were in cahoots with the contractor and shouldn't be trusted. This resulted in contentious relationships between the two. From the contractor's standpoint, we needed a cohesive position, not being caught in the middle between Army factions.

At no time was this more evident than during a meeting in May 1999 with the chief of the TRADOC Comanche office. The Comanche incorporated an advanced helmet display design that incorporated world conformal symbology, referred to as "contact analog." Through the use of three-axis helmet tracking, wide field of view, and binocular vision, this approach overlaid navigation and targeting information in its place in the real world. The pilot simply had to look in the direction of interest and an icon would appear in the helmet showing the object. Previous designs incorporated screen-referenced symbology—essentially a replicate of the instrument panel—with no directional information. Symbology trials with operational pilots outside of TRADOC indicated the new design provided dramatically improved situational awareness, both in how well the pilot understood his situation in the world and how fast he could acquire targeting information.

The Army TRADOC pilots strongly objected to the design and wanted to stay with the screen-referenced approach, which was most similar to the symbology philosophy found in the legacy AH-64 Apache aircraft. They wrote a whitepaper for their commander, who then visited West Palm Beach to meet with Comanche program pilots. Two test pilots—the chief contractor pilot and the chief Army test pilot—arrived for the meeting. Six TRADOC pilots were present. It was well known that the Army test pilot supported the contact analog symbology approach. Before the meeting started, the TRADOC commander, a colonel, directed the Army test pilot to leave the room. This left the contractor pilot alone to address the issues. Ultimately, most of the concerns were due to a lack of understanding by TRADOC pilots of how the design worked. (Not all of them had even seen it operate in the simulator.) The meeting was concluded with the TRADOC commander admitting that he was the victim of a misunderstanding, and the design was continued into flight test. Ironically, military jet fighter aircraft fielded after the Comanche was canceled incorporate helmet-mounted displays with contact analog symbology.

Although not specifically part of the combined testing process, full mission simulators were used extensively in evaluating the crew interface to mission systems. The combined test team (CTT) pilots all participated in simulation activities, including validation of new drops of software. Familiarity with the aircraft was essential to assessing whether new versions of software would work in the flight environment. The simulation environment was where TRADOC pilots were best able to make operational assessments. The only mistake the Army made in this process was that the pilots they sent to the simulation activities in Stratford, Connecticut, came from Fort Rucker, not West Palm Beach. This exacerbated the communication problem—TRADOC pilots observing the test program did not know what the others were doing in the simulator and vice versa. The test pilots (Army and contractor) participated in all the activities and were comfortable communicating both the strengths and weaknesses of the aircraft design.

In the end, the Army test pilots were best informed and most articulate, and presented the best day-to-day guidance to not only flight test, but also the development program.

The TRADOC pilots, however, were most influential in changing requirements. TRADOC seemed to resist designs that evolved from and improved upon legacy systems, such as contact analog symbology. Ironically, though, it seemed they never encountered a brand new technology they could resist. An example was the insistence on including an image intensifier sensor in the pilotage turret, which already had a forward-looking infrared (FLIR). (The helmet-mounted display also incorporated an image intensifier.) They were enamored with the idea of image fusion, an immature technology that has yet to emerge from the research and development stage. What they did not understand was the cascading impacts such a change would have. Although the image intensifier itself was not heavy, it had to be mounted adjacent to the FLIR, which resulted in a large change in internal structure and outer mold line of the turret, adding considerable weight. This also negatively impacted external visibility from the cockpit. The power and cooling required to accommodate the additional sensor impacted the electrical and environmental control system. Additional processing was required to accommodate the exotic software algorithms required to create image fusion. As a result, it was a very heavy, costly design change.

Although we are not suggesting that TRADOC was the primary cause of the design instability, changing requirements were certainly reflected in the numerous program restructures.

The Boeing Sikorsky Joint Program Office (JPO)

Because the Joint Program Office was the primary interface with the Army and other DoD agencies, the companies minimized their direct contact with the Army to avoid confusion. There were still many ongoing contacts between the companies and the Army, but these were coordinated (usually) by the JPO. Any proposed specification or contract modification was the responsibility of the JPO. The JPO led the contract negotiations with large teams of company people behind them.

Internally, the JPO was the mediator of issues between the companies. Both companies clearly wanted this team to be as productive as possible and to produce the winning LHX design. But we were working with a competitor. There was some initial distrust on both sides as well as company pride in the way we did things and the products we produced. We found some major differences, and that took some time to absorb. JPO led the task of dividing the work between Boeing and Sikorsky. The work was divided into clearly defined areas of responsibility, but almost without exception there needed to be a sharing of information and understanding in all areas. JPO set up the award fee program to keep the companies focused on the most important tasks.

The program was under constant congressional reviews, fact-finding, requirement negotiations, restructuring, proposal preparation, and subsystem design reviews. These tasks were performed by the JPO. Its role included being the guardian of the program and regulatory constraints (e.g., performance, schedule and cost limitations, and security classification guidelines), conveying constraints to the team members, and working the customer side when certain changes needed to be made or challenges were directed at the program. The JPO was also responsible for scheduling, overseeing, conducting, and supporting periodic reviews and audits in an effort to keep the Army, suppliers, and team members informed on issues, including progress made, challenges, and changes. The JPO facilitated and supported conflict resolution and technology trade studies. It maintained configuration control to ensure changes were documented, understood, and agreed to by all parties.

Low observable signature levels were negotiated with the Army by the JPO. Comanche could have been easily pushed to be categorized as a "black" program had the JPO not worked the classification guide to keep the aircraft unclassified. A "black" program status would have killed Comanche just on cost.

The Army expectations for the program were high—virtually all of its future helicopter development funding was tied to the LHX. As such, many in the Army were enthusiastic. But others were jealous of the funding being applied to the program. Similar feelings were held by various member of Congress, both positive and negative. The program was so large that it received a lot of attention in all these quarters. The JPO had to deal with all of this on a daily basis. It continually supported the Army to keep the program on track. It did technical analysis and worked with the Army to present these data to appropriate groups, friend or foe. This included the Undersecretary of the Army plus certain members of the House and Senate who wanted to kill the program. Many times these groups were fed misleading information by people or groups who wanted the program canceled for their own reasons. Much effort was expended to counter these positions.

To illustrate JPO activities, the following events occurred in the first few years of the contract. By May 1991, one month after contract award, the program was under review in Washington as part of the defense budget allocation process. The Office of the Secretary of Defense (OSD) and the House Armed Services Committee reviews required JPO support through June. In July, the Conventional Systems Committee of the Defense Acquisition Board was auditing Comanche's capabilities and its impact on combat success and Army Aviation. The JPO generated Comanche functionality briefings in support of this review.

Design definitions of a number of the major subsystems were also becoming finalized by our team members around July 1991. An, Electro-Optic Sensor System (EOSS) design working group spent August at Martin Marietta to finalize the subsystem's external contours and design requirements. September and October were used to finalize the aircraft's shape, based on the results of radar cross-section (RCS) tests. November was spent on Design Flex Clause negotiations. Preliminary design reviews began in in December.

Martin Marietta's preliminary design review (PDR) was in February 1992. After Secretary of Defense Dick Cheney directed that the program be restructured to a technology program, and added the LONGBOW radar, the JPO spent a month with the Martin Westinghouse LONGBOW Joint Venture creating the interface requirements to put the radar system on Comanche. MS II exit criteria were negotiated in April, and nine new design requirements were added by the user that had to be integrated into the design. Award fee criteria and scoring were negotiated in June. Kaiser's PDR was in July. The program restructure fact-finding went through September and October. The, Cost Analysis Improvement Group (CAIG) assessment of the restructure was in November and December. Contract award was in December 1992.

The following year, 1993, started with a bottom-up review (BUR) in February. RAND, IDA, OSD Program Analysis and Evaluation (OSD-PA&E), and Deputy Chief of Staff, Operations (DCSOPS)/Secretary of the Army Research and Development Group (SARD) required JPO support to familiarize their people with the Comanche system. The technical and cost risks of the bearingless main rotor decision were reviewed. The flight control subsystem, dynamic subsystems, and airframe critical design reviews (CDR) were conducted in June, August, and September of 1993. The full-scale model RCS tests were in July. By September, budget shortfalls led the Army to direct restructuring the program once again, deleting the LONGBOW requirement and slipping the program 3 months to the right to cover the funding shortfall.

In February 1994, the program was directed to include commercial business practices, commercial parts and logistics supply chain approaches, and reduced testing by increased use of simulations. This required an updated analysis of alternatives (AoA) effort, cost benefit analysis, and a RAM/supportability assessment.

As further program restructures were required, it was the JPO's job to formulate these, negotiate them with the Army, explain them to various government entities, and coordinate the Boeing Sikorsky response in a responsible plan with a realistic plan of achievement.

The final restructure in 2002 included what was thought to be a cost saving measure—the restructuring of the JPO to eliminate the separate company program offices. The JPO role was now expanded to direct the work activities within the two companies and contract management of the suppliers, as well as its previous activities of customer interface and activities to keep the program alive and represented appropriately within OSD. Many of the people in the previous JPO organization were removed from the program, and some of their previous activities were reduced or eliminated.

EVMS/CSCS

EVMS/Cost Schedule Control System (CSCS) kept everyone honest by revealing problems before they got out of control. The system could be deceived for a while, but eventually any variances were uncovered. Everyone quickly learned that bad news did not get any better with age, and it was much better to reveal problems early so they could be rectified.

It was always a struggle to stay on schedule and within budget because of unanticipated technical problems and staffing issues. It is also interesting to understand how the budgets were established. Very intensive detailed negotiations were conducted by the Army at an old military facility in Granite City, Illinois, near its headquarters. These negotiations went on for several months, during which each statement of work line item and cost estimate was examined and negotiated by the Army's technical matter subject experts. The result was a further trimming of the proposed budgets. Everyone sighed with relief when the bloodletting was over, and though the budgets were challenging, they were achievable.

After contract award, the budgets were distributed to the team to start work. Lo and behold, the design teams found out that the budgets were actually cut another 10% or so to create a management reserve. The negotiating team had previously been sent to classes and instructed about the importance of "truth in negotiation." This issue went to the compliance officer for review. He met with the team, heard their case, and then deemed that the budget cut was legal. He said that program management and contracts had informed the government that they were taking an additional challenge, which the government contracting officer agreed to, and it was therefore legal. So, once again, the team had to sharpen their pencils and get on with it.

There was a flaw in the system from the IPDT viewpoint. The IPDTs depended on other organizations for technical data and interface information. Dependencies could be broken down into two categories. Some technical groups, such as flight sciences, had their work tasks defined as levels of effort. This meant they did not have discrete deliverables that had to be provided in accordance with a schedule. The IPDTs did their best to negotiate deliverables, but these deliverables were not tracked by the CSCS, so if they were missed, only the IPDT was penalized. The other problem regarded obtaining accurate interface information from other design organizations. Although interface schedules were established by these groups, the actual design work, in some cases, was subcontracted to equipment suppliers, whose contracts were still being negotiated. In fact, some of these subcontracts were deferred to much later in the program. This was the case for a large portion of the armament system and the mission equipment package (MEP) avionics that was managed by Boeing. Their interfaces were based on best available preliminary designs, and significant changes were made later, after detailed analyses were performed. This required the IPDTs to make changes, sometimes missing schedule milestones.

Senior Sikorsky management took a strong interest in keeping the program on track and within budget. It was clear that if we ever got too far off track on these parameters we were opening ourselves up to possible questions as to whether the program could be performed as planned. All levels of both program and functional management reviewed these data at least weekly.

Ray Kurlak joined the company as executive vice president, and he conducted weekly reviews with the IPDTs. Ray's strong management style tended to keep everyone focused. These reviews were particularly torturous. Presenters always felt like they were stark naked when making presentations in this forum, because they couldn't hide anything. The IPDTs usually had budget and schedule problems for a variety of reasons. Designing components to meet challenging requirements was difficult enough. In addition, the IPDTs had to deal with a host of conflicting requirements and organizations with other vested interests. Interfaces kept changing, and resolving overweight conditions took a toll. On top of that, the IPDT leadership was constantly immersed in exercises for the systems integration group and the program manager's office. Frequently, the IPDT leaders were spending more time reporting than doing productive design work.

Running the IPDTs in a very chaotic environment certainly seemed like a no-win, thankless job. Workaholism was rampant, and the keys to survival were dedication and having a personality that could thrive on chaos.

Sikorsky president Gene Buckley had weekly Comanche status meetings in which all pertinent issues were discussed. We were required to present the cost and schedule status. The charts shown in Figs. 24.4 and 24.5 are typical.

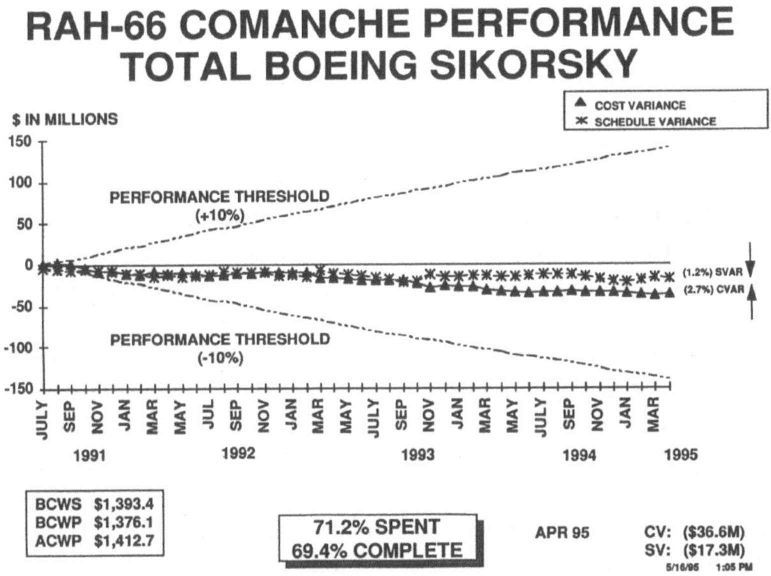

Fig. 24.4 Cost and schedule performance against the plan.
Note: SVAR = schedule variance; CVAR = cost variance; BCWS = budgeted cost for the work scheduled; BCWP = budgeted cost for the work performed; ACWP = actual cost for the work performed; CV = cost variance; SV = schedule variance.

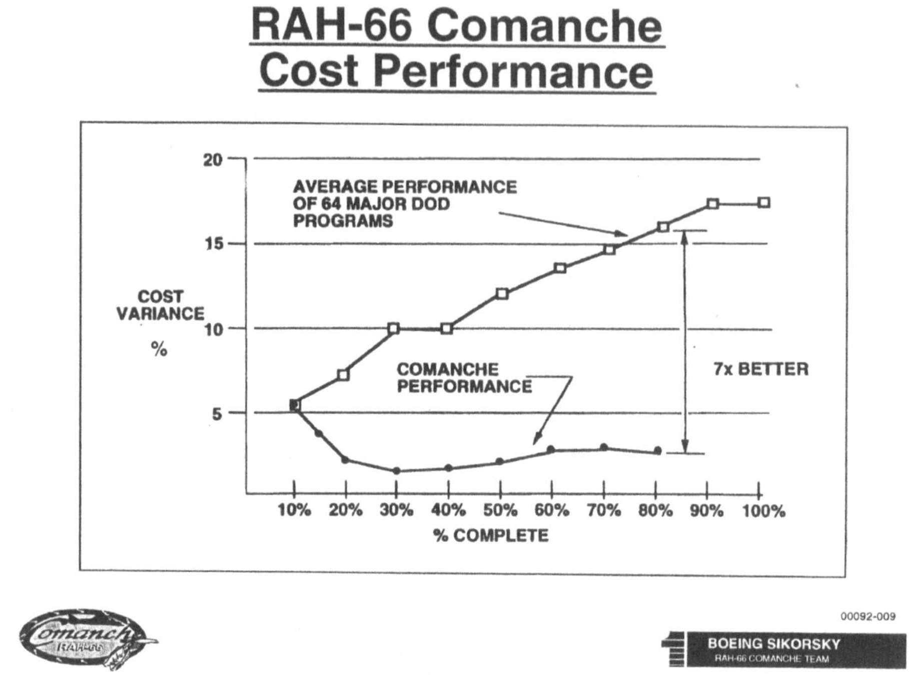

Fig. 24.5 Cost performance versus other DoD programs.

(These particular charts are from a February 1996 presentation to Senator Ted Stevens, R-Alaska.) We obviously wanted to stress that, unlike most DoD programs, we remained on track cost-wise.

Ultimately the CSCS process prevailed, with the IPDTs pulling together all the loose ends to close the design and move on to fabrication. As a top-level result, in the 13 years of the contracted program, and through four major restructures, the program never got above a small variance in either cost or schedule.

However, it must be admitted that with four restructures over 13 years, the team never got more than a few years into any particular plan. Then the whole process started again, and a new plan was developed. Could the cost and schedule discipline hold over a longer period? The team believed it could, but it was never proven.

Senior management generally agreed that although the process was complex and time consuming, it did give management timely data on program progress so that problems could be identified early when there was time to efficiently correct them. The success of CSCS on Comanche led Sikorsky to use a similar system on other new programs, like the S-92.

METRIFICATION

Metrification was accomplished in a fairly straightforward manner with just a few minor problems, which were primarily cultural. Lengths and weight conversions were readily adapted, but it was difficult to think in metrics or develop "feels" for other quantities such as stress and density. In the English system, these quantities were expressed as pounds per square foot and pounds per cubic inch, respectively. The metric equivalent units were pascals and kilograms per cubic meter. The natural sense of how big or small these units were took time to learn. For convenience, some analyses were conducted using English measurements and then converted to metric for formal documentation. Other minor conflicts arose on whether to use hard or soft conversions for certain materials and hardware. Although hardware such as bolts and screws could be procured in metric increments, this was not always the case for other off-the-shelf commodities; for example, aluminum sheet thicknesses were standardized in English units, and when converted to metric resulted in uneven numbers, so the English system was maintained for these conditions.

An LHX inboard profile drawing dated 24 January 1992 shows the fuselage stations in millimeters. The rotor location, normally at something like 300 in., is at 10,000 mm.

The LHX weight empty requirement was usually stated as 7500 lb. At some point, someone embarrassed the Army by pointing out that this was not metric. The weight was then restated as 3402 kg.

A specific metric toolbox was defined, with a limited range of tool sizes. This would simplify the maintainer's tool requirements, but meant that some fasteners were oversized at a cost of increased weight.

However, with the 1993 streamlining initiative directing the Comanche program to implement commercial parts, the requirement for use of the metric measurement system was dropped. With considerable effort, the aircraft and the subsystems designs were converted to English units. No estimate was ever made as to what this "detour" ended up costing the program.

The T801 engine had completed over 500 hours of testing by time of the streamline restructure. It remained in metric units.

CONCLUDING REMARKS

The size and complexity of the Comanche program required the creation of new management systems to operate, control, and track the program. In hindsight, some alterations in these procedures might have been productive. In the end, however, the fact that the Boeing Sikorsky First Team successfully went through five major program restructures and kept the program on track in spite of some difficult issues is a tribute to the management organization and systems that had been set up.

In an article by David Harvey in the June 1981 issue of *Rotor & Wing International* [2], the Boeing Sikorsky program director, Lou Cotton, was asked

> ...whether he thought the government's approach to forcing technology development through ruthless competition was the best way to go. Would the results have been the same if the Comanche had been chosen in some other way? "There's no doubt about the answer," he said. "The state of rotorcraft art would not have advanced to the point it now has without the specific process which LH had to go through. ... If we'd just [gone through the old process] I'm certain we'd now be years behind where we are now. The discipline in the process was tremendous. From that point of view, you'd have to conclude the government was right."

REFERENCES

[1] "Comanche Pilot—Living the Dream," *Army Magazine*, September, 1998, pp. 24–31, by Dennis Steele.

[2] Harvey, D., "First Team's Perspective on the LH Competition," *Rotor & Wing International*, June, 1981, p. 70.

Chapter 25

The SuperTeam

Arthur W. Linden

After Boeing and Sikorsky teamed in 1985 to form the First Team, the two remaining major helicopter companies, McDonnell Douglas and Bell Helicopters, teamed in 1986 to form the SuperTeam (Fig. 25.1). The major emphasis of the LHX program was on the mission equipment package (MEP), so the SuperTeam emphasized that McDonnell Douglas Aircraft would be on the team to develop the MEP. Bell would have responsibility for the air vehicle engineering, and McDonnell Douglas Helicopter would be the

Fig. 25.1 The McDonnell Bell SuperTeam LHX concept (McDonnell Bell SuperTeam Photo).

lead on supportability. They stressed that they were the team with the scout/attack production experience.

THE SUPERTEAM COMPANIES

Like Boeing Sikorsky, the SuperTeam added a number of the leading companies in the industry to their LHX effort. These included Texas Instruments and Hughes Aircraft for the electro-optic sensors and the night vision pilotage system (NVPS); Hughes and Honeywell for the helmet-mounted displays; Texas Instruments, AT&T, Unisys, and Hughes for computer processors; Litton for controls and displays; Hughes and Motorola for communications and navigation; and Northrup and Eaton for aircraft survivable equipment (ASE). General Electric and Honeywell would do the flight controls. The training systems would be developed by a team consisting of McAir Training Systems, Hughes, Loral, Grumman, and EEC.

THE MISSION EQUIPMENT PACKAGE

An extensive article in the 6 March 1989 issue of *Aviation Week & Space Technology* titled "McDonnell Douglas/Bell Team Introduces Its LHX Concept" provided much detail on the SuperTeam's concepts and approaches [1]:

> Hughes Electro-Optical & Data Systems Group and Texas Instruments Defense Systems & Electronics Group have formed a joint venture, called Mission Equipment Team (MET) to design and produce the mission processor, the EOTADS, and the sensors and processing for the pilot's night vision.
>
> The architecture of the mission processor is to be common with that planned for the Air Force advanced tactical fighter and the upgraded version of the Navy A-12 attack aircraft. Officials would also like to see electronic modules to be interchangeable as well.
>
> American Telephone & Telegraph and Unisys Corp.—the two companies developing avionics architecture for the Northrop/McDonnell Douglas ATF team—are working with MET to make the LHX mission processor common with the ATF.

The mission processor was defined as having "...two 32-bit computers that can work at full performance after failure of some modules and can reconfigure for graceful degradation after that."

The electronics would use size E standard electronic modules (SEM-E), as the Boeing Sikorsky team was also using, but the SuperTeam planned to use liquid cooling whereas the First Team had an air-cooled design.

It was planned to have both infrared and radar jammers.

In an article in the March 1990 issue of *Rotor & Wing International*, titled "LHX Infrared Sensing: SuperTeam's Breakthrough," [2] the SuperTeam's concept for infrared (IR) sensing was identified as revolutionary. It described the existing IR sensing technology as using photocathode technology, limited to an IR detector array of 180 detectors, whereas the new system used "photovoltaic" technology. Advantages quoted were the lack of need for a heavy preamplifier and that "...the PV detector can be about 50 times more responsive, and arrays can now be made up of thousands of elements, rather than being limited to 180."

The SuperTeam stressed their scout/attsck helicopter experience over Boeing Sikorsky (Fig. 25.2), and their mission equipment experience, (Fig 25.3) [4].

THE AIR VEHICLE DESIGN

The SuperTeam did a good job of confusing its competition—that would be us—about any details on their aircraft configuration. Illustrations shown publicly were continually changing. (See Figs. 25.4 and 25.5.)

These illustrations were sufficiently vague to prevent us from knowing its design concepts for things like rocket/missile stowage, low observable concepts. realistic empennage designs, and so forth.

The 6 March 1989 issue of *Aviation Week & Space Technology* [2] explained that there were to be 10 weapons stations, 6 being low observable.

> McDonnell Douglas and Bell chose fixed weapon sponsons instead of ones that retract into the fuselage as on the Boeing/Sikorsky design.
>
> To minimize the increase in radar cross section from exposed weapons, the team is looking at conformal carriage or possibly retracting weapons into the sponsons.

The SuperTeam planned to use electric, rather than hydraulic, flight control actuators.

In a major marketing brochure distributed prior to the down-select, the SuperTeam emphasized three areas of the air vehicle—the rotor, the anti-torque/yaw control system, and the gun.

THE IMPROVED 680 BEARINGLESS MAIN ROTOR SYSTEM

The McDonnell Bell team selected what it called a "high agility rotor" derived from the Bell 680 rotor. It included advanced light rotorcraft (ALR) blades. The advantages claimed in its marketing literature were:

- Moderate hinge offset for agility
- Four blades
- Soft in-plane with elastomeric dampers
- Bearingless
- Composite blades and hub
- Fail-safe, on-condition replacement
- Low parts count
- No lubrication required

Fig. 25.2 The SuperTeam stressed their scout/attsck helicopter experience over Boeing Sikorsky [3].

The Bell 680 rotor was initially developed as a four-bladed replacement for the two-bladed teetering rotor that had served Bell very successfully for decades. Quoting the 6 March 1989 issue of *Aviation Week & Space Technology* [1]: "A prototype of the rotor system was first flown on a Bell Model 222 technology demonstrator aircraft in May, 1982.... On January 24, [1989] Bell flew the newest version of the bearingless rotor system on a U.S. Marine Corps AH-1W SuperCobra...."

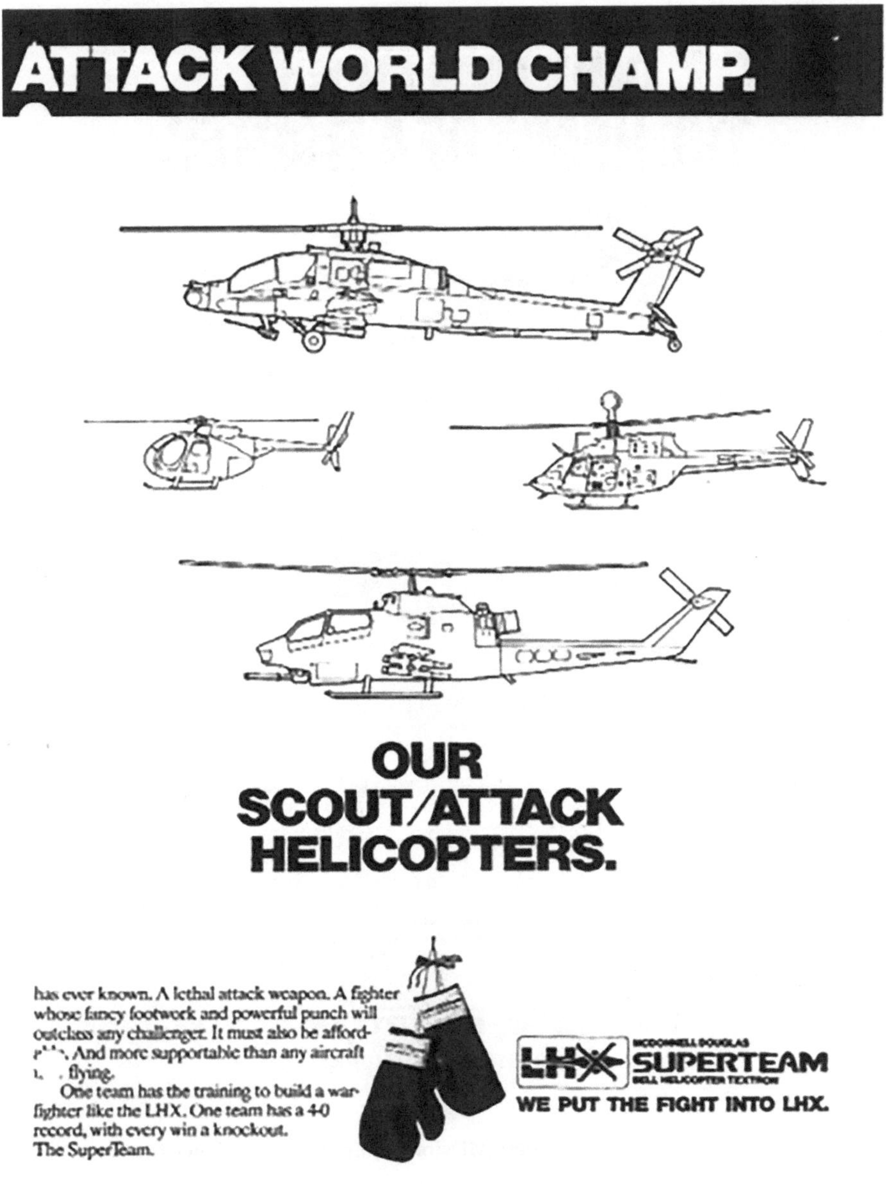

Fig. 25.2 (continued)

According to the article [1]:

> Bell developed a one-piece, glass fiber and epoxy four-bladed yoke that attached the composite rotor blades to the transmission mast….
>
> The yoke is constructed of a series of interleaved, filament-wound glass fiber/epoxy belts with additional plies of glass fiber/epoxy tape at different cross-grain angles added to the mast attachment areas.

Fig. 25.3 The SuperTeam stressed their Mission Equipment experience [4].

Furthering the development story, the article continues [1]:

> Both the 4BW and the LHX system will incorporate several improvements upon the Model 222-680 rotor. The new, larger rotor will use two beam-type yokes, stacked on top of each other and attached at the mast….
>
> The 4BW system will have separate joints, allowing the SuperCobra to maintain its blade folding capability.

Presumably, the LH rotor would also need that feature.

Fig. 25.4 Cover of *Army Aviation*, 31 December 1990 [4].

Fig. 25.5 Trade show handout illustration (McDonnell Bell SuperTeam Photo).

The article quoted Bell's claims for this rotor as:

- Higher agility
- Fifty percent fewer parts
- Fifteen percent rotor system weight reduction
- Lower life cycle costs than Bell's existing two-bladed systems

The article also said that Bell "sought to reduce vibration levels with minimum weight penalties, provide tolerance to high explosive incendiary projectiles up to 23mm and lower manufacturing and maintenance costs" [1].

In its advertising, the SuperTeam stressed that its main rotor was flying, whereas the Boeing Sikorsky First Team's was not (Fig. 25.6).

That point was correct. Both Boeing's and Sikorsky's in-production rotors were fully articulated, a configuration very well suited to their utility and transport missions, but not necessarily for a high agility SCAT aircraft.

THE NOTAR ANTITORQUE/YAW CONTROL SYSTEM

McDonnell Douglas Helicopters had recently flown a new concept for an antitorque/yaw control system. This is what it called the "NOTAR," for no tail rotor. This replaced the conventional single-rotor helicopter's tail rotor with a fan in the tailcone that provided pressurized air to exhaust through slots along the tailcone, creating sideward lift using circulation control (the Coanda effect). In addition, some of the air was exited through a nozzle at the end of the tailcone to provide a way to modulate the sideward thrust and provide yaw control (Figs. 25.7 and 25.8).

The advantages claimed for NOTAR included:

- Superior handling qualities
- Sideward flight equally controllable in either direction
- Reduced drag at high speed
- Low radar and visual signature
- Low acoustic signature
- No strike hazard
- No hazard to ground crew
- Reduced ballistic hazard
- No intermediate or tail rotor gearbox

McDonnell Douglas had been developing the NOTAR since 1975. In 1981, it used a fan from the French Gazelle Fenestron helicopter as its first test hardware. It then tested an improved NOTAR on an OH-6. According to *Helicopter World* [5]: "Pilots who have flown the U.S. Army OH6 NOTAR prototype (including Scott Wilson from The Shepard Press) are full of praise."

Fig. 25.6 They stressed that their rotor was flying while ours was not [6].

But the next step in development did not go well. The Army's 160th Special Operations Force had contracted with McDonnell Douglas to modify some of its fleet of OH-6s to the NOTAR configuration. In fall 1990, McDonnell Douglas demonstrated the NOTAR OH-6 at Fort Campbell, Kentucky. The 160th pilots were unhappy with yaw control and issued a stop work order until the problem could be fixed. In January 1991, McDonnell Douglas proposed a fix, using a yaw axis stability augmentation system (SAS).

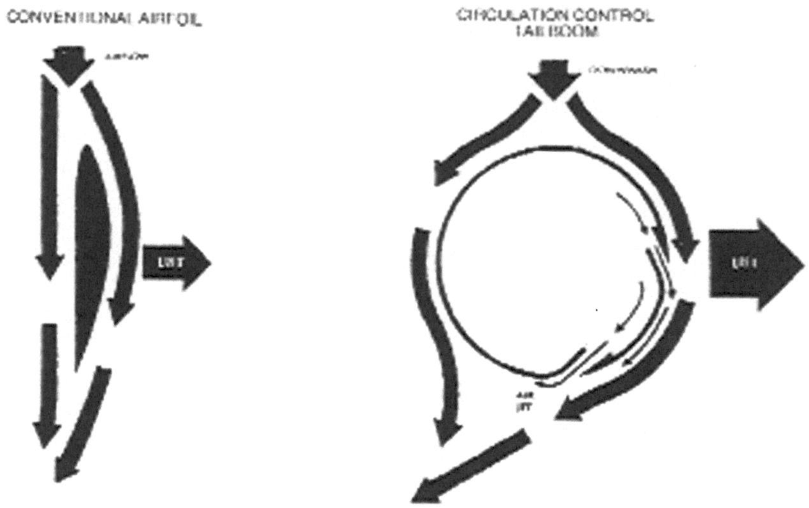

Fig. 25.7 The circulation control concept (McDonnell Bell SuperTeam Photo).

There were also revisions to the LH NOTAR design at this time. As reported in the 4 February 1991 issue of *Aviation Week & Space Technology* [7]:

> The McDonnell Douglas Helicopters and Bell Helicopter Textron team has modified their NOTAR anti torque design for the U.S. Army's Light Helicopter program….
>
> In the most significant change, the primary fan has been moved to the rear of the tail boom from the previous location in the base of the boom

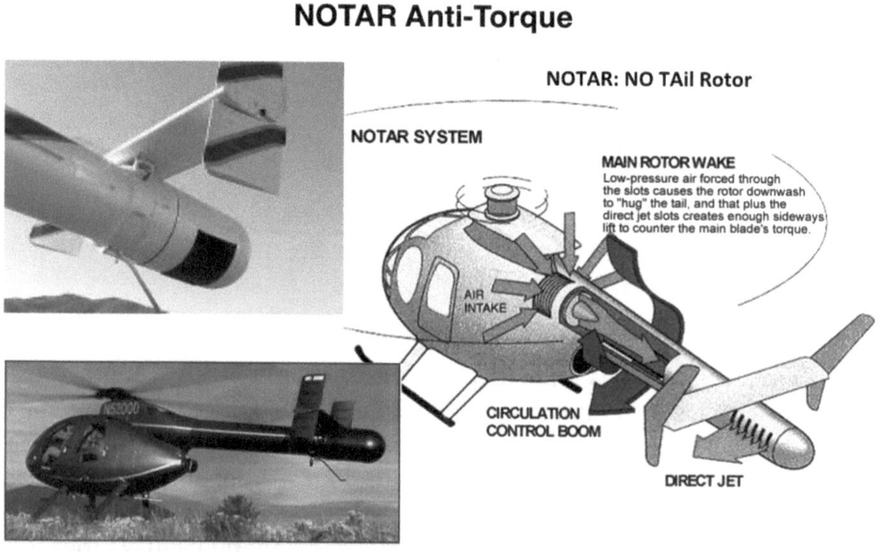

Fig. 25.8 The NOTAR anti-torque cioncept (McDonnell Bell SuperTeam Photo).

> and will provide independently controllable pressurized air to the jet thruster. The 30-in.-dia. fan will be powered by a rotating output shaft from the main transmission and take in outside air from an additional intake on the right side of the boom.
>
> A smaller fan has been added at the base of the tail to supply low pressure air to the Coanda circulation control slots that run the length of the tail. A small duct directs the air from this fan to the slots.

From a February 1991 "Fact Sheet" sent by the SuperTeam to the trade press [7]:

> The slotted tail boom produces roughly ½ the anti-torque required at low speeds and/or high power settings for the small power required by the slot fan. The vertical tail produces the anti-torque required at high speed for the small power required to overcome its drag. The thruster produces yaw control moments as required in all flight regimes. A more optimum thruster/fan match was achieved when the slot requirement was removed [from the] thruster fan [and the] power required was reduced significantly by the second generation design.

But in the *Aviation Week* article [7], a McDonnell Douglas official said that the changes were due to a need for additional space in the tailcone for the LH's mission equipment package, that "...space for the MEP electronics was so critical that engineers adopted the more complex configuration." There was no reference to any need to improve NOTAR's performance.

In the March 1991 issue of *Rotor & Wing International* [8], the SuperTeam program manager, Al Haggerty, is quoted as saying:

> The [*Aviation Week*] story upset us...because we wanted to maintain every inch of competitive advantage right through the final stages of the LH contest. ...Frankly, I'm tired of the negative things I keep hearing about NOTAR. ...It's just flat wrong that the LH contest has somehow been turned into the "battle of the tail rotors." There's much more to the competition than that.

Just weeks before the Army was to downselect to one of the LHX teams, *Inside the Army* reported in its 4 March 1991 issue that the OH-6 Special Operations NOTAR test contract had been placed on hold due to yaw stability problems. In the article, McDonnell Douglas officials said that [9]: "The Delta Force's problem with NOTAR will not carry over to the LH.... As the LH would be built from the ground-up around the NOTAR technology. In the Delta Force's aircraft, NOTAR was added as an afterthought."

McDonnell Douglas also spent considerable effort on the NOTAR's empennage design. NOTAR program manager Jim Van Horn explained in an article in *World Aerospace Technology '90* that using NOTAR eliminates the tail rotor with its stabilizing influence in forward flight. Additionally, the flow field over the NOTAR tailcone [10]: "...changes dramatically when transitioning from powered flight to autorotation. While fixed vertical stabilizers can be made to work adequately within these two different flow fields, moving vertical stabilizers optimize the system in all flight regimes." The experimental OH-6 NOTAR aircraft ended up with a large "H" tail. At the time of the article, they were still evaluating whether this needed to be a movable tail or could be fixed.

The artist's concept paintings for the McDonnell Douglas LH design showed a fixed vertical fin on the aircraft centerline. It was not clear to the Boeing Sikorsky engineers how yaw stability was achieved with this design.

As the Boeing Sikorsky team developed our LH design, the NOTAR concept was evaluated as one option for antitorque/yaw control. An important conclusion was that the NOTAR concept did not scale up to larger sized helicopters linearly; weight and performance penalties grew greater as the aircraft's size increased. Thus, the results from the smaller OH-6 might not be applicable to the larger LH.

At the debrief after the downselect, Al Hagerty was proven correct. All of the emphasis by the two teams on NOTAR vs Fantail, and after the extensive ground and flight tests demonstrations were done, the Army said that the antitorque solution did not pay a major role in the downselect scoring.

THE "CHAIN GUN"

In its marketing brochure [11], the SuperTeam described the chain gun as an "[i]nertially stabilized articulating turret features hydraulically powered, single barrel, 20mm Chain Gun automatic cannon." Advantages listed were as follows [11]:

- Highly Reliable
- Low Maintenance—Operates in a Dirty Environment
- Light Weight—Lighter Than Other Gun Systems
- Selectable Aiming Modes
  - Fixed Forward
  - Target Acquisition System Slaved
  - Crew Line of Sight Slaved
  - In-Flight Boresighting

This was similar to the 30-mm chain gun on the production Apache.

SUPERTEAM FLIGHT DEMONSTRATORS

Like the First Team, the SuperTeam used test aircraft to develop and demonstrate its technologies (Fig. 25.9). An Apache demonstrated digital fly-by-wire flight controls and was planned for future testing of the night vision

Fig. 25.9 Three SuperTeam flight demonstrators [1].

pilotage system; a Bell 222 demonstrated the LH 680 main rotor system; and an OH-6 demonstrated the NOTAR antitorque/yaw control system.

REFERENCES

[1] Michael A. Dornheim, "McDonnell Douglas/Bell Team Introduces Its LHX Concept," *Aviation Week & Space Technology*, 6 March 1989, pp. 56–59.
[2] David Harvey, "LHX Infrared Sensing: SuperTeam's Breakthrough," *Rotor and Wing International*, March 1990, p. 43.
[3] SuperTeam Ad, *Aviation Week & Space Technology*, 14 November 1988, pp. 104–105.
[4] SuperTeam Ad. *Aviation Week & Space Technology*, 3 September 1990, p. 184.
[5] *Helicopter World*, April–June 1989.
[6] SuperTeam Ad, *Aviation Week & Space Technology*, 8 October 1990, p. 35.
[7] "Notar Design for LH Helicopter Modified," *Aviation Week & Space Technology* 4 February 1991, p. 20.
[8] "SuperTeam Reveals NOTAR Change," *Rotor & Wing International*, March 1991, p. 10.
[9] "Army's Elite Special Ops Unit Finds Problems with MAC DAC's 'No Tail Rotor' Helo," *Inside the Army*, Vol. 3 No. 9, 4 March 1991.
[10] James R. van Horn, "NOTAR-tm (no tail rotor) anti-torque system," *World Aerospace Technology—'90*, published by Sterling Publications, London England, pp. 115–116.
[11] "SuprTeam LH—A Warfighter by Design . . ." SuperTeam marketing brochure.

Chapter 26

Marketing

Arthur W. Linden

In addition to developing the LHX/Comanche technology, it was important to market the activities and the progress of the program. There were any number of sources who would rather spend Comanche's funds elsewhere. An active marketing effort was required to counter these. These marketing efforts targeted the Army and Department of Defense (DoD) at all levels, plus Congress and the general public.

Early Technology Demonstrations

As part of the Advanced Rotorcraft Technology Integration (ARTI) program and prior to awarding the LHX contract, Sikorsky and Boeing demonstrated key portions of their overall design. Even in these early days, marketing and public relations were important. Sikorsky prepared an ad titled "Eyes of the Hunter" (see Fig. 26.1), which won a national advertising award for both content and graphics. The illustrations in Fig. 26.1 are courtesy of George Sipprell, who led this effort.

Boeing Sikorsky also had a press release showing the Fantail and Shadow test aircraft in the air together (Fig. 26.2).

Mockup Unveiling

In a strategic marketing move, we decided to reveal our LHX mockup (Fig. 26.5) prior to the competitive downselect. There was much speculation as to what the competitive teams were going to propose. The Army was using generalized simplistic drawings when it illustrated the aircraft. We felt that we

Eyes of the hunter.

It is the 21st Century. A lone pilot-warrior in a rotorcraft of awesome sophistication skims the terrain. His mission: Find the enemy. Neutralize his defenses. Destroy him. Come home safe.

Within the context of AirLand Battle 2000, a demanding challenge is posed for both man and machine. Flying at night in adverse weather against an enemy superior in numbers, with sophisticated weapons of his own. Highly automated controls and displays will have to process data at nanosecond speeds, providing the information needed to see deep. Strike fast. Get in. Get out. All in a matter of seconds.

What level of technology will be required to give the Army pilot of the future what he needs to accomplish his mission? Is a single pilot cockpit viable? Can the next generation of Army rotorcraft be smart, fast, and fierce and at the same time dependable and affordable?

To answer these questions, Sikorsky Aircraft has assembled a team of companies whose combined capabilities cannot be equalled. Sikorsky, Collins Avionics/Rockwell International, Hamilton Standard, Martin Marietta, Norden and Northrop. The members of the Sikorsky ARTI/LHX team have developed and built some of the most advanced military aircraft systems in existence.

This is the team with the facilities to test the single pilot concept and carry the ultimate design to completion. With the expertise to pull together all the vital elements, from flight control and communications to weaponry and data acquisition.

The Sikorsky team stands ready, with the resources, the technologies, and with a history of successfully responding to the challenges of Army aviation.

The Sikorsky Team:
The Army experience ARTI/LHX demands.

UNITED TECHNOLOGIES
SIKORSKY AIRCRAFT

Fig. 26.1 "Eyes of the Hunter" advertisement.

Fig. 26.2 Fantail and Shadow test aircraft.

would have a competitive advantage if we revealed our design prior to the downselect. There would not be sufficient time for the SuperTeam to react to anything it might learn.

The most dramatic moment at the Army Aviation Association of America (AAAA) convention in Orlando, Florida, on 12 April 1990, with many senior Army personnel and the press in attendance, was the unveiling of the Boeing Sikorsky LHX mockup (Figs. 26.3 and 26.4). The mockup was displayed in the armed reconnaissance mission configuration with its weapons bay open and the 20-mm Gatling gun in the firing position.

According to *Flight International*, 25 April 1990 [1]: "The mockup is said to be 'between 80 and 90%' accurate. Certain low observability details remain classified."

In his remarks at the unveiling, Sikorsky president Gene Buckley said: "There is no government, no group of people, no small tribe, that ever looked away from technology and survived economically or militarily into the future."

The *New Haven [CT] Register* of 13 April 1990 commented [2]: "…A McDonnell Douglas spokesman suggested that the First Team's new model is pure PR. We have chosen not to have a public mockup…We can do more by showing the actual thing to the Army, and we won't have to spend an awful lot of money."

Fig. 26.3 Public unveiling of the Boeing Sikorsky LHX mockup at the AAAA Convention, April 1990.

This worked out well for us—with only one design available to the Army, it started using the Boeing Sikorsky design to illustrate the program.

MOCKUP TOUR

After the unveiling, the mockup was shown at a number of First Team member locations—Harris Corp. in Melbourne, Florida; Sikorsky in

Fig. 26.4 The crowd gathers around the mockup.

Fig. 26.5 The Boeing Sikorsky aircraft mockup.

Stratford, Connecticut; Hamilton Standard in Windsor Locks, Connecticut; Boeing in Philadelphia; CAE-Link Corp. in Binghamton, New York; Crystal City in Arlington, Virginia; the American Helicopter Association convention in Washington DC; and Westinghouse in Baltimore, Maryland. The mockup was then returned to its fabricator, Advanced Technologies Inc. of Newport News, Virginia, for modifications in preparation for a second tour that began in August.

On this tour, the mockup was displayed at a number of Army posts around the country (Fig. 26.6). It not only provided a tangible representation of the design, but also drew audiences where many aspects of the design could be

Fig. 26.6 Aircraft 2 on its travels around the country.

presented and discussed. One reaction was that the rank and file Army felt that the tour, accompanied by design engineers, showed that the Boeing Sikorsky team was listening to their needs.

The mockup was next displayed at the American Pavilion in the 1991 Paris air show at LeBourget Airport, June 13–23.

Comanche Portable Cockpit

Later in the program we had the Comanche Portable Cockpit (CPC), a rudimentary simulator with a high fidelity cockpit that allowed the Army to actually fly the aircraft. It also went on tour to get exposure to a broader Army than just those fortunate enough to visit one of our facilities. It could do recon and hit targets. At one venue, it linked into a network of simulators to show the potential for interoperability of Different platforms.

Aviation Week Laurels Award

After the April 1991 competitive downselect, in January 1992 the Boeing Sikorsky RAH-66 First Team won an *Aviation Week & Space Technology* Laurels award. From its January 27, 1992 issue [3]:

> To Boeing Sikorsky RAH-66 Comanche team, including Lou Cotton, team director; Bill Walls, former deputy director; Arthur Linden, Sikorsky program manager; Earl Norman, former Boeing program manager, and Marshall Heard, Boeing Helicopters vice president, for winning the long-fought battle to develop and produce the scout/attack helicopter.

Tom Clancy Visits Sikorsky to Learn About the Comanche

In 1993, author Tom Clancy and his technical advisor visited Sikorsky (Fig. 26.7). He spent a morning learning about the potential warfighting value of the aircraft and receiving a series of extensive technical briefings that described the unique characteristics of the helicopter weapons system. He then flew the RAH-66 motion-based simulator over an emulated battle zone with accompanying threats and targets. After this experience, it quickly became obvious that the aircraft would play an important role in one of Clancy's upcoming novels. Comanche project pilot Rus Stiles relates what happened:

> Usually the way it went was the visitor would be dropped off in the simulator briefing area and I would go over what we were going to do, etc. The entourage often thinned out—and in the case of Clancy, I was eventually left alone with him and his assistant. It got really fun at that point, as I saw his imagination in making up the story. He looked around as if to check to see that no one was looking, and then he bent over the table and looked at me with excitement and said "OK, here's the story…". He proceeded to

Fig. 26.7 Tom Clancy with the Comanche.

> explain how Comanche was going to be used in the book—the setup, plot, action—he became a kid in front of my eyes. When he was done, he said "so…do you think it can do that?" I said, well, let's go try. Then we went in the simulator and did everything he imagined. I'll never forget I got to be with Tom Clancy the storyteller.

The result of this background research effort was a new book, *Debt of Honor*, published in 1994. The book featured the Comanche using its unique survivability and detection features in a limited war with Japan and India. Mr. Clancy showed that he had learned well during his briefings. His words about the Comanche were completely accurate [4]: "Though small and a mere helicopter, the RAH-66A was in some ways the most sophisticated aircraft in the world. It carried in its composite airframe the two most powerful computers ever taken aloft, and one of them was merely a backup in case the first should break." Clancy also showed his understanding of the unique aspects of Comanche's acoustic signature:

> The Sikorsky engineers had long known that most of the noise generated by a helicopter came from the tail rotor's conflict with the main rotor. The one on the RAH-66 was shrouded, and the main rotor had five fairly thin composite blades, resulting in a helicopter with less than a third of the acoustical signature of any other rotary wing aircraft yet built.

Additionally, he demonstrated his knowledge of some of Comanche's weapons options: "The pilot selected Hellfire from his short list of weapons options. The weapons-bay door opened, and he fired twice." And his understanding of the aircraft's 20-mm cannon, which is described as emerging from its "canoe like enclosure."

Tom Clancy was no stranger to Sikorsky products and had featured the MH-53J Pave Low helicopter in a 1989 book, *Clear and Present Danger*, which was about conflict with the drug lords of Colombia. His portrayal of the unique value of the RAH-66 in *Debt of Honor* demonstrated a keen understanding of its warfighting potential.

Clancy also helped our cause at a congressional reception in May 1995. Quoting the *Connecticut Post* of 15 January 1996 [5]: "The effort peaked in May when best-selling author Tom Clancy—who featured the Comanche in a recent novel—wooed key lawmakers at an evening reception in Washington's Library of Congress, featuring high-tech exhibits of the helicopter's capabilities."

THE ROLLOUT CEREMONY

Another big event was the rollout ceremony on 25 May 1995. From the *Sikorsky Weekly*, 30 May 1995 [6]:

> The first RAH-66 Comanche prototype was unveiled…at a roll-out ceremony before a crowd of almost 1,000 employees, company officials and military personnel. …During the roll-out ceremony, Army Chief of Staff Gen. Gordon Sullivan said the Comanche "represents a revolution in the application of technology to the goal of domination on the 21st century battlefield. Comanche is the key to winning the information war, owning the night and ensuring defeat of America's future enemies."

The illustration in Fig. 26.8 was provided by Sikorsky employee Jack Caldwell. The image on the left was a poster provided to the attendees of a

Fig. 26.8 Jack Caldwell's unique rollout memento.

Fig. 26.9 Demonstration of designing the aircraft on the computer. Seated, left to right: Dale Humphries, avionics installation designer, and Secretary of the Army Gordon Sullivan. Rear, left to right, Eugene Buckley, Sikorsky president; Connecticut Senator Joseph I. Lieberman; Ed Renouard, vice president/general manager Boeing Helicopters; Maj. Gen. Dave Robinson, aviation branch chief and commanding general, Aviation Warfighting Center; Art Linden, Sikorsky Comanche program manager; and Lt. Gen. Bud Forester, military deputy to the Assistant Secretary of the Army for Research, Development, and Acquisition.

postrollout celebration. At that event, Jack asked General Sullivan if he would sign his poster. The general graciously agreed, and according to Jack seemed pleased to have been asked. Jack then asked Sikorsky president Gene Buckley to also sign the poster, followed by Wallace Coffey, Chairman of the Comanche Nation. So Jack ended up with a unique and prized memento of his work on the Comanche program.

VIP PRESENTATIONS

As is standard procedure with this type of program, the team hosted many military and civilian leaders to explain the program and get their support. One of the most significant was on 17 May 1993, when General Gordon Sullivan, chief of staff for the U.S. Army, visited the team. Also in attendance were U.S. Senator Joseph Lieberman (D-CT) and senior Army officials (Fig. 26.9).

General Sullivan saw a demonstration of designing the aircraft on computers using three-dimensional electronic technology, a breakthrough

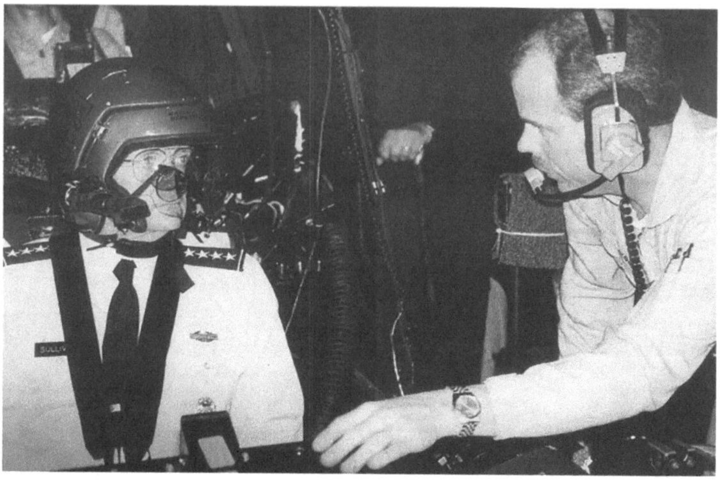

Fig. 26.10 General Sullivan in the simulator with Rus Stiles.

technology at the time. He also viewed the full-sized mockup of the aircraft, had discussions and demonstrations of the mission equipment package, and flew the Comanche simulator, flying a simulated combat mission demonstrating the helicopter's battlefield effectiveness (Fig. 26.10).

General Sullivan then addressed a large group of Sikorsky employees in the Stratford high-bay area. Among other remarks, he said that "...you are, in fact, leading us into the 21st century with a weapons system that I believe will be the hallmark of the early part of that century."

The *Connecticut Post* of 18 May 1993 commented that [7]: "Sen. Lieberman said 'I had the feeling as we walked through the facility today, that not only do we have a distinguished visitor and a great general, but we have a pretty excited customer for the Comanche helicopter.'"

Other important demonstrations followed (Figs. 26.11 and 26.12). Quoting an Associated Press story on 15 March 1994, Secretary of the Army Togo West said [8]:

> Only Comanche has the ability to conduct lethal deep, close, and rear operations and survive while providing critical data, protection, and flexibility to forward-deployed maneuver forces.
>
> Murtha called the Comanche program the Army's most important project and said the Comanche would enhance troops' ability to fight on the ground.
>
> But, he said, even with widespread support—Defense Secretary William Perry has endorsed research funding for the Comanche—the project still will have to compete for funding with dozens of other defense projects.

Fig. 26.11 Presentation on 14 March 1994 to Army Secretary Toga West (third from left in front row) and Rep. John Murtha (R-Pennsylvania), chairman of the House Defense subcommittee of the House Appropriations Committee (second from the right).

DISPLAY AT THE U.S. CAPITOL

On 12 May 1998, the second Comanche aircraft was displayed on the national mall in Washington, DC at the reflecting pool in front of the U.S. Capitol. The event was sponsored by the USO, with the purpose of dedicating

Fig. 26.12 Connecticut Governor John Rowland, Army Comanche Program Manager B. G. Bergantz in the aircraft, and Boeing Sikorsky Program Director Art Linden speaking to the governor.

Fig. 26.13 Souvenir poster from "The Duke" naming ceremony. (Signed by Garth Brooks for Art Linden's daughter Heather, aka Hez, who, not being a country music fan, was not impressed.)

aircraft 2 as "The Duke" in honor of actor John Wayne (Fig. 26.13). Three of his children were in attendance. Speakers at the formal ceremony included Senator Strom Thurmond (R-SC), Speaker of the House Newt Gingrich (R-GA), and Senate Majority Leader Trent Lott (R-MS). According to the *New Haven [CT] Register* of 9 May 1998, Acting Army Secretary Robert Walker said [9]: "Soldiers identify with John Wayne, who portrayed the American soldier as an individual with solid values and high ideals. The Army is honored to have the Comanche prototype No. 2 named 'The Duke.'" Continuing, the article said:

> The event is being organized by the USO as part of a month-long salute to the country's armed forces. They are honoring Wayne because of his long career portraying members of every branch of the armed forces, as well as his support for the USO. He joined a number of USO tours for service members from World War II to Vietnam.

Also included was singer/entertainer Garth Brooks, who is a great fan of John Wayne. "Any time I get a chance to uphold the patriotic ideals that John Wayne stood for, I will do it," said Brooks [9].

The event included a social gathering in the Capitol building that helped build support for the program with the legislators.

While in the Washington area, the aircraft was also displayed at the Pentagon for two days and at Andrews Air Force Base and Fort Meyers.

In December 1999, Secretary of the Army Louis Caldera; Assistant Secretary of the Army for Acquisition, Logistics, and Technology Paul Hoeper; and Army Deputy for Systems Management and Horizontal Technology Major General Joseph Yakovac visited West Palm Beach for a flight demonstration, with both aircraft flying (Fig. 26.14). After the flights, Secretary Caldera joked with pilots Rus Stiles and John Armbrust, asking them, "So how much do you pay us to get to do this job? Very impressive. It's got to be a lot of fun."

Both Secretary Caldera and Mr. Hoeper were also given demonstrations in the Comanche Portable Cockpit simulator (Fig. 26.15). Like others before him, Secretary Caldera declared the Comanche to be the "quarterback of the digital battlefield."

The Comanche aircraft was first shown outside the United States at the 1998 Farnborough air show. A photo of this event was on the cover of the 14 September 1998 issue of *Aviation Week & Space Technology* [10].

In 1999 it was on static display in the U.S. Department of Defense corral at the Paris air show. Interest was high, and the Comanche was considered the primary attraction in the corral. Visitors included U.S. Under Secretary for Acquisition Technology Jacques S. Gansler, Senator James Inhofe (R-OK),

Fig. 26.14 Left to right: Secretary of the Army Louis Caldera, Dean Borgman, and Art Linden on the right, December 1999.

Fig. 26.15 Secretary Caldera (rear seat) and Mr. Hoeper (front seat) in the portable cockpit simulator.

Senator Ben Nighthouse Campbell (R-CO), Deputy Assistant Secretary of the Navy William A. Stussie, and NASA Director Bill Golden. Sikorsky representatives also gave briefings for congressional staffers, other aircraft manufacturers, and numerous press and magazine representatives.

One interesting additional guest was golfer Arnold Palmer (Fig. 26.16), a fixed-wing pilot. "He's interested in flying, and we gave him a briefing in the

Fig. 26.16 Golfer Arnold Palmer in the Comanche cockpit, Paris 1999.

cockpit," said Maureen Fino of Sikorsky's Business Development Office. "He's a really nice guy, very personable."

PRINT ADVERTISING

We did the mandatory print advertising (Fig. 26.17). Some early advertisements stressed the differences between our design and the SuperTeam's

Fig. 26.17 Boeing Sikorsky Print advertisment.
Note: Stressing our technology background.

(Figs. 26.18 and 26.19). We also used the connection with the Comanche Native American nation. Figure 26.20 shows an ad from the 28 February 1993 issue of *Army Aviation* [12].

Of course there were trinkets and giveaways (Figs. 26.21–26.24), and every program has to have shirts (Fig. 26.25).

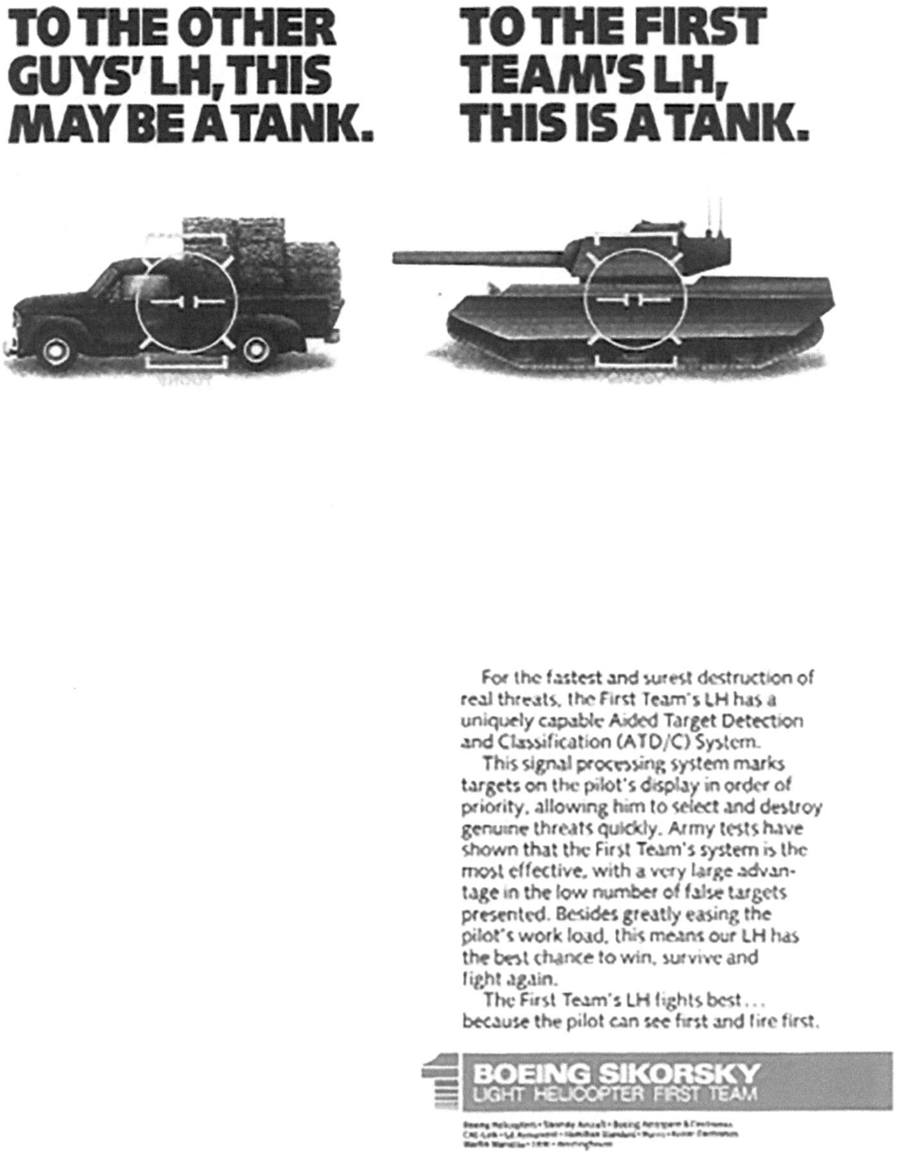

Fig. 26.18 Boeing Sikorsky print advertisment stressing our Aided Target Detection and Classification System.

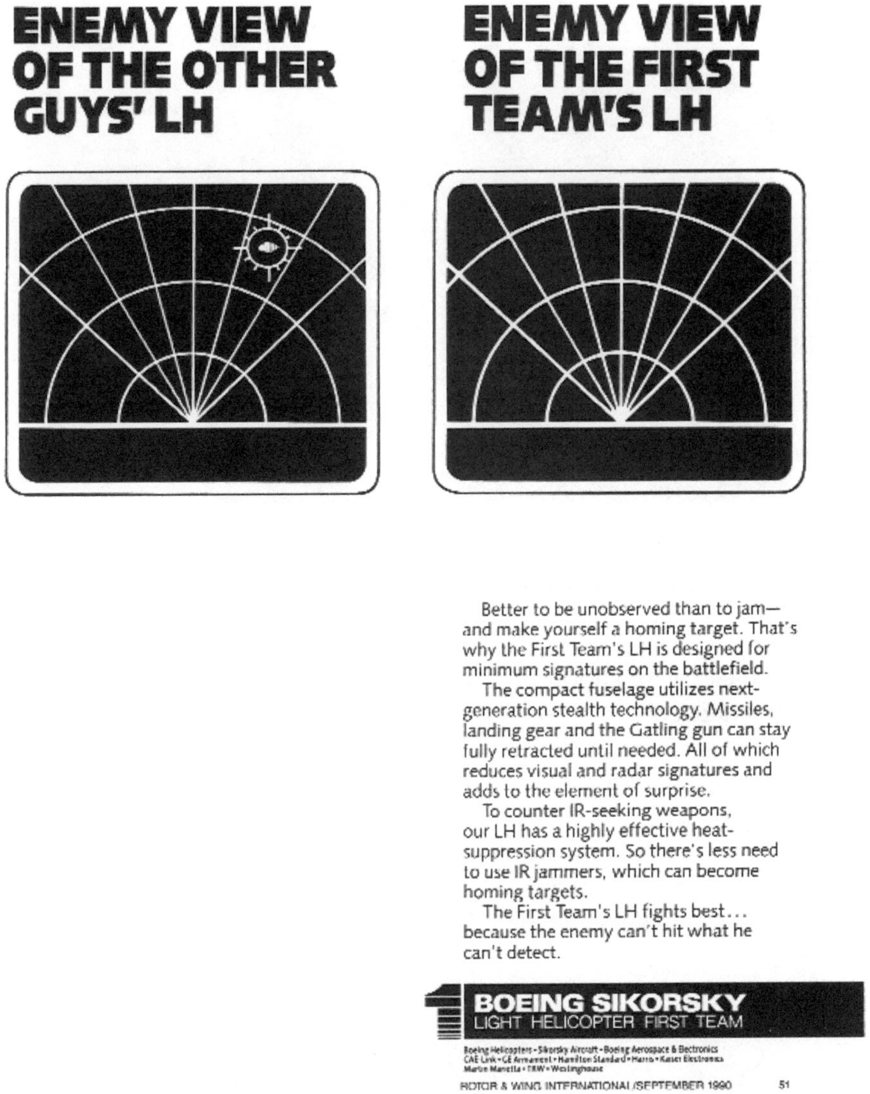

Fig. 26.19 Boeing Sikorsky print advertisment stressing our low observability. (I'd leave the source to the footnote to be consistent with the other ads.) ***Rotor & Wing International*** **September 1990 [11].**

FLIGHT DEMONSTRATIONS

As the flight test program progressed and the flight test envelope was developed, it became clear that formal demonstrations of Comanche's capabilities would be worthwhile. The aircraft was presented to the national media in 1998. Several press days were held, each accompanied by a flight demonstration. In addition, the Discovery Channel did a full show dedicated to the aircraft.

Fig. 26.20 In one ad we stressed our connection with the Comanche Nation.
Note: Boeing Sikorsky print advertisment using our Comanche nation connection.

Most of these were short (10-minute) segments after a regular data flight, usually after data collection was complete for the day. The VIP visitor usually received a program briefing and/or a visit to the telemetry room while the aircraft was conducting its flight test. The test pilots then shifted from data collection to marketing mode. The flight demonstrations displayed the aircraft through nearly its full flight envelope, at quite low altitudes. The unique capabilities and unconventional appearance of the Comanche intrigued both those familiar with helicopters and the general public.

Fig. 26.21 Pins and keychains.

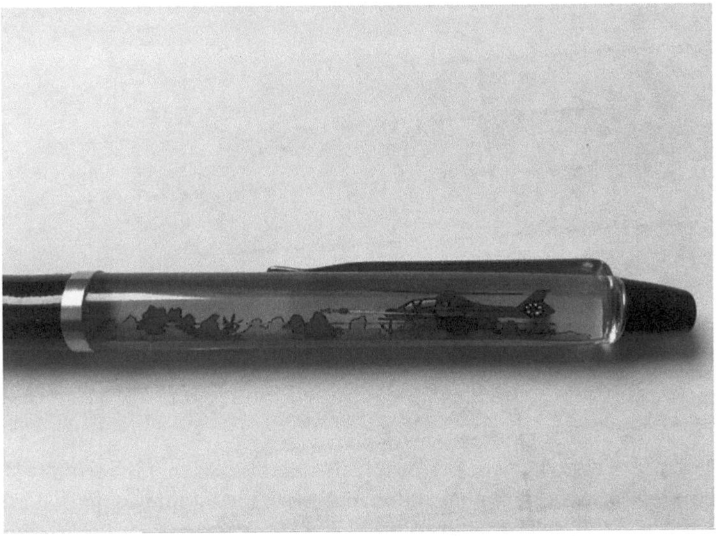

Fig. 26.22 A pen with a floating Comanche model.

Although the Army resisted providing a flight clearance to permit aerobatics, the "split S" (half roll, half loop) was demonstrated on several occasions (Figs. 26.26 and 26.27). The test team wanted to demonstrate the high degree of agility and controllability provided by the bearingless rotor and the fly-by-wire (FBW) control system.

Assistant Secretary of the Army for Research, Development, and Acquisition Gilbert Decker was given a flight demonstration at West Palm

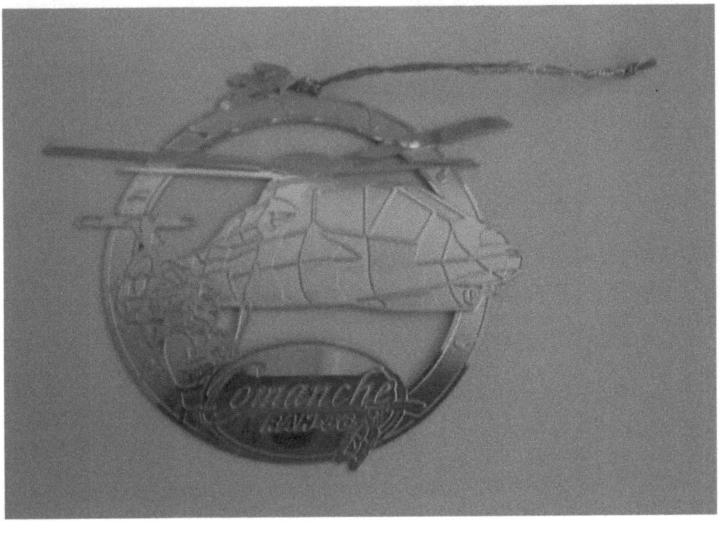

Fig. 26.23 A Christmas tree ornament.

Fig. 26.24 Coffee mugs. The one on the left was unusual in that it normally showed the aircraft in the low observable configuration, but when hot liquid was poured in the image changed to the heavy attack configuration.

Beach on 30 September 1996. The chief of staff of the Army, General Dennis Reimer, was also given a flight demonstration (Fig. 26.28). The general was generous in his praise of the aircraft and the program.

In one demo, a group of congressmen was positioned along the runway to observe a flight demonstration. Quoting the *New Haven [CT] Register* of 29 March 1998 [13]:

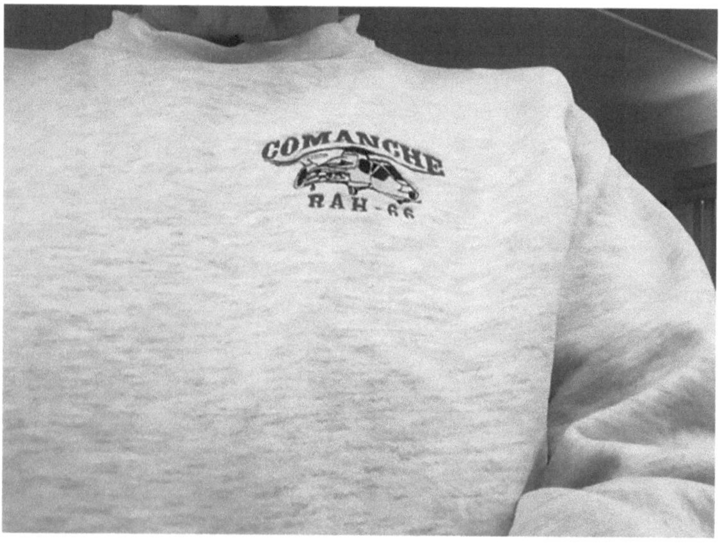

Fig. 26.25 One of the program's shirts.

Fig. 26.26 "Split S."

> ...they ... looked down the runway, listening for the familiar roar of an engine and the pulsing beat of blades slicing through the air. Suddenly the Comanche broke through the sky just above them, its near silent approach mesmerizing the delegation. [According to a Sikorsky spokesman] "One of the key selling points is how quiet it is. We had it come from behind them and fly over their heads—it's pretty dramatic."

Fig. 26.27 "Pushover" maneuver as seen from the aft seat.

In the end, the test team became experts at conducting these demos and found it rewarding to have the opportunity to show what the aircraft was capable of doing.

A very successful 2001 closed out with two days of marketing support, during which an IMAX movie was filmed of the aircraft, clips of which are still included in many aviation documentaries.

As the program progressed into engineering and manufacturing development, these flight demonstrations continued. One of the most important occurred on 15 November 2002, when Deputy Defense Secretary Paul D. Wolfowitz visited West Palm Beach (Fig. 26.29). He saw a Comanche demonstration flight, attended a briefing, flew the simulator, and received a cockpit demonstration.

Aircraft 2 was also shown in a couple of venues, but no aircraft flew anywhere except at West Palm Beach. Additionally, the Army wouldn't let us fly

Army Chief Visits Comanche Flight Test

Army Chief of Staff GEN Dennis Reimer visited Team Comanche at Sikorsky's West Palm Beach, Florida, Experimental Flight Test Facility just before Prototype No. 1 began its upgrade. Reimer expressed support for the program and was thoroughly impressed by a Comanche flight demonstration. Here, (L to R) Army test pilot CW4 John Armbrust and Comanche Team test pilot Rus Stiles (behind Armbrust), discuss the flight with then Comanche PM, BG Jim Snider, Boeing field office director Harry Bendorf, Sikorsky president Gene Buckley, and GEN Reimer. ❑

Fig. 26.28 General Reimer viewing the Comanche.

Fig. 26.29 Deputy Defense Secretary Paul D. Wolfowitz gets a Comanche cockpit demo conducted by Army CW4 Bill Fell.

any pilot except experimental test pilots, and even that list was limited. That may have started creating doubt in the rank and file—what was wrong with the aircraft if no one could fly it? Of course, it finally allowed one non–test pilot warrant officer and General Cody to fly it, but by then it was too late.

On 5 December 2003, Army Lieutenant General Richard A. Cody, Army deputy chief of staff and an experienced aviator, flew the Comanche. According to the January 2004 issue of *National Defense* [14], he "…gave the aircraft a stellar review. …Cody said that the Comanche had 'exceeded his expectation' and was particularly impressed by the rotor system, engine response and control. Cody also reportedly said the aircraft was 'easy to fly' and should be put into the hands of soldiers as soon as possible."

REFERENCES

[1] Julian Moxon, "'Undetectable' First Team LH Unveiled," *Flight International*, 25 April, 1990.

[2] Terry Sacks, "Copter Mockup Turns Heads—Sikorsky, Boeing Unveil LH Model are Army Aviation Meeting," *New Haven [CT] Register*, 13 April, 1990, p. 49.

[3] Laurels, Aviation Week & Space Technology, 12 January, 1991, p. 13.

[4] Jack Ryan, "Limited Partnership. Published by the Penguin Group (USA) Inc.," *Debt of Honor*, The Berkley Publishing Group, Copyright 1994, pp. 822 and 825, ISBN 978-0-425-14758-0.

[5] Lolita C. Baldor, "Copter Funding Scales Hurdles—Comanche Gets Early Support," *Connecticut Post*, 15 January, 1996.

[6] *Sikorsky Weekly*, 30 May 1995.

[7] *Connecticut Post*, 18 May 1993.

[8] Associated Press, 15 March 1994.

[9] Lolita C. Baldor, "Duke Fans Get Friend in High Places—Sikorsky Chopper Has True Grit," *New Haven [CT] Register*, 9 May, 1998, pp. 1 and 5.

[10] "Farnborough Air Show '98," *Aviation Week & Space Technology*, 14 September, 1998.
[11] *Rotor & Wing International*, September 1990, p. 51.
[12] *Army Aviation*, 28 February 1993.
[13] Lolita C. Baldor, "Show'n'tell May Send Comanche Flying," *New Haven [CT] Register*, 29 March 1998, pp. A! and A5.
[14] "Comanche Review Shrouded in Secrecy," *National Defense*, January 2004, p. 14.

Chapter 27

The Competitive Downselect

Arthur W. Linden

In fall 1990, the two competing teams submitted proposals for the full-scale development (FSD) program, detailing the features of their designs and a plan for developing and producing the aircraft. In February 1991, best and final offers (BFOs) were submitted.

On 5 April 1991, the Boeing Sikorsky Joint Venture team was announced as the winner of the competition with an option for engineering and manufacturing development (EMD). (See Figs. 27.2 to 27.4.) This was a \$2.8-billion, 52-month contract for a prototype phase beginning in April 1991. First flight was scheduled for August 1994. A 39-month follow-on EMD phase was planned as a negotiated option in August 1995.

The objectives of the Dem/Val prototype phase were to complete the aircraft design, build prototype aircraft, and conduct flight tests to reduce risk and demonstrate that the system was ready to successfully transition into EMD.

According to the 6 April 1991 issue of *The Washington Post* [1]: "...the Army structured the bidding so that the contractors and Pentagon would share the cost of overruns in the hard-to-predict final development phase of the project. Contractors will be held to their original bids for assembly-line production, however."

On 6 April 1991, *The New York Times* quoted Stephen R. Conver, Assistant Secretary of the Army [2]: "We made a big deal about this in the beginning of getting the contractors to bid realistically, and we told them we weren't going to look at how low their prices were—we were going to look at the realism of their prices."

Fig. 27.1 Aviation Week announcing our LH Win [5].

In those days the paper was still called The Bridgeport [CT] Post, 6 April 1991 [3]: "'The Boeing Sikorsky entry was the preferred one Conver said. 'We had a very good horse race on this. Both companies were very responsive.... Both had realistic pricing.'"

From *New Haven [CT] Register*, 6 April 1991 [4]:

> At Sikorsky, more than 5,000 workers packed into the cavernous plant to hear company President Eugene Buckley announce the news. Many said

Fig. 27.2 Happy employees on victory day.

Fig. 27.3 The Sikorsky factory on victory day, 6 April 1991.

it was the biggest day for the company since it won the contract to build the Army's Black Hawk helicopter 15 years ago....

After seven years of competition the two teams were close on the technical and production measures, said Assistant Secretary of the Army Stephen Conver. But, he said, the Sikorsky-Boeing team won with lower costs and better training, maintenance, reliability and support.

"The winning bid was below the Pentagon's estimate of $8.9 million per helicopter, in 1998 dollars," Conver said.

The Sikorsky-Boeing team had invested $250 million of its own money in developing the helicopter. At Sikorsky, 400 people have been working on the project.

Fig. 27.4 Unofficial Connecticut license plate on Art Linden's car.

From the 15 April 1991 issue of *Aviation Week & Space Technology* [5]:

> The Boeing Sikorsky stub wings—the external fuel armament management system, or EFAMS—were another discriminator among operational suitability factors, [Aviation Branch Chief and Commander of the Army Aviation Center at Fort Rucker, Alabama, Maj. Gen. Rudolph] Ostovich said. Bell and McDonnell Douglas proposed a fixed wing. EFAMS enabled Boeing Sikorsky to offer combinations of additional fuel and armament—up to eight extra Hellfires, versus the four extras required by the Army and offered by Bell and McDonnell Douglas—on a structure that could be removed easily for low-observables armed reconnaissance.
>
> The Army gave the contractors the option of including some radar signature reducing materials and parts in a kit, only to be used when true stealth missions were to be performed. The Boeing Sikorsky team decided to do this, to help meet the 7,500-pound empty weight limit as well as other advantages. The SuperTeam did not.
>
> In their cost proposals, the companies exceeded expectations most dramatically in their operations and support estimate. Last spring, when the Army solicited development proposals, the target was $500 per flying hour for parts used by the Army maintenance personnel at the depot and field unit levels. The target was increased to $702 per flying hour during the summer, after Defense Secretary Richard B. Cheney cut back on the planned acquisition total and production rate. The Boeing Sikorsky offer came in at $422 per flying hour.

We also learned that the SuperTeam reduced the targeting sensor capabilities to reduce its weight empty to the required 7500 lb. This also became a factor in them losing the competition.

The actual scoring of the competitive proposals from the two teams was detailed in the June 1991 issue of *Rotor & Wing International* [6]:

> ...R&WI compiled the following scoring grid based on input from various sources [scores are out of 100]:

Element	First Team	SuperTeam	Scoring Weight
Cost	67.9	45.8	20%
Technical	74.9	74.1	35%
RAM/ILS	82.4	62.6	17.5%
Manprint/training	80.0	64.0	17.5%
Producibility	69.8	71.6	10.0%
Average	75.0	64.62	

> ...Sources say Army evaluators "were surprised" at the amount of the disparity between the two teams.
>
> One senior Army official involved in the LH selection said the McDonnell/Bell Helicopter SuperTeam appeared not to stress "management of the program" enough. "Technically they were fine," he explained. "There was just a whole different philosophy in the approach. The winners took the Army's requirements much more seriously." Other sources said [the] SuperTeam was inclined to offer more technological risk in their LH version.

This was great news for the First Team. The reaction was obviously different from the SuperTeam. According to the 6 April 1991 issue of *The Arizona Republic* [7]:

> The company [McDonnall Douglas] had been optimistic about winning the contract, [president] Tom Gunn said, but reality quickly set in.
>
> Army officials in Washington D.C. said the edge went to Boeing-Sikorsky because of its helicopter's abilities to be maintained more easily on the battlefield.
>
> The Army also said price was a factor. With McDonnell Douglas putting a unit cost of each helicopter at $8.9 million and the Boeing-Sikorsky helicopter at $8.5 million.
>
> The most notable difference between the two proposed helicopters was the tail rotor.
>
> McDonnell Douglas proposed its revolutionary "no tail rotor" or NOTAR design that replaces conventional tail rotors with thrust created by a jet of air formed by air pressure from a ducted fan embedded in the tailboom. The Boeing-Sikorsky team proposed a "fantail" built into the rear of the aircraft.
>
> Gunn said there was no indication that the uniqueness of the NOTAR system was a factor in the Army's decision.

The article also mentioned that at the time, McDonnell Douglas employed 90 people on the project, and the company had invested "more than $200 million of their own money."

In an 15 April 1991 article in *Aviation Week & Space Technology* [5], Army program manager Brigadier General Ronald Andreson verified that the anti-torque design "was not a discriminator" in the competition. The article also quoted Assistant Secretary Conver that

> ...the Army had considered but rejected the idea of trying to merge elements of the two designs. "We looked at that pretty hard . . . There are obviously a lot of things that are attractive and unique about both of the designs. ... It is theoretically possible to do it [combine designs], but from a legal, contractual and technical point of view, the impediments to doing that are almost overwhelming." [Without any firm knowledge of the specifics, we believe this is in reference to the NOTAR. A strong faction in the Army was sold on NOTAR and was therefore disappointed that the SuperTeam lost the competition. At that time they had not accepted our argument that the NOTAR came with a 250-lb empty weight penalty. Eventually they did understand this.]
>
> The Army has "great concern" about the impact of the LH loss on Bell and McDonnell Douglas, Conver said, and it hopes neither company will be driven out of the helicopter business. There are not enough programs to keep all four major U.S. helicopter manufacturers healthy, he said, but the Army will do "everything we can"—including help the companies try to sell military helicopters abroad—"to support the industrial base and keep it form eroding further."

John F. McDonnell, chairman and chief executive officer of McDonnell Douglas Corporation, congratulated Boeing Sikorsky and the Army "for the outstanding way the competition was carried out," and he said that his company "is in the helicopter business for the long haul" and will continue to actively propose and build the AH-64 Apache and its upgraded versions.

After seven years of intense competition, we won the LH competition.

REFERENCES

[1] Steven Pearlstein, "Boeing, Sikorsky to Build New Army Helicopter," *Washington Post*, 6 April 1991, p. A4.

[2] Richard W. Stevenson, "Boeing-Sikorsky Gets Copter Pact," *New York Times*, 6 April 1991, p. 27 and 38.

[3] "Sikorsky wins LH contract—Area, Boeing split prize," *Connecticut Post*, 6 April 1991, p. 1 and 2.

[4] Tamara Lytle and Pamala Coyle, "Sikorsky lands big one—Stratford firm, Boeing to build copters in deal that could bring home $34 billion," *New Haven [CT] Register*, 6 April 1991, p. 1 and 2.

[5] David F. Bond, "Cost, Supportability Key To Boeing Sikorsky LH Award," *Aviation Week & Space Technology*, 15 April 1991, p. 18 and 19.

[6] "LH Team Scoring Analyzed," *Rotor & Wing International*, June 1991, p. 10.

[7] Ken Western, "McDonnell Douglas loses $34 billion copter deal," *The Arizona Republic*, 6 April 1991, p. 1.

Chapter 28

The First Restructure: The Dem/Val Prototype Program, April 1991

Arthur W. Linden

Although the contracted Comanche program lasted 13 years, it never stayed on one contract for more than 4 years. Contract restructures were primarily due to reductions in funding, but the Office of the Secretary of Defense (OSD) also had issues with technical and schedule risk. This led to a great deal of turbulence in the program. There were five contract restructures before we eventually reached Milestone II (MS II).

The first restructure occurred before we even got under contract. It converted the proposed full-scale development program to a "Dem/Val prototype" program before the contract was signed in April 1991.

The government's acquisition process identified key milestones that a program must achieve as it goes forward. Milestone I approved the program for demonstration/validation. Milestone II approved it for engineering and manufacturing development (EMD). The initial full-scale development contract (the program we proposed) combined MS I and II. This basically eliminated the typical Dem/Val technology maturation/risk reduction phase. The rationale was that the necessary technology development efforts were matured during the earlier Dem/Val program and other Army-funded technology maturation and risk reduction efforts during the earlier phase.

But funding reductions derailed a number of these technology maturation efforts, and integration of the technologies was never funded. OSD believed that we needed more technology maturation before proceeding with an ambitious EMD schedule.

At this time there were still many people who were against the program. Some of these were people who felt that the Defense Department's budget was too big, and the way to reduce it was to cancel large programs. Some in the Army coveted the LH funds for their own favorite programs. Others felt the technical risks with the LH were simply too high.

In June 1990, before the two teams had even submitted their full-scale development proposals, according to the 14 June 1990 edition of the *Bridgeport [CT] Post* [1], Les Aspen (D-WI), House Armed Services Committee chairman, received a petition signed by 100 House members calling for the scout and attack helicopter program to be stopped before it went into production.

April 1991—Demonstration/Validation Prototyping Program

Program Features

Four prototype aircraft

Static test aircraft for MEP testing

Propulsion system test bed (PSTB)

Production quantity reduced from 2096 to 1292 aircraft

From 216/year to 120/year

Aircraft empty weight

7500 lb

Schedule

52 months for Dem/Val plus 39 months for EMD, 91 months total

First flight in 40 months, August 1994

IOC in December 1998

Program Cost

Cost plus incentive fee plus award fee contract

$2.8 billion for Dem/Val plus $2.5 billion for EMD, $5.3 billion total

$33 billion projected for the full production run of 1292 aircraft

In August 1990, both the House and the Senate called for deferral of the full-scale development program in favor of prototyping.

At this time there also was a concern developing that the LH basic mission might no longer be required due to the ending of the cold war. In a 23 August 1990 letter to Representative John P. Murtha (D-PA), Secretary of Defense Richard Cheney said:

> The collapse of the Warsaw Pact does not alter the need for the Light Helicopter (LH). The types of scenarios likely to be encountered in the future, such as Operation Just Cause and the current Mid-East crisis, require a versatile and more flexible force. Land combat forces need to respond quickly and decisively to unforeseen global requirement. Helicopters need to be lethal, yet survivable. They must be versatile, with the ability to self-deploy to hot spots. These are all attributes of the LH not found in the current helicopter fleet.

In the summer of 1990, Secretary Cheney conducted a major aircraft review of the program. This led to a decision to continue LH demonstration/validation, get more technology maturation, and add prototyping " … to verify prior to entering full-scale development that the characteristics that we want to achieve in the light helicopter will in fact be achievable with reasonable risk and reasonable cost" [2]. It deferred the EMD phase and extended the prototype phase by an additional two years. The number of aircraft was reduced from 2096 at a rate of 216 per year to 1292 at 120 per year, reducing the total program cost by $8 billion. A $120-million reduction in FY91 funding was achieved.

From *Aviation Week & Space Technology*, 3 September 1990 [3]:

> Supporters of the program cited this concession—and the fact that approval of the development program itself will be delayed until the end of Fiscal 1994—in suggesting that Cheney's action will defuse congressional opposition to LH for the time being.
>
> [Army LH program manager, Major General Ronald K.] Andreson noted that the Army wanted earlier to prototype LH—to minimize the risk of technical, schedule or cost problems during development—but decided that it could not afford to. Under the earlier plan, both industry teams would have built and flown prototypes.

The full-scale development program now reverted to a continuation of demonstration and validation, with prototypes added. It was now a 52-month Dem/Val program, with a 39-month EMD phase as a negotiated option. The objectives were to complete the aircraft design, define the mission equipment package, build prototype aircraft, and conduct flight testing to reduce risk and demonstrate that the system was ready to successfully transition into EMD. Four prototype aircraft were to be fabricated, plus a static test article for mission equipment package (MEP) testing and a propulsion system test bed (PSTB) for testing the dynamic components. First flight was scheduled for August 1994.

The unapproved future EMD program option would move the program toward production with development of production tooling as well as fabrication of two additional aircraft for development testing. The first production aircraft was scheduled to be delivered in 1998. Initial operational capability aircraft also were to be introduced in 1998.

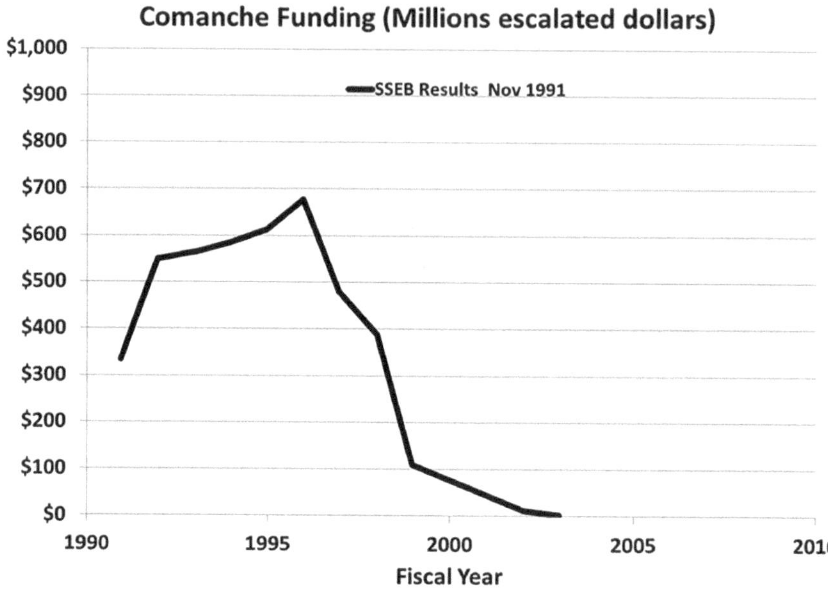

Fig. 28.1 Comanche funding in 1991 (escalated dollars).

The decision to revert to prototype aircraft rather than going directly to the production configuration had major impacts on the program that would be felt as the program moved forward. On the one hand, it gave us the latitude to postpone development of difficult or immature technologies. On the other hand, it meant that our very visible test aircraft were not going to be fully representative of the production configuration. Weight reduction opportunities were postponed, adding future costs to convert the prototype design into the weight-compliant production design. Some suppliers stopped work, and software work was tied to flight controls only.

Another issue with this approach that set the program on a difficult path was that requirements are typically fluid through a prototype phase. At MS II, the start of EMD, they become fixed. By restructuring the program to a prototype phase, there were many opportunities for making modifications to the technical requirements, further slowing the program down. If we would have gone directly into full-scale development like the original contract plan, requirements growth would not have been approved.

There were two different interpretations of these events. Some people in DoD and Congress believed this was the end of LH as a production program. By reverting back to a Dem/Val program with no commitment to EMD, they were satisfied that they had at least slowed it down to where it would be hard for the Army to ever get to production. Others, particularly in the Army and our Boeing Sikorsky team, thought this was a tactical retreat and that the program would overcome this setback and return stronger than ever.

We took solace in the fact that, even though we were now only a Dem/Val prototype program, the revised requirements included additional technical features for later development, which certainly implied that there was something after Dem/Val.

DoD took no strong public position either way. Boeing Sikorsky did everything we could to maintain the track to production. Senior Army leadership continued their full support for the production program.

In early 1991, the Army was very positive about the revised program. According to the *Bridgeport [CT] Post* of 25 February 1991 [4]:

> Army Secretary Michael Stone told the House Armed Services Committee last week that he's confident the program is on schedule and will meet the procurement cost estimate of $7.5 million per aircraft.
>
> Army chief of staff Carl E. Vuono was at the armed services hearing [and said] that the LH is the system he most would have liked to have in Saudi Arabia because of its lethality, maneuverability and deployability.

In its 15 April 1991 issue, *Aviation Week & Space Technology* stated that one of the program's strongest supporters was the Army Chief of Staff, General Gordon L. Sullivan [5]. He declared that the Comanche would be the Army's "quarterback of the digitized battlefield" and the "flagship of the 21st-century army."

By the time of the contract signing in April, the design weight empty remained at 7500 lb, but the unit production cost had grown from the original $7.5 million to $8.9 million due to the quantity reduction and other program changes. Total program cost was estimated to be $33 billion.

The contract was structured as a cost plus incentive fee type with an award fee. The value for both the Dem/Val and EMD phases was $5.3 billion. Figure 28.1 shows the funding profile over the contract duration. In subsequent chapters we will show how this changed with each program restructure.

The original plan, when the total number of aircraft to be produced was 4500, was to have both Boeing and Sikorsky fully tooled up for production and compete against each other for the various production lots. With this smaller 1292-aircraft program, that concept was eliminated.

In Assistant Secretary Conver's comments at the downselect press conference, there was a perhaps little-noticed comment on the Army's funding problems: " … the Army's procurement budget has dwindled from a level of over $14 billion a short two year ago to our present position in the '92 budget of $8 billion …". He presented this as a positive for the LH:

> … therefore when we make a commitment to pursue a large program like the LH, you know that we are really and truly committed to it, and indeed we are committed to the LH. It represents our efforts to be serious about modernizing the Army, and in this instance, the aviation branch of the Army.

This restructured approach could reasonably be stated as one of the key reasons the program development took so long and eventually became too costly. There was no shortage of potential pitfalls ahead. A headline in the 8 April 1991 issue of *Inside the Army* said, "Industry Privately Says Program Still Underfunded." Army budget fluctuations would continue yearly and present one of the major problems for the program.

References

[1] Paul Gargaro, "LH leader quits post at Sikorsky," *Bridgeport [CT] Post*, 14 June 1990, p. C7 and C8.
[2] This comes from a DoD press conference on 8 April 1991, with Assistant Secretary of the Army for Research, Development, and Acquisition Stephen K. Conver and others announcing the Comanche downselect.
[3] David F. Brown, "Defense Dept. Will Build Prototypes of LH Helicopter," *Aviation Week & Space Technology*, 3 September 1990, p. 37.
[4] Alica Lipowicz, "Army backing of LH program seems assured—Sikorsky—Boeing team in running," *Bridgeport [CT] Post*, 25 February 1991, p. B3.
[5] *Aviation Week & Space Technology*, 15 April 1991.
[6] "Industry Privately Says Program Still Underfunded," *Inside the Army*, 8 April 1991.

Chapter 29

The Second Program Restructure: The Technology Development Program, December 1992

Arthur W. Linden

December 1992 Program Restructure

Program Funding

Reduced by $100 million to $300 million per year for next five years

Schedule Impact

47-month program extension

Dem/Val extended from 52 to 78 months

Engineering, manufacturing, and development (EMD) extended from 39 to 60 months

(11 years, 6 months total vs 7 years, 7 months)

First flight delayed one year to August 1995

Initial operating capability (IOC) was December 1998; now January 2003

Program Changes

Three prototype aircraft vs four

Static test article eliminated

Additional Requirements

LONGBOW radar provisions

High-frequency radio

Infrared (IR) jammer

Radiofrequency (RF) jammer

Armor

Combat kits in empty weight

Design Impact

Maximum allowable weight increased from 7500 lb to 7950 lb

Required 12% growth engine

Program Cost

$6.7 million

The second restructure was even worse than the first. The recent dissolution of the Soviet Union led to large Department of Defense (DoD) budget cuts. Yearly funding was reduced by $100 million to $300 million per year through 1997, as shown in Figure 29.1.

This was the time when the Navy's A-12 program had just been terminated well into its development process. Some felt these reductions were necessary to ensure that Comanche didn't fall to the same fate.

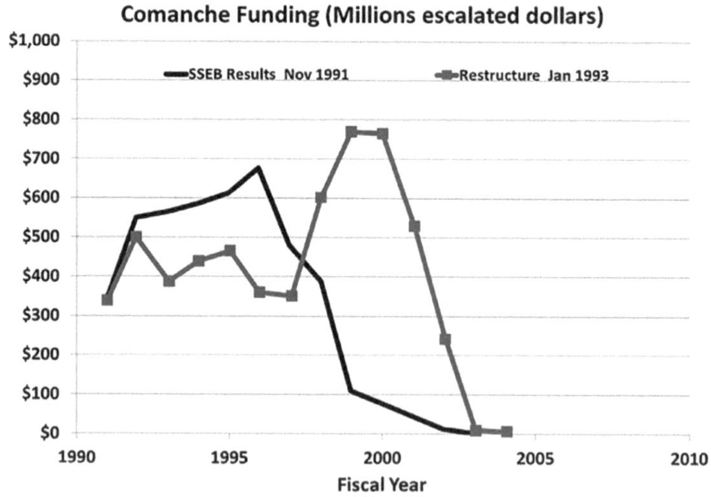

Fig. 29.1 Restructure funding profile vs original Dem/Val prototype Source Selection Evaluation Board (SSEB).

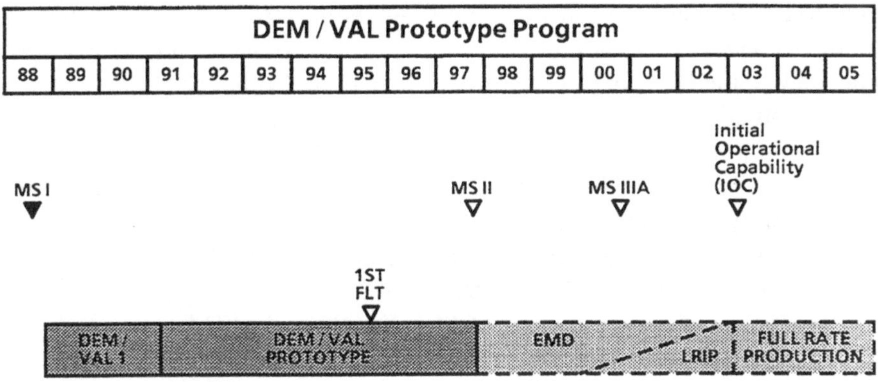

Fig. 29.2 Revised program schedule.

The program had to be revised to fit the new funding profile. Additionally, a DoD Cost Analysis Improvement Group (CAIG) identified cost and schedule risks in the areas of airframe, propulsion systems, LONGBOW radar, mission equipment package (MEP), supportability, and software, and recommended that more money should be applied to the Comanche program.

With the revised program, the prototype aircraft quantity dropped from four to three, and the static test bed was canceled. First flight was pushed from August 1994 to August 1995. Some new systems were added to try to get more technology maturation into this phase. These included infrared and radio frequency jammers, the LONGBOW radar, a high-frequency radio, and miscellaneous other changes.

The restructured program was approved with the Dem/Val phase stretched over two years from 1995 to 1997, from 52 to 78 months as shown in Figure 29.2. Although not authorized, EMD was reintroduced, now defined as a 60-month, rather than 39-month, effort. Thus, the 91-month program was now stretched to 138 months, an extension of almost four years. Dem/Val would be complete in September 1997, and the initial operating capability (IOC) would slip from December 1998 to January 2003.

In addition, the acquisition process now also incorporated a technology readiness level (TRL) gate assessment philosophy. The exit criterion for the Dem/Val phase to transition to EMD was a TRL level of 6 prior to MS II. This required engineering feasibility demonstrations in a relevant environment using system/subsystem models or prototypes.

A TRL 7 was the exit criteria for Milestone III (MS III) to go to low-rate initial production (LRIP). This required demonstrations in a realistic environment with prototypes integrated with supporting elements for field testing.

Adding the LONGBOW radar was an addition of approximately 500 lb and \$1.0 million to the unit fly-away cost. The radar, although still unproven on the Apache, was viewed by the government as a lower risk solution to threat detection and classification than the Comanche's Aided Target Detection System (ATDC) system. The Office of the Secretary of Defense (OSD) believed that the LONGBOW was required for the Comanche to be viable.

Our simulations did not support the addition of the radar.

The weight-empty limit was increased to 7950 lb, and the engine power was increased by 12% to accommodate LONGBOW. This then resulted in changes to the other dynamic components.

Boeing Sikorsky needed to submit another proposal. This was a full cost proposal, not a delta to the April 1991 contract.

In an article on Comanche in its 16 March 1992 issue, "U.S. Army Revamps Comanche Plans, Considers Next Bid for More Funds," *Aviation Week & Space Technology* said [1]:

> The U.S. Army is committed as ever to development and production of the RAH-66 helicopter, but it may wait a couple of years before trying to win back Defense Dept. support for the program.
>
> The timing of an Army attempt to carry Comanche beyond the current prototyping effort is "an open issue," Stephen K. Conver, assistant secretary for research, development and acquisition, said.
>
> ... Most Comanche supporters presumed that the Army would try this year, as Fiscal 1994-95 plans are developed, to provide funds in the 1998–99 for restoring the development program and starting production. Conver said he thinks this would be the wrong approach. It would do better to wait two years, until the Fiscal 1996–2001 plan is prepared. Comanche prototype development can proceed as planned until then, and the helicopter's advantages will be understood better when the Army tries to make the case for it, he said.

The article also quoted Army Chief of Staff Gordon R. Sullivan as saying "I'm convinced the Comanche is going to sell itself... . It's a gangbuster aircraft."

On 17 May 1992 General Sullivan visited Boeing Sikorsky to review the Comanche program. In his remarks to the team he stated:

> This gets the United States Army into the 21st century. The future of the United States Army, certainly the future of Army Aviation, is represented right here... . What a weapons system ... a state-of-the-art system that leaps ahead, and keeps us in froht of everyone else in the world.
>
> The United States Army needs Comanche. We are anxious to have the Comanche on the flight line, out there working every day with the men and women of the United States Army, Active Guard and Reserves.

This restructure did not look good for the program. It gave us 11.5 years to develop the system, including 6 years to get the EMD phase reinstated. It provided almost 2 additional years to prove out all critical components, including avionics, the growth engine, and the added LONGBOW radar.

Some felt that we probably needed that time to transition all the technologies to production solutions.

Although EMD was not recognized, the fact that LONGBOW, HF radio, and jammers were added indicated that OSD expected the production program to be reinstated, but the position was that the program needed to demonstrate technology maturity before that decision could be made.

The restructured program was now defined as:

- Maximum weight empty of 7950 lb
- 500 ft/min vertical rate of climb (VROC) at maximum rated power (MRP) (4000 ft/95°F)
- 170-kt dash speed
- 1344 shp engines
- Propulsion system test bed (PSTB)
- LONGBOW radar with a 12% increase in range
- MEP risk reduction demonstrations (off-the-shelf communication and navigation on prototype aircraft)
- Target acquisition system (TAS) with no backup pilotage
- Forward-looking infrared (FLIR) only night pilotage
- Gun system integration design work
- Aided target detection and classification (ATD&C)
- Electromagnetic environmental effects (E^3) 200 V/m demonstrated on flight-critical hardware only

The following items were deferred to EMD:

- Flight testing of the growth engines
- Weight reduction technologies
- Integrated communications, navigation, and identification avionics (ICNIA) development and integration
- FLIR image intensification
- Aircraft survivability equipment (ASE)
- Low radar cross-section (RCS) antenna development and testing
- Armament other than the gun, plus development of a lightweight gun
- Sensor fusion

Although supportability was worth approximately 40% of the SSEB evaluation criteria for the original 1991 downselect, much of this development was now deferred to EMD. This included reliability, integrated logistics support (ILS), and training activities. MANPRINT and maintainability activities remained mostly unchanged. Achieving the overall reliability, maintenance man-hours per flight hour (MMH/FH), and design to operation and support cost (DTOSC) goals became risk issues.

Reliability analysis was tailored to safety of flight items and prototype aircraft-critical components. Software reliability analysis was limited to three computer software configuration items (CSCIs)—the crew interface management (CIM), aircraft systems management (ASM), and mission management and control (MMC). Deferring corrective action development and verification to EMD added risk to meeting reliability and diagnostics growth requirements. However, the number of systems being demonstrated for diagnostics capability increased from two with the 1991 proposal to six with the restructure.

Aircraft systems would not be representative of the production approach. They were neither weight compliant, electromagnetic compliant, nor reliability compliant, and would use rudimentary working software to support only flight testing. Funding was taken from many of the First Team major suppliers: the Lockheed Martin target acquisition system (TAS) effort was reduced by $46 million, and the TRW ICNIA friend or foe (IFF) avionics work was unfunded, as was the radar cross-section (RCS) material development and material weight reduction efforts.

Some suppliers left the team; some used their own internal funds to continue to work their subsystems. Major suppliers were going directly to the Army to complain about the funding they were getting.

In January 1992, the program completed its preliminary design review (PDR). According to an internal document, the Comanche program had spent $2.55 billion by that time, from 1983 through 1992. This included $1.8 billion spent by the government, $496 million by Boeing Sikorsky, and $250 million by LHTEC.

Each restructure opened the door for the user to add the most recent widget to the aircraft. In some cases this happened just to get the user's support for the program. It was a result of people doing what they felt was required to keep the program alive, but in the end this turbulence continued to delay the program and contributed greatly to the program's eventual termination.

REFERENCE

[1] AW&ST, "U.S. Army Revamps Comanche Plans, Considers Next Bid for More Funds," *Aviation Week & Space Technology*, McGraw Hill Inc, New York, NY, 16 March 1992.

Chapter 30

THE THIRD PROGRAM RESTRUCTURE: THE STREAMLINED PROGRAM, DECEMBER 1994

ARTHUR W. LINDEN

Next we got "streamlined."

December 1994 Streamlined

Program Program Funding

Reduced by an additional $424 million

$78 million in FY94 and $19 million in FY95

Schedule Impact

Three-month additional program slip

First flight delayed by one year to 1996 (was April 1995)

Program Changes

Two prototype aircraft vs three

Six early operational capability (EOC) aircraft planned for future engineering, manufacturing, and development (EMD)

"Streamline" ideas (see text)

Additional Requirements

Integrated communications, navigation, and identification avionics (ICNIA) communications

Fifth-percentile female

Integrated mission support system (IMSS)

Air bags

Use of beryllium-aluminum armor

Image-intensified charge-coupled device TV camera

Reduced Requirements

Eliminate gun

Eliminate electro-optical target acquisition and designation system (EOTADS)

LONGBOW development delayed

At this point in time, Boeing Sikorsky had spent 47.9% of the available funding for the prototype-only program, and 86% of the aircraft design and 90% of the tooling design were complete. Sixty percent of the tools were fabricated. Also completed were:

- 3500 hours of second-generation forward-looking infrared (FLIR) testing
- 100 hours of Fantail flight testing
- 9300 hours of wind tunnel testing
- 4500 hours of radar cross-section (RCS) testing

The reliability of the system architecture had been demonstrated with no failures experienced during FLIR bench testing. Seven mission computer modules had been tested without failure during 1000 thermal cycles. The Comanche player station was at Fort Rucker, Alabama, allowing Comanche to participate in battle lab and Louisiana Maneuvers Task Force exercises. Ten maintainability assessments had been conducted by a joint contractor/Army team.

Then additional Army funding shortfalls were identified in November 1993. Comanche would be further reduced by $78 million in FY94 and $19 million in FY95.

In February 1994, the Office of the Secretary of Defense (OSD) directed the Army to "streamline" the program and stop development of the three prototypes and limit the work on LONGBOW. OSD's perception was that Team Comanche had scrubbed the overall development effort to an absolute "bare bones" program with the prototype program. Additional funds would be allocated to the advancement of the technologies then in development. The bottom line was that OSD was not sure the funding was supporting the activities that would lead to reinstating EMD.

In total, $424 million was removed from our funding. This did have the advantage of smoothing out a very high funding profile in 1998 to 2001 (Fig. 30.1).

"Streamlining" was a cultural change concept to look to commercial products and product development best practices and standards and adapt those that would lead to program schedule and cost reductions. It eliminated non-value-added laws, regulations, specifications, policies, and traditions, and also reduced government oversight.

Commercial development was not subject to the DoD 5000 series acquisition laws, regulations, specifications, and decision gates that were believed to be at least part of the schedule and cost forcing function. Changing the government oversight would be a cultural change that the Army struggled to support.

A combined government/industry process action team (PAT) was created to explore streamlining ideas. These included:

- Acquisition reform such as use of commercial practices
- Military standards and specifications eliminated, tailored, or specified for guidance
- Elimination of reports that need to be redone in EMD
- In-process review (IPR) briefing
- More simulation to reduce flight testing requirements
- Reduced government oversight
- Data submittal in contractor format
- Contractor-supplied logistics support during the first few years of fielding
- Delaying incorporation of LONGBOW
- Elimination of the gun and EOTADS

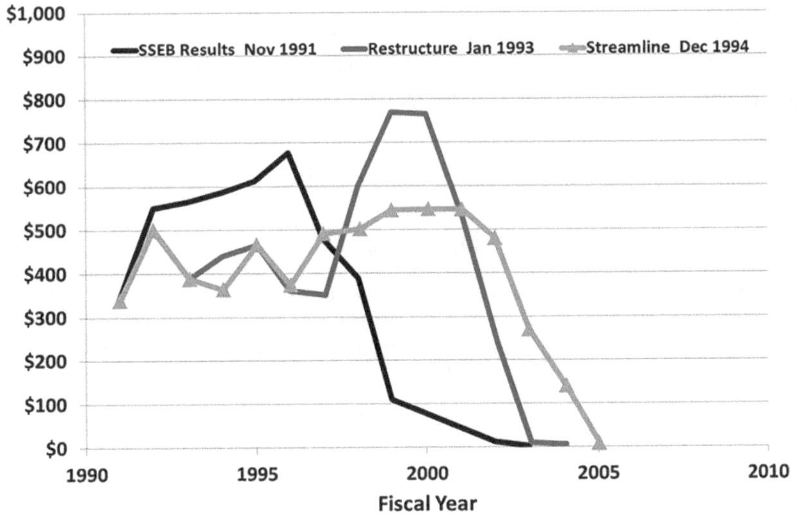

Fig. 30.1 Streamlined funding profile vs the restructure program.

The Comanche's full capabilities and performance were to be maintained, as was all the Manpower and Personnel Integration (MANPRINT), maintainability, reliability, and producibility influences on the design. The combined test team approach was to continue. Non-value-added tasks were identified and eliminated.

Along with the funding reductions and stretchout, we also had a new requirements document. The new requirements reflected changes in the world situation, including experiences in Operations Desert Shield and Desert Storm, the demise of the Soviet Union, and new technology advancements. ICNIA was put back into the program and a fifth-percentile female requirement was added, as were an integrated mission support system (IMSS), air bags, use of beryllium-aluminum, and an image-intensified charge-coupled device TV camera.

Major General DeWitt "T" Irby, the Army's Aviation Program executive officer, was quoted in the 12 April 1993 issue of *Aviation Week & Space Technology* as saying [1], "' . . . the 1994 Army budget cuts are not likely to kill any Army aviation programs, but warned against further Comanche delays. The program has been slipped too far already,' he said."

We were authorized to begin implementation of some of the streamlining initiatives immediately and to stop the development of three prototypes and to limit support of LONGBOW development.

Streamlining initiatives savings estimates were a $72-million cost avoidance, but a $250-million Research, Development, Test and Evaluation (RDT&E) shortfall still remained.

Dr. Paul Kaminsky, under secretary of defense, directed the Army to proceed with Boeing Sikorsky on the streamlined program.

Then another new concern was introduced. An article in the 27 June 1994 issue of *Aviation Week & Space Technology* [2] said, "Critics claim that unmanned aerial vehicles like the Army's Hunter or joint service Pier 2 UAV could do the reconnaissance role more cheaply and with less danger to aircrews."

All this activity did not bode well for the future of the program.

REFERENCES

[1] David A. Fulghum, "Army Orders One-Third Cut of Comanche Program Cost," *Aviation Week & Space Technology*, AW&ST, 12 April 1993, p. 60.

[2] David Fulghum, "Comanche Production Slashed in First Five Years," *Aviation Week & Space Technology*, AW&ST, 27 June 1994.

Chapter 31

The Fourth Program Restructure: Early Operational Capability (EOC), May 1995

Arthur W. Linden

May 1995 Early Operational Capability (EOC) Program

Program Funding

Reduced by $2 billion from FY96 to FY01

Funding added in later years

Schedule Impact

Program extended four more years to 15 years, 9 months

Milestone II delayed to 2001–2002

Initial Operational Capability (IOC) moved from 2003 to 2006

Changes to Program

All eight Dem/Val and engineering, manufacturing, and development (EMD) aircraft to be initial Army Comanche fleet

Reintroduced a plan for production

Production rate reduced from 120 to 72 aircraft per year

Additional Requirements

LONGBOW to be Pre-Planned Product Improvement (P^3I)

Standard Advanced Dewar Assembly (SADA)

Link 16

satellite communications (SATCOM)

Incorporated LO kits

Deleted Requirements

See text.

Design Impact

We increased to 8265 lb.

Program Cost

$7.8 billion

Colonel Robert Birmingham took Army command of the Comanche program in July 2000 as it entered the EMD program phase. After program cancellation, he wrote a book. In it he lists five program restructures, rather than the six described here. He combined what we're calling restructures four and five into one because they were a redefinition of the program without altering the program funding. We are calling these separate restructures because they were major changes to the Boeing Sikorsky program statement of work.

Later in 1994 there was another funding reduction, this time of $2 billion. The new funding profile is shown in Figure 31.1. The LONGBOW Apache development was overrunning, and many that felt LONGBOW and Comanche could not both survive based on the Army Aviation–approved budgets. There

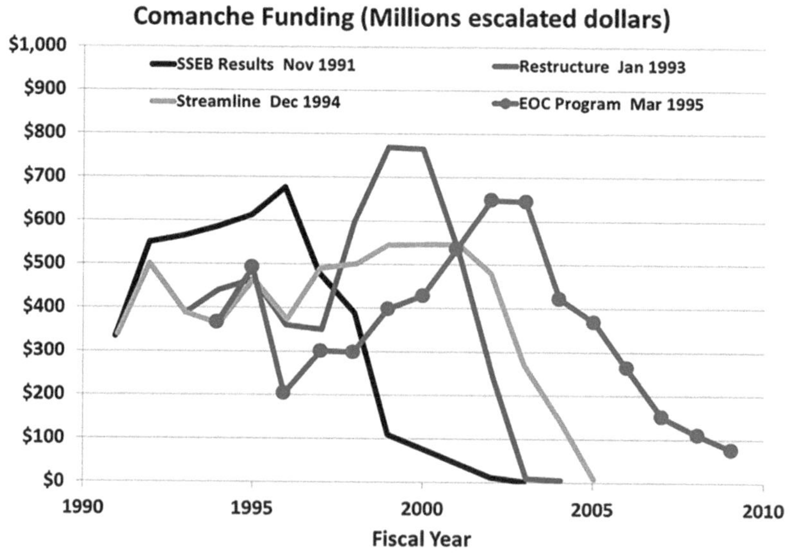

Fig. 31.1 EOC funding profile vs the streamlined program.

were directives within the Office of the Secretary of Defense (OSD) to cancel or again restructure Comanche. An article in the 14 October 1994 issue of *Aviation Week & Space Technology* stated [1]: "The Pentagon's top civilians have made a preliminary decision to kill production of both the U.S. Army's RAH-66 Comanche helicopter and the Air Force-run tri-service Standoff Attack Missile as a budget-cutting measure, but they will not announce the plan until after the November elections to avoid political backlash."

Secretary of Defense John Deutch directed the Army to develop alternatives that would lead to the termination of the Comanche program.

This set off alarm bells within the Army and the companies. The companies developed contact plans at many levels including the CEOs of Boeing and United Technologies Corporation (UTC) contacting Secretary of Defense Perry. The Army similarly worked its chain of command to keep the program viable.

The Army program manager, Brigadier General Orlin Mullen, had instituted many of the streamlining initiatives of the previous restructure. Eventually his efforts won over Under Secretary of Defense Dr. Paul Kaminski, who convinced Secretary of Defense John Deutch to restructure the program into a technology-based project.

The Army at all levels really believed in the armed reconnaissance requirement and the Comanche's ability to provide it; however, constrained budgets had now threatened the cancellation of the prototypes, an integral part of the original strategy. Without flying prototypes to demonstrate the Comanche's value, the program would probably be axed.

The next Comanche program manager, Brigadier General Jim Snider, felt that the only way Comanche could survive was to provide some reconnaissance operational capability quickly, in FY01, followed by the armed reconnaissance/attack capability in FY04. His approach was called the early operational capability (EOC) program. Hopefully this would get the Comanche back on the Army's priority list. He really saved the program by providing the Army with something the original strategy would have provided if not for budget reductions: sufficient aircraft to demonstrate at least some of the Comanche's capability.

And then, to make our lives even more complicated, in a 28 February 1995 memo approving these program changes, Gilbert Decker, Army Aviation executive, added: "This acquisition strategy must have the flexibility to execute an accelerated program if given additional funding in subsequent years."

Boeing Sikorsky continued to market the program to shore up support. Mockups and simulators were used to demonstrate the Comanche's capabilities at many OSD activities.

In substantially reducing program cost, the EOC program obviously also reduced the aircraft's complexity and capability. It had:

- No armament systems development
- No backup pilotage system
- No moving target indication

- No onboard inert-gas generation system (OBIGGS)
- Limited fault detection/fault isolation (FD/FI)
- No reconfiguration and degraded mode operations
- No radar cross-section (RCS) treatment
- No tactics expert functions
- No situational awareness functions
- No active countermeasure sensors
- No LCD helmet-mounted display (HMD)

And we had some additions. These were image intensification (I^2) on the nose, Standard Advanced Dewar Assembly (SADA), Link 16, satellite communications (SATCOM), and Low Observable (LO) kits. All were added to weight empty, which increased risk.

And the concern with unmanned aerial vehicles (UAVs) was addressed. The revised program added UAV control and a tactical data link (TDL). The concept was that a UAV would provide the over-the-hill view, keeping the Comanche platform less vulnerable to exposure to threat assets. The addition of a UAV control requirement added the need for a common data link (CDL) system with an antenna that provided high data rates for rapid download of UAV flight information and intelligence data and images of the battle area.

Brigadier General Snider felt that UAV control was important and needed to be on the platform early.

This change in the contract did not remove the radar from the Comanche configuration definition or any other system on the aircraft. Operationally the radar made little sense for the reconnaissance platform, but it was in the testing phase as part of the Apache-D program to demonstrate the radar's capability to find and target ground threat systems. The real question was how often the Comanche would have a companion UAV and how often the Comanche would need the radar. These questions were never answered.

An accurate track of configuration control suffered during this bare-bones EOC program. The evolutionary acquisition approach did not have a clear definition of the follow-on block upgrades to reach the full production configuration.

The armed reconnaissance/attack configuration would start following MS II in FY02, and IOC/MS III was planned for FY06.

Something new was also added to our process here. The EOC contracting used what was called "alpha acquisition," where contractor and government personnel are included in the process from the beginning:

- Proposal written, audited, and negotiated concurrently vs sequentially
- Requirement development
- Statement of Work (SOW)
- Schedule
- Budget

- Weekly meeting
- Technical review team on-site during proposal review
- Joint technical and cost reviews
- Written responses to questions
- Fact-finding results documented and published
- Pen and ink changes
- Issues cleaned up in 60 days
- Debrief team members

The alpha acquisition contracting cycle was 108 days. This saved us 6 months over the more conventional approach.

The contracting strategy was to modify the existing contract into three phases—phase I represented the actuals spent up to FY95, phase II represented undefinitized contracting actions (UCAs) up to FY97 to accommodate the modified contract needs, and phase III starting in FY97 through FY05 to complete the EOC contract.

Commercial processes and procedures as well as parts were acceptable. Cost as an independent variable (CAIV) analysis was to be used to control weight and cost growth. At first glance this would be 5 years of technology maturation with an early initial operational test and evaluation (IOT&E) of initial capacities. EMD would follow MS II to fully develop the production configuration. Although OSD approved the EOC acquisition approach in March 1995, Boeing Sikorsky didn't start work on the EOC proposal until January 1996.

With a 21 March 1995 acquisition decision memorandum by Paul Kaminski, the Army approved the program. The EOC contact was signed in December 1996. It was again a cost-plus-award fee contract with a 10% fee based on quarterly performance reviews.

Reference

[1] David A. Fulghum, "New Defense Cuts Target Comanche, TSSAM," *Aviation Week & Space Technology*, October 1994.

Chapter 32

THE FIFTH PROGRAM RESTRUCTURE: THE PREPRODUCTION PROTOTYPE PROGRAM, DECEMBER 1998

ARTHUR W. LINDEN

December 1998 Preproduction Prototype Program

Program Funding

No change

Schedule Impact

Milestone II moved forward to 2000

Program Changes

Early operational capability (EOC) aircraft canceled

Six preproduction aircraft

Eight initial operational test and evaluation (IOT&E) aircraft added

Mission equipment package (MEP) development accelerated

Additional Requirements

LONGBOW radar back in

Program Cost

$8.3 billion

As time moved forward, various Army and government officials went on the record with their thoughts on the program. Generally, the Army continued to express the need for the Comanche's capabilities, but all parties were struggling with how to fit the program funding into the Army's overall budgets.

A few quotes from the press:

From a 6 January 1997 *Aviation Week & Space Technology* article titled "Comanche Avoids 1998 Budget Cut" [1]:

> The Army's RAH-66, at least for the moment, appears to have avoided a recommendation for cancellation by the Pentagon's Defense Resources Board.
>
> The program had been "put on the table" for cancellation before Christmas to help compensate for a last-minute $600 million cut to get the Fiscal 1998 budget in shape for submission to the White House in late January, according to military and aerospace officials. But the program was subsequently removed from consideration. "Even if the Resource Board made a recommendation [for cancellation] it does not mean it will be endorsed by Army leadership," noted an Army official.
>
> Instead, the Army last week signed a multiyear $1.7-billion contract with Boeing and Sikorsky for continued development of the low-observable, scout/attack helicopter. The agreement runs from 1997 through September, 2002, when the sixth Comanche is to be delivered.
>
> Army officials both in the Comanche program office and Pentagon staff confirmed that the Comanche program was considered for cancellation, along with other major programs, and that it has avoided the ax. They caution that budget pressures might cause a reconsideration.
>
> There is little doubt that the Army's dilemma is real. Officials of the Joint Chiefs of Staff say that the Army can maintain 10 active divisions or the Comanche program, but not both.

In March 1997, Lieutenant General Ellis D. "Don" Parker, Retired, was quoted in *Army* magazine [2]:

> Several Army development programs are caught in the budget squeeze, but the Army's new combat helicopter, the Comanche, typifies the conflict between need and resources. Few programs have been as thoroughly examined in every detail from the start by people of solid experience, good will and genuine concern for our long-term national security, but Comanche is still the butt of too much skepticism and misinformation, even within the Army itself. It has been stretched out, delayed, restructured, sliced and diced year after year. Nevertheless, few programs enjoy such robust justification. I know. I was there at the at start.

From *Inside the Pentagon*, 3 April 1997 [3]:

> Senior Defense Department officials say the ongoing Quadrennial Defense Review is likely to either cancel or severely cut back the Army's RAH-66 Comanche helicopter program. The dim prospects for the armed reconnaissance helo have evolved out of a number of factors including Defense Secretary William Cohen's reported desire to take serious

> reductions in tactical aircraft programs and the Army's own signal recently that it would prefer to see the $45 billion Comanche program take a budget hit than to endure reductions in another of its high-profile efforts, the Crusader self-propelled artillery program
>
> Army Vice Chief of Staff Gen. Ronald Griffith recently signaled in a meeting of the Joint Requirements Oversight Council that when the time comes to take cuts on procurement programs under the QDR, Crusader takes priority over Comanche.

From *Defense Daily*, 6 November 1997 [4]:

> In November 1997 a DoD across-the-board 2.75% funding cut in the FY 1998 Defense Authorization Bill caused the delay of the first flight of aircraft #2 from September 1998 to March 1999. In 1998 aircraft #2 was intended to be a backup to the flight testing of aircraft #1, before it took on its primary role of developing the MEP. The program felt it was most important to not delay any MEP development, so that the overall MEP schedule was not impacted.

From *Helicopter News*, 5 December 1997, in an article titled "Army Aviation SITREP" [5]:

> But the Boeing/Sikorsky ...RAH-66 Comanche [said an anonymous Army official] remains the "absolute number one thing we do around here" because of "what it will bring to the digital battlefield." The aircraft "uses about ten percent of its systems and processing capability for actual flying, and has ninety left over for battlefield commanders," he said. "That's about one hundred eighty degrees from what's happened in the past with a new helicopter."

Also from *Helicopter News*, 5 December 1997, in a separate article titled "National Defense Panel Bears Down on Comanche" [6]:

> The National Defense Panel (NDP) called into question the procurement level for the cornerstone of the Army's aviation modernization program in its report "Transforming Defense: National Security in the 21st Century."... the panel questioned the number of Boeing[BA]/Sikorsky[UTX] RAH-66 Comanche armed scout helicopters to be purchased in the future.

Major General Ronald Adams, Army Deputy Chief of Staff for Operations and Plans for Force Development, was quoted in *Defense Daily*, 15 January 1998 [7]:

> As for the...Comanche scout helicopter program, Adams stressed that the program must move ahead. "Comanche has got to stay on track...we have got to field it and not just experiment with it," he said. Adams also hinted that the Army might actually have a requirement for more Comanches as it moves closer to the Army After Next (AAN).

From an internal Boeing Sikorsky memo:

> At a February 26, 1998 meeting of the House National Security Committee (HNSC) "Congressman Norman Dicks said he was a big supporter of Comanche. Army Chief of Staff Gen Reimer responded that the Comanche was the Quarterback of the Digital Battlefield, self-deployable, easy to maintain, and a great technology platform. Congressman Duke Cunningham also mentioned the Comanche by saying that he understood that the program was mismatched (helicopter ahead of the MEP and the MEP not in synch with Div 00 and Corps 04. He asked if [Congress] could help."

From *Army Times*, 13 April 1998 [8]:

> 1997's Quadrennial Defense Review...suggest[s] that the Army cannot afford both Comanche and the Crusader self-propelled howitzer.

From the *New Haven Register*, 13 May 1998 [9]:

> At a display of Comanche Aircraft #2 at the U.S. Capital building in Washington D.C., Army Chief of Staff General Dennis Reimer said: 'We call it the quarterback of the digital battlefield. It's very much part of our future. It means a lot to the Army.'

Also in 1998, General Reimer urged Congress to fully support the Comanche, calling it the Army's highest research and development program.

Next, the new Army Chief of Staff, General Erik Shinseki, wanted to transform the Army into a lighter, more lethal, expeditionary form, the Future Combat System (FCS). Comanche was to be an integral part of the transformation. Major General James Snider, then Army Program Executive Officer for Aviation, and Brigadier General Joseph Bergantz, then Army Comanche Program Manager, restructured the program within existing funding, moving MS II forward by two years to 2000 and adding the LONGBOW radar back into the design. The early operational capability (EOC) aircraft were canceled. This program would now be the preproduction prototype (PPP) program. The revised program included six PPP aircraft and eight initial operational test and evaluation (IOT&E) aircraft. The aircraft would be delivered to the Army two years later, but they would be in close to full production configuration, not just the reconnaissance configuration. MEP development was accelerated. Initial Operational Capability (IOC) would remain at December 2006.

The program would align the aircraft delivery with the Army Digitization Plan for the first digitized division (FDD) and create a single aircraft tooling, facilitation, and flight test approach to eliminate potential duplication and throwaway cost. It would create a single training program for pilots (one configuration), a single set of production technical manuals, and a single set of support equipment, and provide field evaluation of production representative aircraft. It allowed the contractors to develop tooling and enter production earlier.

The new PPP aircraft would include situational awareness, radar/forward-looking infrared (FLIR), Aided Target Detection and Classification (ATD&C) target detection and classification, moving target indication, search-on-the-move, and full fault detection/fault isolation functions. They also would include full low observables and nuclear, biological, and chemical (NBC) compliance, and full Force XXI communication capabilities, weapons, and image intensification (I^2) TV pilotage.

There were 40 modifications and 35 additions to the operational requirements document (ORD) associated with this restructure. They were to be incorporated using the cost as an independent variable (CAIV) tradeoff process or delayed for a preplanned product improvement (P^3I) block upgrade. A majority of the weight reduction initiatives would also be part of the P^3I effort.

Due to the ORD-directed equipment changes that increased the aircraft total weight by around 470 lb, the rotor diameter was increased by 1 ft to 40 ft. (The original design included sufficient space between the main rotor and the aircraft tail, anticipating this need.) The total changes to regain the 500 fpm capability at takeoff gross weight also required the use of 100% maximum rated power (MRP) and anhedral blade tips. To regain the terrain avoidance maneuver capability, a chord increase of 11% would be required.

Total program cost was now $8.3 billion.

It took Boeing Sikorsky three program planning iterations to develop a program definition that fit the funding profile.

On 27 July 1998, the program was approved by the Under Secretary of Defense for Acquisition and Technology, and the acquisition decision memorandum was signed. General Bergantz "was pleased with the program restructure and felt that this finally put the program back in motion towards production and away from a demonstration and validation program only."

On the downside, in its 14 December 1998 issue, *Inside the Army* had an article titled "Top Pentagon Cost Analysts Warn Comanche Is Underfunded, Overweight" [10]:

> A group of Pentagon cost analysts has warned that the Army's Comanche helicopter program faces a serious funding shortfall and a weight problem that may necessitate a "nearly complete redesign" of the aircraft, according to a memorandum reviewed by inside the Army.
>
> Comanche proponents, however, strongly disagree…
>
> The most serious conclusion drawn by the [CAIG] group focuses on funding. According to the group's estimates, Comanche is underfunded by $151.3 million between fiscal years 1999 and 2006.

> These money problems could cause severe damage to the program's structure, warns the CAIG.

Also, the MEP was described as "relatively immature."

PPP was awarded in April 2000 and included authorization to proceed with EMD. There was a $62-million FY99 plus-up to increase the flight test hours, start work on the earlier deferred target acquisition system (TAS) functions, and build an additional Comanche portable cockpit to support experiments and demonstrations to integrate Comanche into the digitized force. It included updating the current Comanche portable cockpit (CPC) and Comanche players station (CPS) to training devices, starting fire control radar work, and funded high resolution LCD display integration into the helmet integrated display and sighting system (HIDSS). The contract was for $3.15 billion.

REFERENCES

[1] David A. Fulghum and Anthony L. Velocci, "Comanche Avoids 1998 Budget Cut," *Aviation Week & Space Technology*, 6 January 1997, p. 30.

[2] Lt. Gen. Ellis D., "Why Army Modernization is Vital to Our Future," *Army*, March 1997, pp 15–17, {Don} Parker, U.S. Army retired.

[3] "Storm Clouds Seen Over Army's Comanche Helo in Quadrennial Review," *Inside the Pentagon*, 3 April 1997, p. 1

[4] Brian Shannon, "Army: Funding Cuts Will Delay First Flight of Second Comanche," *Defense Daily*, 6 November 1997.

[5] "Army Aviation SITREP," *Helicopter News*, 5 December 1997, Vol 23 No. 25.

[6] "National Defense Panel Bears Down on Comanche," *Helicopter News*, 5 December 1997, Vol 23 No. 25

[7] Brian Shannon, "Army Looking to Modernize Black Hawk Fleet Beginning in FY '02," *Defense Daily*, 15 January 1998, Vol. 198 No. 10.

[8] George I. Seffers, "Comanche Faced Cloudy Future—Helicopter Competes for Funding with Crusader," *Army Times*, 13 April 1998.

[9] Lolita C. Baldor, "Newt, Garth Brooks Help Christen Copter," *New Haven Register*, 13 May 1998, p. A6.

[10] "Top Pentagon Cost Analysts Warn Comanche Is Underfunded, Overweight," *Inside the Army*, 14 December 1998, pp 1–3.

Chapter 33

The Engineering, Manufacturing, and Development (EMD) Proposal, 2000

Arthur W. Linden

EMD Proposal

Program Features

Five EMD aircraft, eight Initial Operational Capablility (IOC) aircraft

Acquisition of 1292 aircraft reduced to 1213 at 72/year

$34-billion production program

Additional Requirements

Mission range increased by 25% using existing internal fuel tank

Added auxiliary tanks within the weapons bays

Design Impact

Mission fuel required increased by 400 lb

Rotor diameter increased from 39 ft to 40 ft

Aircraft empty weight increased from 8952 lb to 9305 lb [1]

Schedule Impact

Low Rate Initial Production (LRIP) in 2005

IOC in 2006

Full rate production in 2010

Last delivery in 2023

Program Cost

$3.1 billion

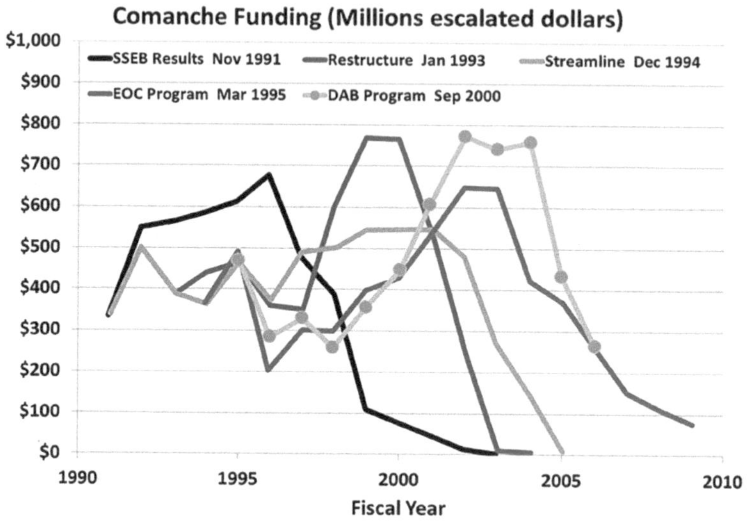

Fig. 33.1 Defense Acquisition Board (DAB) funding vs. Early Operational Capablility (EOC).

As the Dem/Val program was winding down and the proposal preparation for the EMD program was beginning, the Army program manager, Brigadier General Joseph Bergantz, gave an interview to the Sikorsky public relations department in December 1999 (Fig. 33.2). This was reported in the December 1999 issue of *Inside Sikorsky.* The title of the article was "BG Bergantz Says 'Best Program in the Army' Is Ready for Next Phase" [2].

The following are some of Bergantz's comments:

> He expressed optimism about Comanche and said the program enjoys solid support within the Army and U.S. Department of Defense....

Fig. 33.2 Brigadier General Bergantz discussing Comanche's future, December 1999.

> It's key in that getting permission to go into this last phase of development [EMD] is always a strong signal that the program is on good footing and is ready to go forward....
>
> The Army has what they call a "One to N list" in terms of priorities. The first 10 items had to do with readiness and quality of life. The first modernization program that's on that list is number 11, and that's the Comanche program. It's the number one modernization program in the Army's priorities....
>
> The Army has six operational patterns . . . Comanche does all six of those, but its forte is probably bringing information dominance....
>
> It's quite an exciting program when you think about how much technology has changed between Apache and what the Comanche can bring to the table....
>
> It does tie up a lot of the Army's money over the next 15 to 20 years, so there's going to be a whole heck of a lot of attention on something of that magnitude. I think we're ready to handle all that.

Continuing our story with more press quotes:
From the Association of the U.S. Army's "Defense Report" [3]:

> One of the most successful warfighting weapon systems coming on the scene is the RAH-66 Comanche helicopter. It will meet the goals of reducing U.S. casualties, limiting collateral damage, operating during periods of inclement weather, and much more. It is a lethal weapons system....
>
> Comanche is on the leading edge of technology growth for the 21st century. There is no capability in the force today that provides the armed reconnaissance requirements for Army XXI [Army 21]. Comanche fills that void, and will thus be a key system in Army After Next warfighting concepts. It is essential the Comanche development and production program stay on schedule and the production program stays funded to support fielding of the Comanche.

From *Rotor & Wing's Helicopter News*, 17 December 1999 [4]:

> "Congress, frankly, is frustrated with the Army's approach to the Comanche," says John Barnes, military aviation expert with the Senate Armed Services Committee, "They say that it's the their number one aviation program. Yet they consistently underfund it. When you do that for too long, people tend not to take you very seriously."

From the *Wall Street Journal*, 2 March 2000 [5]: "The Army is expected to drastically scale back its purchase of Comanche helicopters after the leaders of the Senate Armed Services Committee warned that the branch's ambitious modernization plans wouldn't be funded without some serious sacrifices."

John McWethy from *ABC World News Tonight* said the following on 5 April 2000 [6]:

> Army leaders are gambling that asking for huge expenditures on the new helicopter will not end up jeopardizing another program, the critically important effort to transform the Army into a lighter, more mobile land force. The announcement signals a flat refusal by the Army to bend to pressure for deep cuts in the Comanche program from Congress and Defense Secretary Cohen's office, the savings to be used for building a lighter force. Army Chief of Staff Eric Shinseki must have decided after scouting the political battlefield that retreat was not in the Army's best interest.

An important technical issue was what specific requirements did the program have to prove to move into EMD. In a 21 March 1995 memo to Gilbert F. Decker, Dr. Paul Kaminski identified the exit criteria for Milestone II. These were:

- 350 fpm rate of climb
- A night forward-looking infrared (FLIR) recognition range of the operational requirements document (ORD) requirement minus 0.4 km.
- A nose-on radar cross-section (RCS) signature of the ORD value plus 6 db, as measured in a pole test
- An infrared signature of the ORD value plus 3 watts/steradian.
- Ballistic vulnerability demonstrated by damage tolerance on fatigue test specimens
- Demonstration of interactive electronic technical manual capability and capability to download aircraft data to Unit Level Logistics System-Aviation

The vertical rate of climb (VROC) was proven with an aircraft test. The pilotage FLIR was tested with a brassboard bench test and the targeting FLIR with a test at Lockheed Martin mounting the FLIR on a 150-ft tower and directing it at vehicles on the ground. Radar signature was proven with a full-scale model of the aircraft. Infrared (IR) signature received a full-scale test at Sikorsky's Florida facility. Ballistic tolerance was proven with live fire tests on the main rotor flexbeam, fuel tank, composite panels, Fantail drive shaft, and main rotor mast. Demonstrations of the Portable Intelligent Maintenance Aid (PIMA), interactive electronic technical manuals (IETMs), and Unit-Level Logistics System (ULLS)-A were done on the prototype aircraft. All tests were conducted successfully.

Quoting an article in the internal Sikorsky magazine, *Inside Sikorsky*, from April 2000 [7]:

> On April 4, the Department of Defense Acquisition Board granted COMANCHE Milestone II approval. The board certified that the

COMANCHE program is properly planned and administered and has successfully completed all demonstration/validation phase program requirements. Once the EMD contract is formally signed, the manufacture of 13 additional aircraft can begin.

"Also on April 4, the U.S. Army released its long-awaited 2000 Aviation Force Modernization Plan which calls for fielding COMANCHE on schedule starting in 2006.

The plan calls for full procurement of 1,213 aircraft, worth about $34 billion over the full production cycle. The plan cited COMANCHE as the linchpin of the Army's aviation modernization efforts.

"Reconnaissance/security represents the single greatest deficiency in Army aviation" the report noted. "The Army remains completely committed to the COMANCHE, a variant of which may be considered as the possible long-term solution for the attack helicopter beyond Apache Longbow."

When briefing news reporters about the modernization plan, Army officials reiterated the service's support of COMANCHE:

"This plan provides a significant departure from the past because it aligns the aviation strategy with the Army vision. It places the COMANCHE as the centerpiece of Army aviation modernization," said BG Craig D. Hackett, the Army's Director of Requirements for the Assistant Deputy Chief of Staff for Operations and Plans—Force Development.

Under Secretary of the Army, Lieutenant General Gansler, in signing the acquisition decision memorandum authorizing EMD, said that Comanche was the cornerstone of the Objective Force and "without Comanche, FCS cannot survive."

The contract for EMD was signed on 1 June 2000 (Figs. 33.3 and 33.4). The contract value was $3.1 billion. The program was to manufacture and deliver five new Comanche helicopters for EMD testing and deliver eight new aircraft to the Army for operational testing and evaluation. Testing was to include qualifying the fully integrated digital mission equipment package and advanced cockpit management systems incorporating navigation, communication, and target acquisition capabilities. Low-rate initial production was scheduled to begin in FY05, with a ramp up to planned full production, 72 per year, with Lot 6 in FY10. Final delivery was slated for 2023. The first U.S. Army Comanche unit was projected to be operationally equipped in 2006.

The *Huntsville [Alabama] Times*, on 2 June 2000, quoted Major General Al Sullivan, commander of the Army Aviation and Missile Command [8]: "Today is a great day for soldiers. What you have here is something that makes a soldier's life easier."

The *New Haven [Connecticut] Register* on the same day quoted Darrell Harrison, the Army's deputy program manager, saying that the signing of the EMD contract " …is a pretty bold step for the Comanche. 'Once the program gets into the [engineering and manufacturing development phase], it is likely to go into full production'" [9].

Fig. 33.3 EMD signing ceremony, 1 June 2000. Seated, left to right: Samuel Huffstetler, Army Contracting Office; Robert Lamina, Boeing Sikorsky contracts director. Observing, from left: Arthur Linden, Boeing Sikorsky program director; Roger Krone, VP, general manager, Boeing Helicopters; Dean Borgman, Sikorsky president.

Fig. 33.4 The full JPO team at contract signing.

This was the first time in 20 years that an Army aircraft program had reached the EMD phase. But the Army did have some concerns. Lieutenant General Paul Kern's congratulatory letter to Boeing Sikorsky expressed a concern about the status of mission equipment package (MEP) development (Fig. 33.5). This seemed unusual for a congratulatory letter.

17 APR 2000

Mr. Arthur Linden
Boeing Sikorsky Joint Program Office
International Plaza II, Suite 645
Philadelphia, Pennsylvania 19113

Dear Mr. Linden:

Congratulations on successfully transitioning the Comanche into Engineering and Manufacturing Development. Your team's performance during the Product Definition and Risk Reduction phase resulted in a well executed program that met cost schedule and performance goals while successfully dealing with a variety of challenges. The effort paid off.

As you move into Engineering and Manufacturing Development, I would like to re-emphasize the importance of integrating the Mission Equipment Package (MEP) into the maneuver force. Successful MEP integration is the key to demonstrating the real value of the Comanche into the future force during operational testing. I challenge you to keep that goal in mind as you continue the Comanche development.

I am very proud of the Comanche Team's accomplishments and what they are doing for our soldiers. Please extend my sincere appreciation to all program participants. I have great expectations for the future of the Comanche program.

Paul J. Kern
Lieutenant General, GS
Military Deputy for
Assistant Secretary of the
Army (Acquisition, Logistics
and Technology)

Fig. 33.5 Congratulatory letter from Lieutenant General Kern.

Fig. 33.6 Accepting the UTC Board of Director Trophy, 21 May 2001. From left: Bill Harper, JPO technical director; Rick Isaacs, JPO planning director; George Sipprell, Sikorsky weapons systems integration manager; Jayne Kissam, JPO finance director; and Rich Gallagher, Sikorsky air vehicle integration manager.

UTC BOARD OF DIRECTORS TROPHY

On 21 May 2001, the Sikorsky Comanche team was awarded the United Technologies Corporation Board of Directors Trophy for the EMD contract win. This corporate trophy is given out once per year and "...recognizes success in completion and qualification of a significant new product family or system." A core of five employees was selected to represent the program (Fig. 33.6).

In his remarks, Sikorsky president Dean Borgman said: "Comanche is a pillar of our company's bright future, and UTC justifiably recognizes the significance of the transition to EMD. My compliments to the entire Sikorsky Comanche team."

REFERENCES

[1] *Flight International*, 26 May 1999.

[2] "BG Bergantz Says 'Best Program in The Army' Is Ready For Next Phase," *Inside Sikorsky*, December 1999.

[3] "RAH-66 Comanche Helicopter - Key to Army Warfighting," Association of the U.S. Army, "Defense Report", April 1999/DR 99-5.

[4] "Comanche May Fall Victim to New Army Vision," *Rotor & Wing's Helicopter News*, 17 December 1999, p. 3.

[5] Greg Jaffe, "Spending Limitations May Lead Army to Scale Back Helicopter Purshes," *Wall Street Journal*, 2 March 2000.

[6] McWethy, J., "US Army Saying It Needs New Fleet of Comanche Helicopters To Replace Unreliable Ones, Anchor: Anderson Cooper, reported and copyrighted by Burrelle's Information Services," *ABC World News Tonight*, 5 April 2000.

[7] "Army Behind COMANCHE As It Enters EMD Phase," *Inside Sikorsky*, April 2000, p. 8.

[8] Shelby G. Spires, "$3.1B contract lifts Comanche - Army's next armed scout helicopter enters advanced testing phase," *Huntsville [Alabama] Times*, 2 June 2000.

[9] "Sikorsky Wins Big - $3.1 billion Army deal to add hundreds of jobs," *New Haven [Connecticut] Register*, 2 June 2000, p. 1.

Chapter 34

THE APACHE

ARTHUR W. LINDEN

The Comanche relationship with the Apache program took many swings during the over 20 years of the program. In the beginning the two aircraft were to work together, with the Comanche as a replacement for the Army's light helicopters and designed for reconnaissance and light attack. The Apache's role was heavy attack. Simple enough.

But, as with many Defense Department programs, there was always a concern about funding. In the 1987 budget debates, the Army reduced the Black Hawk and Apache planned buys to help fund the Comanche program.

Hughes Aircraft, the original developer of the Apache, was worried about this way back in the 1980s. From *Howard's Whirlybirds—Howard Hughes' Amazing Pioneering Helicopter Exploits* by Donald J. Porter, page 198 [1]:

> The possibility of no more Apache program, meager commercial helicopter sales, and only a chance at building the LHX, meant that sales could drop off to a trifling fraction of what they once were. The situation would be reminiscent of the days immediately following the loss of the OH-6A contract years earlier. The company was so worried that it pressured Congress to use some of the LHX funding to extend Apache production beyond fiscal 1988. Seeking to head off the debacle, the company was funded to build several hundred additional Apaches. In May 1988, Congress voted to fund development of an advanced-technology Apache, planned for delivery starting in 1992.

Although Comanche was designed for light attack and Apache for heavy, over the years as the Comanche requirements continued to grow and its weight increased, the differences between the two aircraft started to fade. Some Army proponents of Comanche were stating that it was more or less self-evident that the Comanche would replace the Apache. With the External Fuel-Armament Management Subsystem (EFAMS) installed, we could carry 14 Hellfires. The Apache could only carry two more. We could carry nearly as many rockets as they could. We had a gun that was arguably going to be more accurate with a higher rate of fire. We got the LONGBOW radar, had a better sensor, and had air-to-air capability. We could operate around radar and infrared (IR) threats and survive in nuclear, biological, and chemical (NBC) environments. We had superior communications capability, which worked best when communicating between Comanches. The Apache was not similarly equipped.

Boeing Buys McDonnell Douglas

Then things really got complicated when Boeing bought McDonnell Douglas. Now the Apache program was owned by one of our team. We stopped any implication that Comanche was a potential replacement for the Apache.

The reaction of the Boeing senior leadership on the Comanche program was interesting. Initially, being totally involved with and dedicated to the Comanche, they were not at all enthused about the Apache. Their conclusion was that the Apache was now dead. They thought Comanche was so superior that Boeing would put all its chips on the Comanche and let the Apache wind down and be retired. This was quickly dispelled by the views of the Boeing folks outside the Comanche program.

Sikorsky was also speculating about the impact of the Apache now being part of Boeing. One view was that Boeing would now favor the Apache because it had 100% of that program but only 50% of Comanche. But it was pointed out that the Apache was contracted differently than the Comanche. Boeing was contracted only for the air vehicle. Forty percent of Apache was contracted directly with the mission equipment package (MEP) contractors, so it was somewhat more comparable.

Boeing soon arranged for an Apache to fly to West Palm Beach for a "family photo" of the two aircraft together (Fig. 34.1).

Boeing said:

> The U.S. Army Aviation's one-two punch met and flew together for the first time last month, giving the Army, the helicopter's manufacturers and the world a glimpse of the future. . . . The meeting of Oct. 24 came less than three months after the Boeing-McDonnell Douglas merger brought the Apache Longbow and Comanche together under the Boeing banner.

The external turbulence continued. Back in 1997, according to the 6 January 1997 issue of *Aviation Week & Space Technology*, in an article "Comanche Avoids 1998 Budget Cut," it was reported that [2]:

Fig. 34.1 ***Army Aviation*, 31 December 1998 (Boeing/*Army Aviation* Magazine Photo).**

… a JCS [Joint Chief of Staff] official said, "My belief is that [eventually] they will choose to continue Apache Longbow production at a modest rate and give up the Comanche low radar cross section …".

One way to save money would be to abandon the low-observable Comanche airframe but to continue development of advanced sensors and other mission payloads that would be installed on the existing AH-64 Apache helicopter.

This argument would come up periodically. We found it amusing in that the development of the Comanche airframe was virtually complete, and it was getting rave reviews. The primary remaining task was the development of the mission equipment. How moving that effort over to the Apache program would save money was never quite clear.

REFERENCES

[1] Porter, D. J., *Howard's Whirlybirds—Howard Hughes' Amazing Pioneering Helicopter Exploits*, Fonthill Media, United Kingdom, 2003.

[2] David A. Fulghum and Anthony Velocci, Je. "Comanche Avoids 1998 Budget Cut," *Aviation Week & Space Technology*, 6 January 1997, p. 30.

Chapter 35

The Sixth Program Restructure: Block Upgrades, 2002

Arthur W. Linden

2002 Block Upgrades Program

Program Features

Develop Comanche capability in blocks:

I. Training

II. Armed Reconnaissance

III. Attack

Production quantity reduced from 1213 to 650

Aircraft Empty Weight

9975 lb

Schedule

First engineering, manufacturing, and development (EMD) aircraft flight, 2004

Initial operational test and evaluation (IOT&E), 2007

Initial Operational Capability (IOC), 2009

Low Rate Initial Production (LRIP) delayed from FY2005 to FY2009 through 2011

Production would end in 2019.

Program Cost

EMD program cost grew from $3.1 billion to 6.5 billion.
Production program was reduced from $39.3 billion to $26.9 billion (due to lower quantity).

At this point, the authors want to point out that we were gradually leaving the program, but we did keep in contact with those people still working on Comanche and we actively followed program progress. These last chapters are our best understanding of what was going on.

As the EMD program commenced, aircraft 1 continued in flight test, and it completed its flight testing at the end of 2001. It had flown 311 flights for a total of 377.5 hours. It had tested the more powerful Light Helicopter Turbine Engine Company (LHTEC) T800-LHT-801 engines, with a 17% power increase to 1563 hp. It was then taken off flight status to be used as a backup for aircraft 2. At that time, aircraft 2 had flown 103.5 hours and would continue in flight test, concentrating on the MEP [1].

The Comanche flight test program had proceeded with remarkably few major technical issues. This was the result of the extensive detailed development program conducted prior to the first flight. The only major technical issue that came up with the air vehicle was the configuration of the tail. The initial design had low control power and was determined to be not acceptable. The airflow was being washed out by the turbulence coming off the main rotor hub. In December 2000, a revised tail was flown with a simulated version of the LONGBOW radar (which had recently been added to the Comanche's requirements), a new main rotor pylon, an updated rotor hub fairing, and vertical tail endplates on the horizontal tail. These modifications were concluded to be successful.

In spring 2002, the Lockheed Martin night vision pilotage subsystem was delivered for integration into the aircraft. It included the helmet displays, commercial processors, and functional cockpit displays. The target acquisition subsystem and other MEP elements were to be incorporated later [1].

Aircraft 2 made its first flight with new MEP software and the more powerful engines on 23 May 2002. The MEP software provided the foundation for the aircraft operating system and would support the pilotage, vehicle systems reporting and health, and the digital map function [2].

In 2001, post 9/11, we actually benefitted from an environment of a country going to war, although it wasn't quite sure where. We received enormous support from all the Army branches, Congress, the public, and especially the Army chief of staff.

Management Changes

Around the time of the EMD proposal and contract, things were changing with the management at Sikorsky, with the Boeing Sikorsky Comanche team, and with the Army program office.

The Comanche program director, Art Linden of Sikorsky, was retiring at the end of his four-year stint as director. Per the agreement between the two companies, the lead for the next four years would be Boeing's. Boeing chose Chuck Allen, who had been the deputy director under Art, to be the new director.

On the Army side, Brigadier General Joe Bergantz was promoted to Major General and replaced Major General Snider as the Army's program executive officer for aviation. Colonel Robert Birmingham was brought on to replace Bergantz as the Army Comanche program director.

Thus we had two new senior Comanche leaders, Boeing's Chuck Allen and the Army's Colonel Birmingham. Allen had been the deputy Comanche program director, but his overall background was in fixed wing aircraft. Bob Birmingham had worked on the Comanche engine program, but was not familiar with the details of the aircraft program.

From the Sikorsky side, other program management changes were being made that probably were more impactful than originally recognized.

The Sikorsky Comanche management team had always been highly engineering oriented. All the Sikorsky program directors except one had been engineers. Since the beginning, Sikorsky's Joint Program Office (JPO) directors and deputy directors were all engineers. Sikorsky's engineering department was always heavily involved with Comanche and was led by an active engineering vice president, Dr. Ken Rosen. Although none of this was a program requirement, it did lead to a very strong team to resolve technical issues and move this highly complex system through development.

All this changed with the new Sikorsky EMD management team. At Sikorsky, there were now no engineers in any position above the program engineering manager. The new Sikorsky engineering manager was brought in from outside the program with little knowledge of the program history and issues. He had no senior technical leadership, no "gray beards" to consult with and seek guidance from on technical issues.

The JPO technical director, Bill Harper, left the program in 2001.

Although a change in management is a common practice as a program transitions to its next phase of development, typically there is a transition period during which the new management team is brought up to speed on the technical aspects and technologies of the program along with technical and program issues and risks. This did not happen with Comanche. The old management team was essentially terminated in 2000, and the new management team started work the next day.

Changes had also been made at the senior management level at Sikorsky. Company president Gene Buckley and engineering vice president Dr. Ken Rosen, both strong Comanche supporters, had retired. They were replaced with people from outside the company, people who again were not familiar with the program and some of its unique management approaches.

In one fundamental change in the way we operated, the new senior company leadership did not buy into the integrated product development team approach. We were accused of running a separate engineering organization. In one intense meeting, the new Sikorsky engineering vice president claimed that Comanche was being run "...like you were a separate corporation." He told the team that this would end immediately.

He was also a much more hands-off engineering leader, very different from what we were used to.

With all of these changes, our new engineering manager had a challenging job with his new Comanche assignment.

At the request of the Army in 2000, the JPO had moved from Philadelphia to Huntsville, Alabama, to be close to the Army program office.

Then, in 2001, Chuck Allen made major changes to the JPO organization. The two companies had always had their own organizations with the JPO operating separately, usually in another location. This kept all the parties focused on their own responsibilities. But Chuck, presumably with Colonel Birmingham's concurrence, felt that this was not working and moved the JPO back to Connecticut. He also made a fundamental change in the role of the JPO. The unique aspect of the Comanche program had been the teaming of industry, the Army science and technology (S&T) community, other military S&T programs, and the Army acquisition office to address the warfighter's combat needs and issues. The role of the JPO included being the guardian of the program and regulatory constraints, plus scheduling, overseeing, conducting, and supporting periodic reviews and audits, all in an effort to keep the team members informed and to keep the program sold. JPO facilitated and supported conflict resolution and technology trade studies to reach the end objective, while also maintaining configuration control. It was the JPO's responsibility to address independent studies and efforts by external agencies and individuals to influence the program.

Over all these past years it was not the role of the JPO to direct the work of Boeing and Sikorsky. Chuck changed that. The new JPO organization would retain the responsibility of guardian of the program, but now it would also directly manage the work done by the two companies.

ANOTHER RESTRUCTURE

Early in the EMD program, doubts started to occur about the risks in the program plan. This started with the Army Program Management Organization (PMO) staff discussing the program with the newly installed Colonel Birmingham. He then brought up the subject with Chuck Allen. After numerous analyses were conducted, both concluded that the program's cost and schedule metrics indicated that the program could not be completed as planned. In 2001, Colonel Birmingham (who prided himself on "telling the unvarnished truth," according to his Comanche book, *The Last Comanche—An Army Program Manager's Story* [3]) stated that the program was not executable. Quoted in the 30 November 2002 issue of *Army Aviation*, he stated [4]: "Due to fiscal constraints coupled with an extremely ambitious development schedule, the MS II program was a noble effort but a long shot to succeed." In the Spring 2002 issue of *Vertiflite*, he said: "The Army and industry's resolve and commitment to get Comanche fielded often led to a degree of ingenuous confidence and optimism which kept the program supported and 'moving forward,' while unknowingly pushing the program envelope in terms of acceptable risk" [1].

Another restructure would be required.

He gave a number of specific reasons for this, but the major one—as mentioned previously—was that the new management team felt the contracted plan was simply too optimistic. They felt that it had little cushion for the inevitable unknown-unknowns. Other reasons mentioned were that the software plan was unrealistic, the task was underestimated, and the companies could not acquire the software engineers needed to do the job in the time scheduled. (Software for the program was initially estimated at 800,000 lines of code. Over the years this had grown to 4 million lines and was still growing.)

There was necessary technology maturation work that had yet to be done.

And there was aircraft weight. Like most development aircraft programs, the weight had grown. The initial plan for an empty weight of 7500 lb now had a goal of 9975 lb. Additional weight reduction work was needed to get to that weight, and the effort was thought to be above what had been estimated.

Figure 35.1 shows a chart used by Colonel Birmingham to explain the critical issues at that time to Army leadership.

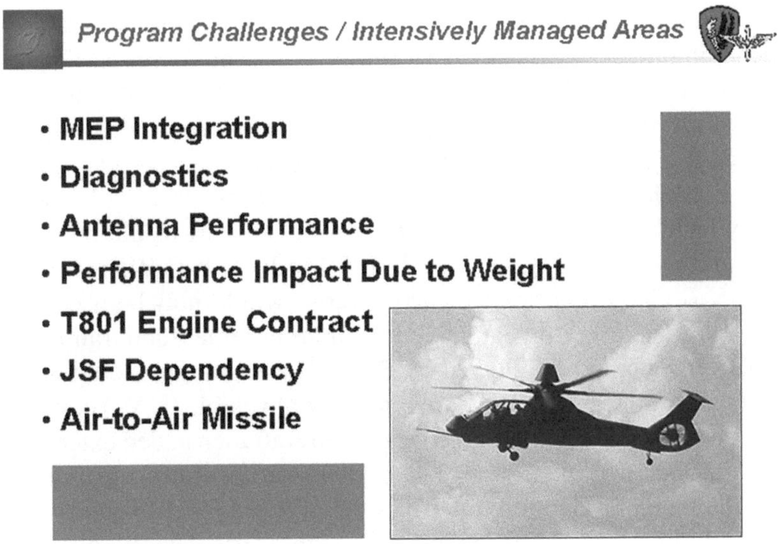

Fig. 35.1 Program challenges/intensively managed areas.

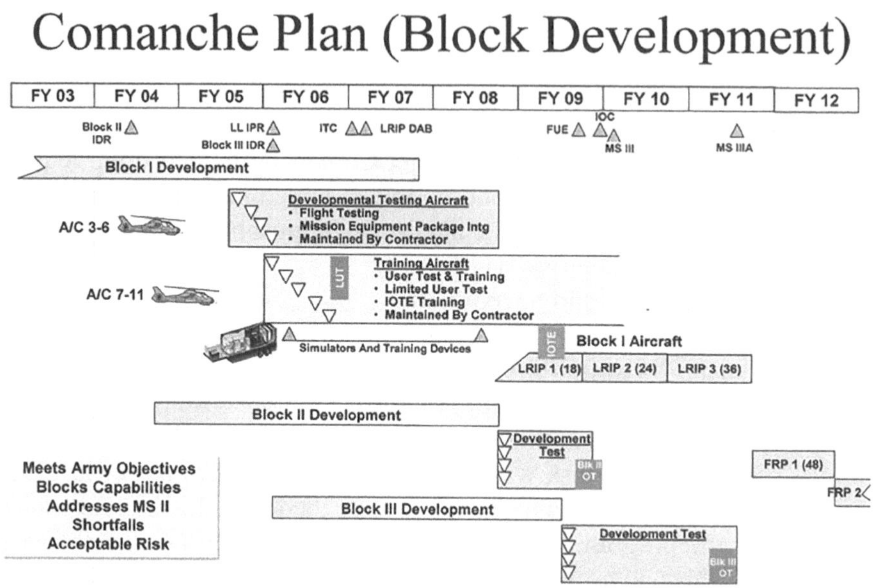

Fig. 35.2 Comanche plan (block development).

A new development plan was created. The program would be slowed and would be "…in progressive block upgrades to increase combat effectiveness over time, utilizing revolutionary new technologies and meet[ing] the Army's transformation goals," according to an article in the 31 March 2002 issue of *Army Aviation* magazine, written by Colonel Birmingham [5]. These blocks would be training, armed reconnaissance, and attack versions (Fig. 35.2).

The projected development cost was now twice the previously allocated Comanche funding. On top of the $3.1 billion approved for the EMD program, $3.4 billion in additional funding was required. The team felt that asking for more near-term money was not realistic. They proposed instead staying with the approved Comanche funding profile, but using early-year production money to complete the required additional development. Initial Operational Capability (IOC) and production fielding of the aircraft would be delayed to fit the funding available. Low Rate Initial Production (LRIP) would move from FY05 to FY09. Four of the early aircraft would now be used for additional developmental testing, and five for limited user test and training (Block I aircraft.) This would be followed by Block II aircraft that would update the computers and add Link-16 and unmanned aerial vehicle (UAV) control to get the full reconnaissance capability. Block III aircraft then added external stores, air-to-air missiles, and microclimate cooling to get the attack capability. It also included the weight reduction initiatives to get to the 9975 lb empty weight.

The development would be extended by eight years. The full incorporation of the MEP would not occur until 2012.

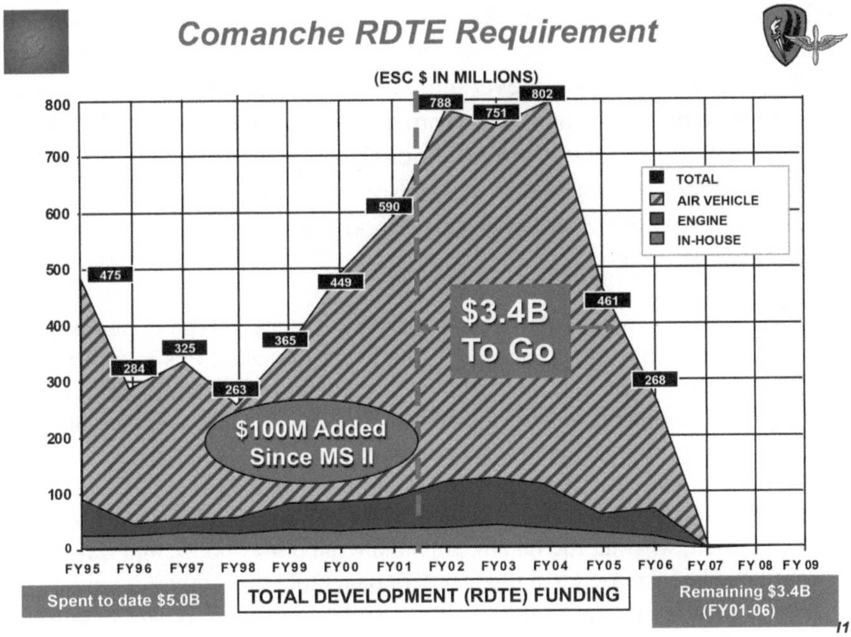

Fig. 35.3 2002 Comanche funding profile.

This evolutionary acquisition approach did promise resolution of all the issues. It dramatically reduced the technical and schedule risks, and allowed for technology insertion. It outlined the total development cost to get to a full production-compliant configuration (Fig. 35.3).

An article in the 3 June 2002 issue of *Aviation Week & Space Technology* explained [6]:

> About half the added [$3.4 billion] money is for more flight testing and the acceleration and implementation of capabilities, such as Enhanced Position Location Reporting System connectivity. Another 25% is for items such as software development tools or the ability to control a UAV from the aircraft. The remainder is to cover cost growth.

According to the 28 October 2002 issue of *Aviation Week & Space Technology*, the quantity was reduced from 1213 to 650, with a peak annual build rate of 60 aircraft per year [7]. The average production cost grew to $32.2 million per aircraft from $24.1 million. The total production bill was reduced from $39.3 billion to $26.9 billion due to the smaller production quantity. About $3 billion of that was required to pay for the expanded development program. Production would run until 2019.

The restructured program also included the Army plan to develop an unmanned adjunct to Comanche and to upgrade the AH-64D Apaches.

In the 31 March 2002 issue of *Army Aviation*, Colonel Birmingham implied that maybe the weight issue was not fully resolved in the proposed program [5]: "The aircraft design will leverage lighter-weight and higher performing technologies, which will increase Comanche's performance envelope over time. Blocking strategies will provide our future platforms more opportunity to decrease system weight as this technology becomes available."

The 28 October 2002 *Aviation Week & Space Technology* article discussed the aircraft weight issue [7]:

> In the coming months, the Army plans to initiate a total system power study to determine what changes it may want to make to its propulsion plans, Birmingham said. He stressed that the current baseline engine, T800-LHT-802, which features a 100-hp. increase over the original -801, meets all requirements for the immediate future. But, by 2006 the Army may want to increase shaft horsepower limits of the engine and transmission. But first the service wants to determine how much engine life would be degraded, whether the fantail can handle the power boost, and if rotor speed can be increased. An independent panel told the Army it would be more cost-effective to forgo engine life in favor of more power than to spend a lot of money on reducing the aircraft's weight, Birmingham said.

Perhaps the biggest issue with the new program was the delay in getting Comanche into the field; the program was not consistent with the Army modernization plans. Additionally, this restructure consumed a large percentage, over 40%, of Army Aviation's available budget for many years.

The proposed plan was controversial within the Army PMO organization. According to Colonel Birminham's book [3], Major General Bergantz was not in favor of a restructure at that time. He proposed waiting a couple of years to let the program continue on the contracted plan while the issues were worked out. Colonel Birmingham did not agree. He felt that the program was going to be canceled if he did not take action, so he went forward to Army leadership with his proposed plan.

According to an article in the 3 June 2002 issue of *Aviation Week & Space Technology* [7]:

> Service officials also believe that the final word hasn't been spoken about Comanche's fielding plan. "I don't accept that longer schedule," said the Army's acquisition chief Claude Bolton. "I accept that is where we are today, but I will not accept that it is how we do the program." He added "if there is an opportunity between now and when we field this aircraft to bring that [in service date] back to the left, I intend to do it."

The Army's internal Systems Acquisition Review Council signed off on the new plan in June 2002. On 14 November 2002, the contract was signed.

Colonel Birmingham and Frank Wallace of the Army program office were awarded the AAAA Robert M. Leich award for developing this new plan, according to the March/April 2003 edition of *Army Aviation* [8].

Meanwhile, Boeing Sikorsky kept the program moving forward. On 15 September 2003, the team formally opened the RAH-66 Comanche production facility in Bridgeport, Connecticut. The keynote speaker was Lieutenant General John Riggs, director of the Army's Objective Force Task Force. Also in attendance was Connecticut Governor John G. Rowland and U.S. Representative Rosa L. DeLauro [9]. Bridgeport mayor Joseph P. Ganim had previously said that the economic spinoff from the facility would "propel Sikorsky to the top of the city's list of taxpayers" [10].

In June 2003, Colonel Birmingham was replaced as the Army program director by Colonel Michael E. Cantor.

In November 2003, the program director's position rotated back to Sikorsky, with Sikorsky's Mike Blake appointed to the position. This was earlier than the usual four-year rotation. Mike had served as the deputy program director under Chuck Allen since November 2001 (Fig. 35.4).

In the 23 February 2004 issue of *Aviation Week & Space Technology*, a letter from Colonel Cantor and Mike Blake was published responding to a previous article assessing the status of several Pentagon development projects [12]. This letter was cleared by Army PAO. The letter said:

> The program is within cost, on schedule and meeting all key performance parameters. Four aircraft are in production and a fifth will join the line shortly. In 2005, the first engineering and manufacturing development aircraft will be in flight test. Contrary to your article, Comanche is below its target weight. The other technical challenges cited (communications and navigation, radar cross section, software requirements and the 20mm gun) are par for the course for a development program, and all are being actively managed. Bottom line: the program is performing well.
>
> Lt. Gen Richard Cody, the first general to fly in Comanche, recently told the Assn. of the U.S. Army Aviation Symposium that it is "the best

Ending his tenure on a high note, Comanche Program Director **Chuck Allen** was awarded the Bronze Order of St. Michael on Nov. 21, 2003, his last day on the job.

In the photo, Mike Blake (left), president of AAAA's Connecticut Chapter, presents Allen with the OSM as they stand near RAH-66 Aircraft #4 on the production line at the Comanche Major and Final Assembly facility in Bridgeport, Conn. Blake, the former Comanche program deputy director, has been named director.

Allen took over leadership of the Comanche Program Office in July 2000. He oversaw the successful 2002 program restructure that resulted in a new $6.7 Engineering, Manufacturing and Development contract with the Army. In 2003 the Boeing Sikorsky team began major assembly of the first EMD aircraft.

Allen has taken a new position within The Boeing Company as vice president and program manager of the firm's Orbital Space Program in Huntsville, Ala.

Fig. 35.4 Chuck Allen receiving the AAAA award (*Army Aviation* Magazine Photo).

aircraft that we've ever built." We remain confident Comanche will provide future warfighters leap-ahead technologies and transformational capabilities.

REFERENCES

[1] "Boeing-Sikorsky RAH-66 Comanche," *Vertiflite*, Spring 2002, pp. 19–20.

[2] Comanche #2 Flies With New Software and Engines, Sikorsky press release, 23 May 2002, pp. 19–20.

[3] Birmingham, R., "Defence Helicopter," *The Last Comanche—An Army Program Manager's Story*, Shephard Press, vol. 25, No 3, June/July 2006, pp. 24–26.

[4] Col. Robert P. Birmingham, "Comanche Program Gets Real, "*Army Aviation*, 30 November 2002, pp. 10, 12, 14, and 28.

[5] Birmingham, R., "RAH-66 Comanche: Adjusting for the Warfighter," *Army Aviation*, 31 March 2002, p. 10 and 30.

[6] Robert Wall, "Army Sets Comanche Replan; Other Helo Woes Remain," *Aviation Week & Space Technology*, 3 June 2002, pp. 60, 61 and 62.

[7] Robert Wall, "New Comanche Plan Gets Green Light," *Aviation Week & Space Technology*, 28 October 2002, pp. 27–28.

[8] Robert M. Leich Award, "Sponsored by Northrop Grumman ESSS Comanche Program Office," *Army Aviation*, Redstone Arsenal, AL, March/April 2003, p. 42.

[9] *Connecticut Post*, 16 September 2003.

[10] Marian Gail Brown "Comanche era takes off - Sikorsky adds Bridgeport jobs, $15m plant, new business," *Connecticut Post*, 13 March 2002, pp. 1 and A13.

[11] "Comanche Program Director Chuck Allen was awarded the Bronze Order," *Army Aviation*, 21 November 2003, p. 61.

[12] Cantor, M. E., and Blake, M., "Comanche Performing Well," *Aviation Week & Space Technology*, 23 February 2004, pp. 7–8.

Chapter 36

Program Cancellation

Arthur W. Linden

Just because we had finally passed Milestone II in 2000 and entered into the engineering, manufacturing, and development (EMD) phase did not mean we had clear sailing. On 28 February 2000, the *Washington Times* reported [1]:

> Gen. Eric Shinseki, Army chief of staff, announced a major Army transformation in October. Since then, he has urged his staff at a whirlwind pace to create the first lighter, more mobile brigades at Ft. Lewis, Wash, and scrub the five-year budget plan to find the billions of dollars in needed savings.
>
> … The idea is to take an essentially Cold War Army designed to fight a great land battle in Europe and convert it into swifter brigades able to respond to peace enforcement missions in a matter of days, not weeks.

One thing under consideration was to reduce the total planned Comanche buy from 1213 to 650. Back to the *Washington Times* article [1]: "It remains to be seen whether the chief of staff will support such a reduction, the confidential congressional report says. 'The Army is holding out the expectation that by the time the 625th Comanche is produced, a 'Comanche Plus' program will begin to replace the aging Apache.'"

And still later in that article [1]:

> The congressional analysis sums up the Army's dilemma: "Over the last few years, the [Senate Armed Services Committee] has endeavored to convince the Army to deal with the increasing strategic irrelevance of a heavy Cold War–structured force. Gen. Shinseki is to be applauded and generally supported for undertaking that transformation. However, as with any undertaking of this size and importance, there are and will be considerable areas of controversy and dispute."

Moving ahead more than two years, on 30 August 2002, Army Lieutenant General John Riggs went to Florida to see a flight demonstration with "an entourage of congressional staffers and reporters who had flown from Washington with him," according to the 31 August 2002 edition of *The Washington Post* [2]. But Lieutenant General Riggs had to announce that: "The airplane is hard down and will not be flying today." An engine malfunction had grounded the aircraft. The article went on to say that the Comanche " ... has become synonymous with Pentagon ineptitude at procuring new weapons systems."

The *New York Times* reported on 15 April 2002 that [3]:

> Defense Secretary Donald H. Rumsfeld convened a meeting with the three service secretaries today to tell them they needed to cut major weapons systems to finance new "transformational technologies," including space surveillance systems and unpiloted weapons, officials said. Big-ticket programs potentially on the block, beginning with the 2004 budget, include the Air Force's next-generation tactical jet and two Army weapons, the Comanche helicopter and the Crusader artillery system, officials said.

By then, Army Chief of Staff General Eric Shinseki had been replaced by General Peter J. Schoomaker. General Schoomaker appointed a review team to discuss the future of Army aviation including Comanche. Brigadier General James O. Thurman led the team. This was all done in secret and was well underway before its existence was revealed. Major General Joe Bergantz, previous Army Comanche program manager and now the Army executive officer for aviation, volunteered to participate but was declined. Lieutenant General John Riggs, Army deputy chief of staff, was excluded. The Army's Comanche program office was also not part of the review team and was not allowed to have any contact with any member of the panel [4]. The team meetings were so secretive the members were sequestered and had to sign nondisclosure agreements.

Some believe their task was to kill the Comanche program. The board was assembled before Colonel Birmingham's proposal for the revised program was even submitted, although his plans had been discussed with various senior people.

By December 2002, the existence of a review board was made public.

The wars in Iraq and Afghanistan had put the Army in the position where replacement aircraft were needed immediately. In contrast, the new Comanche restructure kept the aircraft in a continual development phase for another nine years.

One view is that there were few people on the board who could be considered as Comanche advocates. There seemed to be more anti-Comanche factions than there were supporters. Outside of aviation, Comanche was not popular or appreciated. Inside aviation, the support was low because the program no longer supported the Army's modernization initiatives.

We felt that the technologies and capabilities of the Comanche were not completely understood by all the team members, and in some case they were misrepresented.

The board's chief technology officer was closely involved with Army Special Operations. He didn't want Comanche fielded because he didn't want the low observable technology in the general aviation population. He felt that they couldn't protect the technology if it was in the field. He also saw the Comanche as a drain on funds and saw no need for the Comanche in the special operations community.

The Training and Doctrine Command (TRADOC) representative from the Comanche Systems Manager Office told the termination board that the Comanche capabilities did not mesh well with the projected capabilities of the Army's Future Combat System (FCS), and to change the Comanche to meet those needs would require still another restructure following the one the program was currently proposing. He made it very clear to the termination board that the Comanche was unacceptable for combat operations.

The Army project manager for aviation systems was reported as telling the review team that the Comanche electronics were obsolete.

And a representative from the aviation acquisition community recommended reinvesting the $14.6 billion in the Comanche program to refurbish 498 older aircraft and purchase 796 newer models that fit the FCS needs.

In this sixth restructure, the initial claim was that no new money was needed, only reprogrammed production money to Research, Development, Test and Evaluation (RDT&E). In his book, Colonel Birmingham says that was the strategy in 2000 and 2001. But by 2002 and 2003, the proposals by the suppliers and the companies did require the Army to find more money.

CANCELLATION

On 23 February 2004, the Defense Department stunned both the Army Comanche team and the contractors with an announcement that it was terminating the Comanche program. This was after it had recently canceled the Army's Crusader mobile artillery system. The Army website said that it ended the program based on experience during combat in Afghanistan and Iraq, and a high-level drive to make the branch better at fighting counterinsurgency and unconventional warfare campaigns. Army Chief of Staff General Schoomaker said:

> If you take a look at when Comanche was envisioned, starting in 1983, and you take a look at the threat that we faced at that time and the kind of battlefield that we envisioned, Comanche made a lot of sense. But it makes less sense today as we go forward. [5]

A press conference was held on February 23, the subject being "Briefing on the Restructure and Revitalization of Army Aviation." Participating were:

- Les Brownlee, acting secretary of the Army
- Claude M. Bolton, Jr., assistant secretary of the Army for acquisition, logistics, and technology
- General Peter J. Schoomaker, chief of staff, U.S. Army
- Lieutenant General Richard A. Cody, deputy chief of staff, U.S. Army
- Lieutenant General H. Steven Blum, chief of the National Guard
- Lieutenant General Ron Helmly, chief of the Army Reserve
- Brigadier General E. J. Sinclair, commanding general of Fort Rucker and army aviation branch chief

In his opening remarks, Mr. Brownlee said that "...the restructure and revitalization being announced was the result of lessons learned and experiences gained by the Army's recent two-and-a-half years of combat in the global war on terror as well as the operational environments envisioned in the foreseeable future" [6]. The study that led to this revised plan concluded that "... we should upgrade, modernize and rebuild our attack, utility and cargo helicopter fleets, and replace our light observation and scout/attack helicopters as rapidly as possible." (This last objective was, of course, the original reason for the Comanche program when it began 20 years earlier.)

Mr. Brownlee went on to explain [6]:

> We've examined closely our resourcing plans for aviation and concluded that some of the capabilities those funds would provide are no longer consistent with the changed operational environment. Therefore, General Schoomaker and I have recommended that the Comanche helicopter program be terminated and those resources reallocated to restructuring and revitalizing Army aviation.

At the press conference, numerous reasons were presented for the cancellation of the program. These included [6]:

- About 40% of the Army's aviation budget up to 2011 was devoted to Comanche. "We're at war, and need those funds for current aviation programs."
- "Comanche was built to go deep. It was built to be a low observable helicopter ... which it is. It was built to be a highly responsive reconnaissance aircraft ... that ... could be the see-first, understand-first and the action agent to act first on the modern battlefield. And we're sure that it could do that. But the operational environment has changed. We're seeing a proliferation of MANPADS, IR missile systems, more sophisticated air-defense systems, as well as, in the joint fires arena, we have now new types of

capabilities to deal with the radar threat environment that 13 or 14 years ago we did not have in the joint force. And so that has changed."

- There was $14 billion in the Comanche program through 2011, but it only bought 121 aircraft. With the new plan, including all aviation funding, not just that from Comanche, the Army would buy 800 new aircraft and upgrade 400–1400 more. "70 percent of the current fleet we have we'll be able to either upgrade, recapitalize and buy new, compared to trading off 121 Comanches. And that's what this whole program's about."
- The Block III Apache "which we will now be able to do in some numbers, will have all of the capabilities that we would have built into the Comanche with the exception of one, and that's the low observability. And if you look at the operational environment in which we're now operating and the one we think we'll be operating in the future, we think that is not where we should put our focus on."
- Emerging technologies, primarily unmanned aircraft, were becoming capable of doing some of Comanche's design role simpler and cheaper.

Figure 36.1 was used at the press conference and had been used a few days earlier at the White House in a briefing to President George W. Bush. The briefers there were Mr. Brownlee and Lieutenant General Cody.

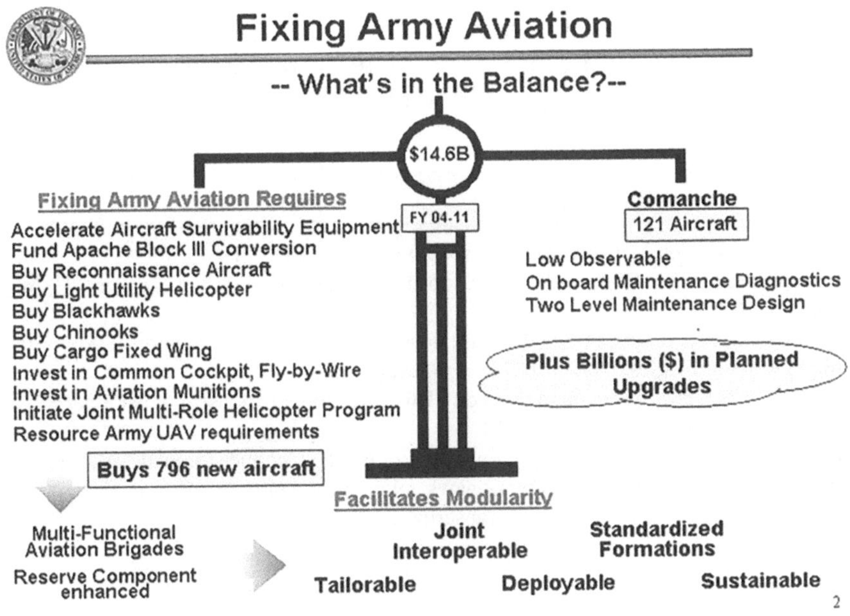

Fig. 36.1 Army Chart: Fixing Army Aviation.

Other, maybe secondary, issues were also at play:

- Slipping the program schedule with the last restructure resulted in the loss of support for the warfighter. The aircraft fielding schedule was no longer aligned with any Army modernization plan or strategy.
- None of the mission equipment package (MEP) had yet been demonstrated on an aircraft.
- The recent replacement of Army Chief of Staff General Eric Shinseki, a strong Comanche supporter, by General Schoomaker, an Armor Division general and a strong supporter of the Special Operations Forces (who were anti-Comanche), put less emphasis on the need for Comanche.

Comanche had started out as a $33-billion program to procure 4500 aircraft, with an initial operational capability (IOC) date of 1998. It became a $26.9-billion program to acquire 650 aircraft with an IOC date of 2009. The final program restructure doubled the remaining development cost and extended the development schedule by eight additional years. The average aircraft unit production cost had grown from $7.5 million ($11.8 million in 2004 dollars) to $32.2 million. This was due to configuration changes with the additional requirements and weight growth, but more importantly to the reduction in aircraft quantities and the amortization of the added research and development (R&D) costs.

It appears to these authors that the program cancellation was no more complicated than the reason the Army first articulated: It could continue with Comanche and get 121 field-ready aircraft by 2011, or it could redirect the Comanche funding to procure many more current-production aircraft to replace combat losses, plus invest in other needed systems.

REFERENCES

[1] Rowan Scarborough, "Copter Cutbacks Get Army Support—Funds Would Help Restructure Forces," *Washington Times*, 28 February 2000, p. 1.

[2] Vernon Loeb, "Fate of Army Chopper on the Block," *The Washington Post*, 31 August 2002, p. A2.

[3] Thom Shanker and James Dao, "Defense Secretary Wants Cuts in Weapons Systems to Pay for New Technologies," *New York Times*, 15 April 2002.

[4] Comanche Review Shrouded in Secrecy "This is National Defense, not National Review," *National Review*, January 2004, p. 14.

[5] Bradley Perrett, "Lockheed Opportunity as Comanche Chopper Canned," AOL News, 24 February 2004, Reuters. http://aolsvc.news.aol.com/news/article.

[6] U.S. Department of Defense, news transcript, 23 February 2004.

Chapter 37

REBUTTAL AND RETROSPECTIVE

ARTHUR W. LINDEN

As a result of the jolting announcement of program cancellation, the Boeing Sikorsky team issued the following statement:

> We are surprised and disappointed by the Army's announcement today to terminate the RAH-66 COMANCHE program. Five of these advanced technology aircraft are on the production line today, and we are on plan for the program.
>
> While we regret the Army's announcement, we are committed to working closely with our customer and will engage in further discussions to ensure we have a complete understanding of the next steps for COMANCHE. We appreciate the commitment of the 1,300 dedicated aerospace employees working on COMANCHE at Boeing and Sikorsky, and the thousands more at our suppliers across the country.

In 1984, Norman Augustine, former under secretary of the Army and then CEO and chairman of Lockheed, published a tongue-in-cheek list of "laws" of defense system development. The most famous of these was [1]: "In the year 2054, the entire defense budget will purchase just one tactical aircraft. The aircraft will have to be shared between the Air Force and Navy 3½ days each week."

Another Augustine law was that the probability of program cancellation is based on its years in development. With this theory, the final Comanche restructure took the program from a 70% probability of cancellation to 100%.

His best quote for Comanche was: "In terms of their chances of surviving, most programs start out kind of slowly and then sort of taper off."

Over more than 20 years of effort, the air vehicle had developed and proven virtually all of its technology advancements. But the mission equipment package

(MEP) was still not mature, and the final production air vehicle design still required additional work to reduce weight and incorporate delayed systems.

The Office of the Secretary of Defense (OSD) had stepped in and offered technology maturation help through the government labs, but that resulted in dependencies on other programs that didn't remain funded. Dependencies that were slipped and/or canceled due to funding changes had a large impact on the program. These now had to be funded by the program. Their impact on the program is often overlooked.

SATCOM, Link-16, and miniaturized ultra high frequency and very high frequency hardware are a few examples of efforts tied to the Joint Strike Fighter (JSF) program. The JSF program slips made these items too late for Comanche integration, so we had to pick up the development as part of the integrated communications, navigation, and identification avionics (ICNIA) program. Aided/automated targeting algorithm development started as an Army lab tech-based program that had to be picked up and completed within the Comanche program. Radar cross-section reduction material work stopped and suppliers went out of business. Focal plane array suppliers went out of business.

The prototype program took funding away from many of the First Team suppliers. The Lockheed target acquisition system effort was reduced by $46 million. The TRW ICNIA work was unfunded, as was the radar cross-section material development and material weight reduction efforts. Some suppliers left the team; some used their own internal funds to continue to work their subsystems.

Some in the Army program office felt that Boeing Sikorsky made the decision to put the available money toward the prototype airframe at the expense of the mission equipment package and weight reduction initiatives. This would get the aircraft into flight test sooner and improve the visibility of progress, but delaying the MEP had its own issues.

Not surprisingly, technology advanced over the extended time period of the program. This was particularly true in the case of the MEP and electronic processing. Think about where personal computers were in 1980 when the LHX program started. In 1991 we signed up to put the capability of four Cray computers on each side of the Comanche in the mission computers. By 2000, cell phones had almost as much processing power as the Comanche MEP.

We had trouble explaining Comanche's overall program approach, which was to proceed in a very structured manner, solving issues as they came up, rather than rushing into the flight test program. This resulted in few serious technical issues in flight test. This approach was in direct contrast to earlier-generation helicopters that experienced major technical issues during flight test, which were time consuming and expensive to fix. A number of government and consultant analysts used data from those earlier programs to project Comanche's development future, with many additional technical issues to be expected that would be expensive to fix. They ignored Comanche's long successful development approach of solving issues early. They took Comanche's long development history and then added projected future development costs based on the history of those earlier programs. This led to very inflated estimates to complete (ETC), to the detriment of the program.

We also did not agree with some of the reasons given by the Army for program cancellation, but these don't seem to have been major drivers.

COLD WAR WEAPON

One argument was that Comanche was a "cold war" weapon. Although it's true that it was conceived in the early 1980s, when the world was still in the intense cold war, the specific Comanche requirements applied to many other conditions of war.

> As mentioned in chapter 28, on April 23, 1990 when this question came up earlier, then secretary of Defense Richard Cheny wrote a letter to Sen John Murtha (D-PA) saying "The collapse of the Warsaw Pact does not alter the need for the Light helicopter (LH)."

The new battlefield was described as asymmetric warfare, including urban warfare, and the Army was trying to understand how best to address this new reality. The combat doctrine transitioned from a heavily mechanized army to a more air-mobile concept that stressed maneuver warfare. This made the Army aviation assets an even more important integral element of combat operations. The strategy still included the armed/reconnaissance helicopter as an integral element of the ground force maneuver and a valuable combat asset that increased the ground force survivability and operational effectiveness. The centerpiece of this transformation was to be the Comanche.

A key part of this "cold war weapon system" argument was the need for low observables. Some strongly argued that with the cold war over, low observability was no longer required. We countered that the very low levels of observability that we achieved would be a strong advantage in many future wartime scenarios.

In a totally opposite view, an opinion was expressed by one senior general that all this advanced Comanche technology seemed to be of no use on the battlefields of the time. We had a lot of inventions and advanced technology flying, but we would be fighting an enemy who (according to this general) was "one step out of the stone age." He asked "What are we going to with all that capability in Afghanistan?"

INFRARED ISSUE

During the General Schoomaker directed review, the Army Aircraft Survivability Equipment (ASE) office reported that the Comanche was vulnerable to modern infrared (IR) seeker threat systems and needed an IR jammer.

Defeating IR missiles was always one of the strong points of Comanche. We believed we had solved this requirement from our initial design days, and the Army technical community had agreed. Our test engineers and pilots also did not believe this was an issue.

The threats had evolved over time from the Soviet (red) threats to multinational (blue and gray) advanced threat systems. In 2002, the Army had Boeing Sikorsky perform a detailed survivability analysis of the Comanche operating within the acquisition range of the U.S. Stinger missile. The results of this analysis supported the Comanche IR signature levels and that a very small potential missile acquisition envelope resulted in a very low probability of intercept by the missile. Earlier hardware-in-the-loop simulations with the Stinger missile seeker, conducted by Sanders, showed no intercepts.

However, the audience in 2002 receiving the results of the analysis took the position that if the analysis did not show the Comanche completely invulnerable to a Stinger engagement, the aircraft would require IR active countermeasure equipment that would increase the unit cost of the aircraft by $1 million.

The flaw in this conclusion was that even with IR ASE, an aircraft has some probability of being hit by the Stinger missile. The Stinger is only a ballistic threat when first launched and guidance control is not yet available. Complete invulnerability to a ballistic weapon, guided or unguided, is unrealistic. In the first kilometer, the engagement range, it really didn't matter because the missile was not up to a high enough speed to maneuver/guide to hit the helicopter. No matter how much ASE you had, it would make no difference. At that first kilometer, the missile couldn't even acquire the Comanche to allow the missile to be launched.

We felt that this argument was simply a rationalization to help explain the program cancellation.

20/20 Hindsight

After the passage of many years, and with 20/20 hindsight, a few things are clear.

The Comanche was the most advanced rotary wing air vehicle system that government and industry had ever attempted. The original LHX program was undoubtedly ambitious. It was to be the first all-new military helicopter design since the Black Hawk and the Apache. It became a magnet for all the varied technologies that government labs and private industry had been working on since the 1970s. Put all together, there was the promise of a capability vastly superior to anything seen before. Here was an aviation system that could find and identify an enemy under the most adverse conditions without ever being detected by that enemy. Unheard of.

Such a program would be costly, but the early planners felt that the cost was justified for the results promised.

Going back to the beginning, the Comanche program went off track even before it was placed on contract. In 1990, the Army had to reduce the funding available due to DoD budget reductions and other priorities. More importantly, there was an ongoing issue with OSD concerning the lack of technology maturity and the resulting risk for some of the concepts and systems we were proposing. The restructure in 1990 not only reduced funding, but also converted the program from a production program to a prototype program. OSD wanted, and got, a technology demonstrator, not a production program. As we moved forward we—the Army program office and Boeing Sikorsky—did everything we could to keep the production option open. This approach put the program in a situation that was difficult to recover from.

Through the years, the Army program office and the Boeing Sikorsky JPO tried to keep the production program a possibility, with limited results. At the last restructure, it was said that some in the Army thought the aircraft had been converted back to a full production design over the course of the numerous restructures. That was not the case.

As the previous chapters have shown, four restructures reduced or further delayed the Army funding. This might have been acceptable, but each restructure permitted the Army to reevaluate the technology then available and introduce additional requirements. With only the best of intentions, people in power required that these be incorporated into Comanche, believing that this would help keep the program sold. But this all added to cost and schedule issues.

This then started a pattern of aircraft weight growth that moved the aircraft out of the "light" category. This was in direct contrast to previous Army programs. In discussing this with Ray Leoni, Sikorsky Black Hawk engineering manager and later program director, he said: "We were really fortunate during UTTAS [Black Hawk] that no changes to the prime item development spec were permitted by Army project managers, notably Generals Turner and Lauer, after the BED Phase contract was signed."

Additionally, technical optimism came into play. Inventions and technology developments did not occur as rapidly as we had projected. (Maybe OSD was right in its original assessment that we were going too fast in too many areas.) Some of this was due to the funding cuts and the fact that other supporting programs were not coming through, but much of it was simply that we were too optimistic on what was realistically achievable. This resulted in even more aircraft weight growth. In spite of intense effort and numerous redesigns, we could not keep the aircraft weight down.

WHAT MIGHT HAVE BEEN DONE TO SAVE THE PROGRAM

WHY NOT WAIT TWO YEARS?

It is believed that the timing of the last restructure was one reason why the program was canceled. The MEP had not yet been demonstrated, and it was a big schedule and cost risk. User support for the program was low, in part because of all the delays forced by the earlier restructures and concerns about the lack of maturity of the MEP.

Discussions were held within the JPO concerning these cost and schedule risks. There was also the realization that any further replan would not change the first few years of the program. Within the next two years, the MEP would have been on the aircraft and in flight test, and the assumed program risk level would have been considerably less. It was thought that the proposed changes being considered, to an incremental or "spiral" or "evolutionary acquisition" of capabilities, would have been more readily accepted at a later date.

This was also MG Bergntz's recommendation, as discussed in chapter 37.

But that approach did not prevail. Senior program leadership decided that it was best to propose the further program restructure as soon as possible.

WHY NOT AN "A" MODEL?

From the beginning, all involved with the program were enthusiastic about putting all this available new technology together into a system that would be dramatically better than anything the world had ever seen before. As we found out, this didn't work.

What would have happened if we started out with a simple A model using only the technology that was mature enough for production at that time? Maybe this should have been proposed in 1991 when the OSD was concerned about the program's technical risk. Maybe instead of a prototype program, we should have gone with an A model production aircraft. This could have been put into production, with B and C model to follow as other technologies became available. This would have followed the very successful examples of the Black Hawk and Apache programs. (The software in the A model AH-64, for example, is vastly different from today's Apache.) The eventual RAH-66C or D or whatever would have incorporated these changes and evolved into the valued asset we envisioned.

Another alternative at a later date, at the time of the final restructure, might have been to convert the very successful flight test aircraft into a production A model, accepting the aircraft design as it then existed and adding currently available low-risk mission equipment. The aircraft could have been stuffed with available sensors, weapons, and existing mission systems limited only by its available useful load at its demonstrated gross weight. We had enough ballast in the nose to replicate the weight of the Electro-Optical Sensor System (EOSS). The sensor development was pretty far along—we were flying the night vision pilotage system (NVPS) and helmet-mounted display (HMD) prior to the cancellation, so although it may not have been full up, we could have had a night vision and targeting sensor in place of the ballast.

We also could have eliminated, or at least greatly reduced, the costly effort to reduce the aircraft weight back to the spec value. The flight test aircraft was flying very well at its higher weight and received that rave review from the Army deputy chief of staff Lt. Gen Richard Cody (see chapter 26, last paragraph.)

This approach certainly would not have been unprecedented. Other previous aircraft programs entered production well above their desired original empty weight. Even those that entered production at their original weight grew as new needs were defined. In those cases there was only a small impact on the aircraft's real-world performance.

Also, why was LONGBOW on a reconnaissance aircraft when the rules of engagement required target identification, which LONGBOW cannot do? Why two enhanced fuel armament management subsystems (EFAMS) stations on a light attack platform? Removing the pressurized cockpit requirement would have been good. Chemical detectors could have been kitted. Why not make the radar cross-section (RCS) a kit? What's wrong with a 50-caliber gun? Why did they need a gun that was so robust you could fire all 500 rounds without stopping?

We pitched a simplification of the night vision system that had two options: First, return to the forward-looking infrared (FLIR)-only design. (The Army wanted both a FLIR and I^2 camera in the turret so it could do sensor fusion.) Second, remove the NVPS turret altogether and utilize night vision goggles (like virtually every other tactical aircraft in the world except for the Apache). The Army added SATCOM. There were so many things in the design that seemed to be there just to get support from one or another technology group in the Army.

With this approach, subsequent models would contain additional bells and whistles because they would have been developed and demonstrated. This might have been a winner for all parties. From published reports, Comanche was claimed to be the best helicopter ever created.

CONCLUSION

Time was a major cause for the termination. It was easy to get enthusiastic support for the original program with an initial operational capability (IOC) date of 1998. This fell off with each new restructure. The Army user community and some senior Army leaders began to wonder if they were ever going to get the promised capability.

Realistically, the dragging out of the program schedule and overly optimistic technical projections, plus the continual addition of new requirements with no hope of ever getting any funding increases, probably made the program undoable long before it was finally canceled.

REFERENCE

[1] Augustine, N. "The American Institute of Aeronautics and Astronautics," *Augustine Laws*, 6th ed., Reston VA 1983.

Chapter 38

POSTSCRIPT

ARTHUR W. LINDEN

At the time of the cancellation, Boeing Sikorsky program director Mike Blake suggested that the Army consider a low-key mini-program where we would finish the build-out of the nine aircraft in engineering, manufacturing, and development (EMD), and the Army could then evaluate them in the future as new threats emerged. He cited this as a way to protect the decades of work that had been done while cutting new costs by 90%. At that time we had four aircraft on the line, and most materials were in the pipeline for the rest of the nine plus initial spares.

We were instead told to stand down and that all the materials were to be destroyed. The Department of Defense seemed to have a fear of the program rising from the ashes, and did what it could to ensure the program would remain canceled. The composite material was shredded and burned. Millions in components were also destroyed and junked. The materials and documentation were stored in Sikorsky's Bridgeport Seaside facility as the direction was given over a three-year period for its disposal.

The bottom line is that it appears the Army wanted no possible resurrection of the RAH-66 program.

TECHNOLOGY TRANSFER

Comanche pioneered many design elements that were new to military helicopters—fly-by-wire (FBW) with unique trim controllers, multifunction displays, digital map terrain shading, helmet display with contact analog symbology, vehicle automation, rotor agility—all technologies that were mission agnostic. Radar and infrared (IR) signature reductions were very impressive and applicable to any threat environment.

We were ahead of even today's technology in some areas. An example is the current fascination with autonomy. By implementing the centralized avionics vehicle management, we automated virtually all the system management functions. We did it so the pilot could focus on mission management, but making the aircraft "optionally piloted" would have been relatively trivial.

The engineering accomplishments of Comanche have indeed greatly influenced the modern helicopter design process and will continue to influence aviation design and development for years. Comanche pioneered the use of 3-D computer-aided design (CAD) and digital mockups that dramatically improved the design process and created substantial benefits to manufacturing. This technology has now been adopted as a standard practice at Boeing and Sikorsky, and across the entire aerospace industry.

All Army helicopters now have digital displays, and new platforms have FBW, high-speed processing mission computers, software-programmable radios, composite structures, fault-tolerant designs, open architectures, optical networks, and so forth, in part because of the work done by the Comanche team. Our work on low observables has gone on to other aircraft. Some of the other technologies transitioned to other systems include the image intensification (I^2) TV camera, radar electronics unit, radar warning receiver, and antennas.

Comanche's technology advancements were paid for by the government and therefore are government owned. Following termination, a spreadsheet was put together to identify where each of the technologies was going to be transitioned.

Program Executive Office (PEO) Aviation and the Comanche Program Management Office (PMO) had a plan for permanent storage of all the data and files, classified and unclassified. The government's position was that the technology needed to be resident only within the government to ensure protection of distribution as much as possible. The concept was to have one resource where the physical and function models of the Comanche technologies, assemblies, subassemblies, parts, and so on resided to control distribution of the technologies and loss of the technology information to other countries.

The primary location for both hard copy and electronic files, including classified files, was the Redstone Scientific Information Center (RSIC). A documented agreement was established between the Comanche office and RSIC to accept and store the files. Most of the files were transferred in the 2005–2006 time frame while termination activities were underway. There are approximately 20 file cabinets of hard copy files in storage at RSIC.

As part of the technology transfer, from 2006 to 2008, Bill Harper of Boeing Sikorsky and Mike Richey of the Army PMO developed and gave classes at the University of Alabama Huntsville on Comanche technology. Their information came from the RSIC files.

Unfortunately, very few of these files were ever cataloged. RSIC recently closed all but online searches. None of the Comanche data are available online. The hard copy files are still in Huntsville, but there isn't a formal plan to scan them, although we understand that was the goal. Given that the files have been at RSIC going on 20 years with little attention, it doesn't seem likely they will ever be scanned. It is unclear what future RSIC has or what the long-term storage plan will be for the files and who will retain custody.

The Comanche flight test trailers were transferred to Bell along with the cockpit design. Current Army helicopters are still federated systems using individual black boxes, so much of the Comanche mission equipment package (MEP) hardware was not transitioned.

For a number of years, the two prototype aircraft resided at Redstone before going to the Fort Rucker aviation museum. The MEP on those aircraft was made operational at Redstone.

WHERE TO NEXT?

The need for a Comanche-like system did not go away with the termination of the program. The requirements and need still exist 40 years later.

History had taken us from the cold war with Russia to asymmetrical warfare in the Middle East. And now the American military's focus is shifting to China.

The Army is now on its third try to replace Comanche. At least one proposed design for the Future Armed Reconnaissance Aircraft (FARA) looks remarkable like the Comanche, showing many of the same features we developed years ago. Maybe the Comanche was dropped too early and now might be what the Army needs. Even better, other technologies have been developed in the interim that could make the FARA more capable than Comanche was 20 years ago.

Sydney J. Freedberg, Jr., reporting in the 26 February 2020 issue of *Breaking Defense*, described flight demonstrations of three experimental aircraft at Sikorsky's West Palm Beach, Florida, test center, attended by Army Secretary Ryan McCarthy and Senator Tammy Duckworth (D-Illinois) [1]:

> I asked, how can the Army convince Congress and the taxpayers that FVL [the Army's Future Vertical Lift program, including both FARA and the Future Long-Range Assult Aircraft (FLRAA)] won't be a repeat of past aviation acquisition disasters, like the stillborn Armed Aerial Scout and the RAH-66 Comanche, cancelled only after the service invested $7 billion?

Senator Duckworth's take: "The Comanche was [driven] by what a bunch of aviators wanted, not necessarily by looking forward to what the real mission was on the ground."

> "The Comanche helicopter, it was extraordinary what they created; it just cost too much," said Army Secretary Ryan McCarthy. "The Army made a decision—it was right when I entered the Pentagon over a decade ago—[that] they couldn't afford it.
>
> "What's different about this time, we're organized very differently against the problem," McCarthy said. Instead of the longstanding

> bureaucratic divide between combat veterans writing wish-lists of high-tech requirements and acquisition officials trying to actually build something on a budget, the 18-month-old Army Futures Command and its FVL Cross-Functional Team "have brought the requirements organization together with acquisition," he said. "There's much greater clarity."
>
> Pointing down the table, McCarthy went on: "Gen. Rugan and Mr. Mason here, the requirements leader and the acquisition executive responsible for this program, they're attached at the hip. They're going to work very hard to prevent requirements fluctuation from confusing the manufacturers.
>
> "The Army has learned its lesson," Sen. Duckworth said....
>
> Some 16 years after Comanche died, does the Army really have a clearer grasp of what the mission requires, what industry can deliver, and what the budget can afford?

Mr. Freedberg also interviewed Major General (retired) Rudy Ostovich, who helped write the Comanche's initial requirements. He said:

> "I believe Army Aviation leadership has learned its lessons from Comanche regarding FVL, . . . We are now properly organized around the problem."
>
> . . . as over-reaching as the initial requirements were, the Army didn't stick to them. Instead, it kept adding to the wish-list over time, which increased Comanche's capabilities—at least on paper—but also its size and price.

As for those of us who worked on the Comanche program, we end with a quote from Walter P. Chrysler, who worked building locomotives before joining the automobile industry [2]: "There is in manufacturing a creative joy that only poets are supposed to know. Some day I'd like to show a poet how it feels to design and build a railroad locomotive."

We didn't do locomotives, but we did create the world's most advanced helicopter weapons system. We were driven by what we were creating.

On cancellation, some in government and people outside of our industry asked why we were so disappointed that it was canceled. After all, we got paid for our work over all those years.

We brought into reality the idea of breakthrough technologies combined into an almost unbelievable future helicopter weapons system, only to see the program canceled just as it was demonstrating the reality of what we had created. It was much more than just people losing jobs. It was the elimination of what to many of us was the culmination of a life's work in the helicopter industry.

REFERENCE

[1] Freedberg, S. J., Jr., *Breaking Defense*, 26 February 2020.

[2] Charle K. Hyde, "A History of the Chrysler Corporation—Riding the Roller Coaster," published in 2003 by Wayne State University Press, Detroit Michigan, p. 1.

Acknowledgements

This book is a compilation of the recollections of many people at Sikorsky who worked on the Comanche program. The following are the major contributors:

Irv Alansky

Irv was the Sikorsky Manager for the Comanche engineering development simulator (EDS) team. His support started with an initial fixed base simulator developed during the Advanced Rotorcraft Technology Integration (ARTI) program. As the Comanche program progressed, the engineering shifted to a high-fidelity motion facility with a state-of-the-art visual system and eventually to a portable simulator known as the Comanche portable cockpit (CPC). The CPC was used by the Army to support various war flighting exercises and assisted in the development of the tactics, techniques, and procedures to be employed by crew members flying Comanche.

Irv and his team coordinated with the Comanche designers primarily in the area of flight controls, handling qualities, avionics, and crewstation design in defining software tools, aircraft and weapons models, and hardware solutions that prototyped the evolving design and permitted piloted evaluations in simulated battlefield and mission-oriented environments.

William Harper

Bill was the original helicopter configuration designer for the Sikorsky DAS configuration study group. Bill and Mark Alber hold the U.S. design patent for the Comanche helicopter configuration. Bill joined the Boeing Sikorsky Joint Program Office (JPO) in 1987 as the LHX engineering manager. He later became the JPO engineering and manufacturing director. His organization had the technical responsibility for the Comanche weapon system. Its primary roles were requirements control and allocation, and mediation among Sikorsky, Boeing, suppliers, and the Army, plus communication of Comanche technologies and operational benefits to decision makers, support of independent studies and evaluation, and management of compliance with major milestone exit criteria. Bill helped pioneer some of the signature techniques used on Comanche. He was instrumental in the development of the radar and infrared detectability and countermeasure effectiveness methodologies for Comanche that are still being used by Boeing, Sikorsky, and the U.S. Army for helicopter operation effectiveness simulation, and the tradeoff methodology used to establish the 7500-lb/$7.5-million 1991 proposal configuration. After retiring from Sikorsky, from 2006 through 2008 Bill was an instructor at the University of Alabama Huntsville (UAH) for the Comanche Technology Transfer, Professional Development classes.

It is not an exaggeration to say that this book could not have been written in the detail shown without Bill's extensive database and personal memory. Over five years, Bill provided many extensive explanations of technical and programmatic history. He actually provided much more detail than we could include in a reasonably sized book.

Stephen Horling

Steve started in the early 1980s working the initial DAS study contracts. He is an electrical engineer with an MS degree. He worked part time to provide all the reliability, availability, and maintainability (RAM) effort related to the mission equipment package (MEP), electrical systems, flight controls, and avionics. He began working full time with the onset of the Advanced Rotorcraft Technology Integration (ARTI) program. He became RAM lead shortly after the onset of the Boeing Sikorsky joint venture and continued as RAM lead until program cancellation.

Bruce Kay

Bruce was the LHX chief of design prior to our teaming with Boeing and helped write the LHX proposal to the Army. He was responsible for the air vehicle design including the airframe, dynamic systems, and propulsion systems. After contract award, when the program was reorganized into an integrated product development team (IPDT) structure, he managed the forward fuselage and propulsion system IPDTs. All of the major air vehicle system trades were done under his watch. He stayed on the program through initial shakedown flight testing. Bruce authored or was a major contributor to the following chapters: Chapter 4, Low Observables; Chapter 5, Armament; Chapter 11, Main Rotor; Chapter 12, Fantail; Chapter 13, Fuselage; Chapter 14, Propulsion; Chapter 15, Transmission; Chapter 16, Hydraulics; and Chapter 24, Management.

Arthur Linden

Art joined the LHX program in September 1987 as the LHX engineering manager, replacing Bill Kuczynski. He became the Sikorsky LHX program director in June 1990. In 1996 he went to the JPO, first as the Comanche deputy program director and then as the program director. He retired in June 2000, after his four-year rotation in that position. He is the primary author of this book. His three boxes of LHX/Comanche material were the original basis for the book. They included all the issues of the aviation press quoted herein. He authored numerous chapters from this material and served as coordinator and editor to put the full book together.

JOHN OCCHIATO

John had the lead role for handling qualities for the duration of the program. Before LHX got started, he was the chief analyst and part of the advanced design team preparing for the program, helping to create the basic required handling characteristics as well as maneuverability and agility requirements, and helping shape the configuration. He headed up and executed the Army's Maneuverability and Agility Design Sensitivity Study, a precursor to LHX, contracted out of Fort Eustis. When Boeing joined the team, John had the Integrated Product Development Team (IPDT) lead that covered both handling qualities and flight controls. Boeing had strong experience with the fly-by-wire design, but it was very critical that Sikorsky maintain a tight rein on the performance of the control system to get the eventual handling that we felt the program needed.

WILLIAM F. PAUL

Bill Paul, Sikorsky president during the early days of the program, provided the details on establishing our teaming agreement with Boeing.

DR. KENNETH ROSEN

Ken was Sikorsky's director of design and programs between 1986 and 1988, and vice president of research and engineering until his retirement in 2000. He was functionally responsible for all the engineering resources dedicated to the Comanche program. He was a very proactive leader. He was tasked with the program technical overview responsibility by company president Gene Buckley. The process he used to accomplish this involved very detailed weekly reviews of each of the integrated product development teams, where weight, performance, and other technical attribute statuses were reviewed. Ken personally conducted these reviews every Friday for over a decade. Ken contributed to numerous sections of the book and to its editing.

FRANK SCHRADER

Frank joined the Advanced Rotorcraft Technology Integration (ARTI) team as avionics/electrical lead after a stint as a functional section leader in avionics. He helped launch the Shadow program. He later became the mission equipment manager just before the teaming efforts with Boeing. After contract award, he took responsibility for Sikorsky's hardware elements in the mission equipment package (MEP) [helmet integrated display and sighting system (HIDDS), Rotary AirData System (RADS), crew station design, and avionic systems engineers] as deputy MEP manager coordinating with Boeing, and also the electrical subsystems (270-V power generation, auxiliary subsystems, fiber optics, and air vehicle interface control system (AVICS). Frank authored the chapters on MEP (with substantial contributions from Raj Kaushik and Rus Stiles), and the electrical and software chapters (Chapters 17 and 10, respectively).

LORREN "RUS" STILES

Rus joined the Sikorsky Pilot Office in December 1982 and started working with the human factors group on the cockpit for the "Advanced Scout Helicopter (ASH)" shortly thereafter. At the same time, Sikorsky was doing research in conjunction with the Canadian National Research Council (NRC) to study controller configurations for fly-by-wire (FBW) helicopters. By 1983, Rus was actively participating in the piloting duties for an S-76 surrogate aircraft with a palletized FBW system, as well as flying the NRC Bell 205, another FBW surrogate. Rus continued working with these aircraft and simulation studies of FBW control laws and advanced cockpit concepts. Shadow, the follow-on to the S-76 FBW surrogate, which incorporated a full research cockpit on the nose, first flew in January 1986 with Rus at the controls. Rus was later assigned as the chief project pilot for the Sikorsky Advanced Rotorcraft Technology Integration (ARTI) program. After the teaming with Boeing was established, he was assigned as the chief program pilot for the LHX program. He made the first flight of the RAH-66 Comanche prototype in January 1996 and stayed with the program in the chief pilot capacity until its cancellation. Rus authored Chapter 19 on flight test and contributed to the chapters on handling qualities and cockpit (Chapters 6 and 9).

Other people who made additional valuable contributions to the book:

MIKE BLAKE

Mike was the last Boeing Sikorsky Comanche program director. He provided valuable information on the final years of the program when most of our authors had left the program.

JACK CALDWELL

Jack provided his unique memento of the Comanche rollout ceremony—a commemorative photograph signed by the Army Chief of Staff General

Gordon Sullivan, the Chief of the Comanche Nation Wallace Coffey, and the President of Sikorsky Eugene Buckley.

TERRY DONAHUE

Terry worked for many years on the program. He provided his files, primarily on programmatic issues.

TOM SHEEHY

Tom joined the Sikorsky Comanche team in 1991 as the director of design and development reporting to Art Linden. In this position he directed and controlled the aircraft design and development activities including basic airframe systems and much of the mission equipment package. In 1996 he was named director of the Sikorsky Project Office when Art was named Comanche JPO deputy program director He provided valuable assistance and guidance in the preparation of the manuscript of this book.

GEORGE SIPPRELL

George was one of the longest duration members of the LHX/Comanche team. He joined the program in 1981 when it was still in the conceptual design stages and did not leave until 2002. He was heavily involved early in the Shadow program and went on to a number of senior technical positions. He provided much Comanche information from his files and numerous quotes for the book.

FRANK COLUCCI

Additionally, professional author Frank Colucci provided guidance and editing of key chapters.

INDEX